Cases on Telecommunications and Networking

Mehdi Khosrow-Pour, D.B.A.
Editor-in-Chief, Journal of Cases on Information Technology

IDEA GROUP PUBLISHING
Hershey • London • Melbourne • Singapore

Acquisitions Editor:	Michelle Potter
Development Editor:	Kristin Roth
Senior Managing Editor:	Amanda Appicello
Managing Editor:	Jennifer Neidig
Typesetter:	Marko Primorac
Cover Design:	Lisa Tosheff
Printed at:	Integrated Book Technology

Published in the United States of America by
    Idea Group Publishing (an imprint of Idea Group Inc.)
    701 E. Chocolate Avenue, Suite 200
    Hershey PA 17033
    Tel: 717-533-8845
    Fax: 717-533-8661
    E-mail: cust@idea-group.com
    Web site: http://www.idea-group.com

and in the United Kingdom by
    Idea Group Publishing (an imprint of Idea Group Inc.)
    3 Henrietta Street
    Covent Garden
    London WC2E 8LU
    Tel: 44 20 7240 0856
    Fax: 44 20 7379 0609
    Web site: http://www.eurospanonline.com

Product or company names used in this book are for identification purposes only. Inclusion of the names of the products or companies does not indicate a claim of ownership by IGI of the trademark or registered trademark.

Library of Congress Cataloging-in-Publication Data

Cases on telecommunication and networking / Mehdi Khosrow-Pour, editor.
    p. cm.
    Summary: "This book presents a wide range of the most current issues related to the planning, design, maintenance, and management of telecommunications and networking technologies and applications in organizations"—Provided by publisher.
    Includes bibliographical references and index.
    ISBN 1-59904-417-X — ISBN 1-59904-418-8 (softcover) — ISBN 1-59904-419-6 (ebook)
    1. Telecommunication systems—Case studies. 2. Computer networks—Case studies. I. Khosrowpour, Mehdi, 1951-
    TK5102.5.C325 2006
    004.068—dc22   2006008425

British Cataloguing in Publication Data
A Cataloguing in Publication record for this book is available from the British Library.

The views expressed in this book are those of the authors, but not necessarily of the publisher.

Cases on Information Technology Series

ISSN: 1537-9337

Series Editor
Mehdi Khosrow-Pour, D.B.A.
Editor-in-Chief, *Journal of Cases on Information Technology*

Cases on Telecommunications and Networking

Table of Contents

Daniel Robey, Georgia State University, USA
Leigh Jin, San Francisco State University, USA

This case describes the successful launch of a free voicemail service across the entire U.S. Although initially successful, iTalk later foundered in its efforts to generate revenues with services that customers would be willing to buy. Technical problems delayed deployment of new services, and relationships between iTalk's engineers and operations personnel became strained. Facing these and other problems, iTalk was not in a strong position to go public or to raise additional funds from private investors. Worse, iTalk had become potential prey for takeover attempts.

Andrew Borchers, Lawrence Technological University, USA
Mark Demski, Lawrence Technological University, USA

As a response to strong competitive pressures, the U.S. automotive industry has actively employed electronic data interchange in communications between suppliers and car makers for many years. This case reviews the recent development of ANX®, a COIN (Community of Interest Network) intended to provide industry-wide connectivity between car makers, dealers and Tier suppliers. The authors identify technical and business challenges to the success of ANX®.

A fire destroyed a facility that served as both office and computer server room for a College of Business located in the United States. The fire also caused significant smoke damage to the office building where the computer facility was located. This case, written from the point of view of the chairperson of the College Technology Committee, discusses the issues faced by the college as they resumed operations and planned for rebuilding their information technology operations.

This case examines policy processes for the introduction of technology-mediated learning at universities and colleges. It is based on the results of a two-year research project to investigate policy issues that arise with the implementation of telelearning technology in universities and colleges.

This case study explores a metrics program to track the development of a new version of a telecommunications company's voicemail product. The study addresses the adoption of the program, its components, structure, and implementation, and its impact on product quality and timeliness. The study also discusses the aftermath of the program and its place in the larger organizational culture.

This case compares and contrasts the issues faced by firms in today's telecommunications environment with an actual telecommuting case study of Trade Reporting and Data Exchange, Inc. (T.R.A.D.E.), a software engineering company located in San Mateo, CA.

The challenges faced by organizations in developing countries in getting reliable, high-speed Internet access to support their mission critical Web-enabled information systems are highlighted in this case. The case prescribes various measures to optimally use the constrained bandwidth available from service providers. The challenges in defining and monitoring appropriate service level agreements with the service providers are discussed.

This case describes a province-wide network of Community Access Internet sites that was supported during the summers of 1996 and 1997 by Wire Nova Scotia (WiNS), a government-funded program to provide staffing, training and technical support for these centers. Remote management enabled the efficient and low-cost operation of a program involving 67 sites.

This case discusses Digifone's (an Irish telecom company) evaluation of an open XML vocabulary as part of the mediation and billing solution for their next generation (3G) mobile services. The case argues that the key to exploiting 3G technology is inter-industry agreement on measurement standards to effectively manage the value of the associated services.

This case study describes the process of integrating the library, computing and telecommunications services in a University with all the problems and challenges that were experienced during the project.

Ten years of efforts in introducing the state-of-the-art information and communication technologies (ICT) and development of ICT infrastructure on the national level are described in this case. The aim of the project was to build Internet in Croatia and to foster its leverage in the broad range of activities of public interest in the society as a whole. The prime target group was academic and research community, as a vehicle for the overall development in the society.

This case introduces one company that has decided to experiment with the telecommuting arrangement. Through the eyes of one teleworker, many of the benefits and challenges of telecommuting are explored.

This case presents experience from design and implementation of a university information system at the Brno University of Technology. The newly built system is expected to provide the students and staff with better tools for communication within the university's independent faculties, departments, and central administration. An object-oriented metasystem approach was used by the IT Department to describe the services offered by the university information system and to generate needed program code for run-time operation of the system.

Chapter XIV

Methods for how information technology tools are currently cutting cost and adding value for NASA Langley internal and external customers are presented in this case. Three components from a larger strategic WWW framework are highlighted: Geographic Information Systems (GIS), Integrated Computing Environment (ICE), and LANTERN (Langley's intranet).

This case examines the role and implications of deregulation in the telecommunications sector on an IT-based services organization, Globe Telecom, in the Philippines. Globe has continued to succeed despite the competition against the Philippine Long Distance Telephone Company, which at one time controlled over 90% of the telephone lines in the Philippines. Globe has been able to do this through strategic partnerships, mergers, and acquisitions. Furthermore, Globe has developed into a leading wireless provider by its effective use of modern information technology.

This case study illustrates the use of a non-traditional approach to determine the requirements for the Naval Air Systems Team Wide-Area Network (NAVWAN). It is considered to be non-traditional because the case data enable the use of stakeholder analysis and SWOT (strengths, weaknesses, opportunities, threats) assessments to determine the requirements instead of asking functional proponents about function and data requirements.

The impact of information technology on the pharmaceutical industry as it responds to new FDA guidelines is presented in this case. One such guideline is that all New Drug Applications (NDA) be submitted in electronic (paperless) format. Pharmacies must now take steps to assure that its use of information technology will allow it to not only meet FDA guidelines, but achieve its corporate goals of improved efficiency and reduced operating costs.

Chapter XVIII

Norwel Equipment Co. Limited Partnership (L.P.) is a Louisiana business retailer of construction equipment specializing in John Deere heavy-equipment and has secured exclusive John Deere rights for most of the State of Louisiana. This case illustrates business and technology issues facing Norwel.

Chapter XIX

This case study presents an experience report on an enterprise modelling and application integration project for a young company, starting in the telecommunications business area. The company positions itself as a broadband application provider for the SME market. Whereas its original information infrastructure consisted of a number of stand-alone business and operational support system (BSS/OSS) applications, the project's aim was to define and implement an enterprise layer, serving as an integration layer on top of which these existing BSS/OSSs would function independently and in parallel.

The case study describes the design process for a collaborative engineering workspace at the University of Technology Aachen, Germany, under development within a research project considering distributed chemical engineering as an example. Current solutions and challenges as well as future work are outlined, including the lessons learned from the study.

This case traces the dynamic evolution/revolution of an e-commerce entity from concept through first-round venture financing. It details the critical thought processes and decisions that made this enterprise a key player in the explosive field of supply chain logistics. It also provides a highly valuable view of lessons learned and closes with key discussion points that may be used in the classroom in order to provoke thoughtful and meaningful discussion of important business issues.

The case study explores the similarities between businesses using Internet technologies to "add value" to their products and services, and the reasons academics use Internet technologies to assist in traditional classroom delivery. This case examines benefits derived by faculty and students when using the Internet to supplement four different subjects at Victoria University, Australia.

This case deals with the experience of a school district in the design and implementation of a wide area network. The problems faced by the school district that made the WAN a necessity are enumerated. The choice of hardware and the software is explained within the context of the needs of the school district. Finally the benefits accruing to the school district are identified, and the cost of the overall system is determined.

Preface

During the past two decades, technological development related to telecommunication technologies has allowed organizations of all types and size to be able to develop effective networking applications in support of information management. Furthermore, telecommunication technologies combined with computer technology have created the foundation of modern information technology which has affected all aspects of societal and organizational functions in our modern world. *Cases on Telecommunications and Networking*, part of Idea Group Inc.'s *Cases on Information Technology Series*, presents a wide range of the most current issues related to the planning, design, maintenance, and management of telecommunications and networking technologies and applications in organizations. Real-life cases included in this volume clearly illustrate challenges and solutions associated to the effective utilization and management of telecommunications and networking technologies in modern organizations worldwide.

The cases included in this volume cover a wide variety of topics, such as virtual e-businesses, automotive industry networks, software metrics at a telecommunications company, telecommuting managerial issues, defining service level agreements based on metrics, mobile licenses and customer value, integrating library, telecommunications and computing services in a university, nation-wide ICT infrastructure introduction, the telecommuting life, the application of an object-oriented metasystem, IT and large organizations, deregulation in the telecommunications sector on an IT-based services organization, implementing a wide-area network at a Naval Air Station, IT in the pharmaceutical industry, Internet upgrades at a construction equipment retailer, architecture implementation for an young company's integration project, collaborative engineering shared workspace, the evolution of an e-commerce entity, Internet technologies used as a supplement to traditional classroom subject delivery, and the implementation of a wide area network.

As telecommunications and networking technologies become an influential force driving modern information technologies, such as electronic commerce and government and wireless and mobile commerce, management will need to stay up-to-date with current technologies, innovations, and solutions related to the utilization and management of telecommunications and networking technologies in organizations. The best source of knowledge regarding these issues is found in documented, real-life case studies. *Cases on Telecommunications and Networking* will provide practitioners, educators and students with important examples of telecommunications and networking systems implementation successes and failures. An outstanding collection of current real-life situations associated with the effective utilization of telecommunications and networking systems, with lessons learned included in each case, this publication should be very instrumental for those learning about the issues and challenges in the field of telecommunications and networking systems.

Note to Professors: Teaching notes for cases included in this publication are available to those professors who decide to adopt the book for their college course. Contact cases@idea-group.com for additional information regarding teaching notes and to learn about other volumes of case books in the IGI *Cases on Information Technology Series.*

ACKNOWLEDGMENTS

Putting together a publication of this magnitude requires the cooperation and assistance of many professionals with much expertise. I would like to take this opportunity to express my gratitude to all the authors of cases included in this volume. Many thanks also to all the editorial assistance provided by the Idea Group Inc. editors during the development of these books, particularly all the valuable and timely efforts of Mr. Andrew Bundy and Ms. Michelle Potter. Finally, I would like to dedicate this book to all my colleagues and former students who taught me a lot during my years in academia.

A special thank you to the Editorial Advisory Board: Annie Becker, Florida Institute of Technology, USA; Stephen Burgess, Victoria University, Australia; Juergen Seitz, University of Cooperative Education, Germany; Subhasish Dasgupta, George Washington University, USA; and Barbara Klein, University of Michigan, Dearborn, USA.

Mehdi Khosrow-Pour, D.B.A.
Editor-in-Chief
Cases on Information Technology Series
http://www.idea-group.com/bookseries/details.asp?id=18

Chapter I

iTalk:
Managing the Virtual E-Business

Daniel Robey, Georgia State University, USA

Leigh Jin, San Francisco State University, USA

EXECUTIVE SUMMARY

iTalk was founded in January 1999 and one year later successfully launched a free voicemail service across the entire U.S. The service was made possible by the development of proprietary software that interfaced iTalk's Web and phone servers with the switched networks of the established U.S. telephone companies. iTalk immediately experienced phenomenal Web site traffic, rivaling established telecom providers such as AT&T. Although initially successful, iTalk later foundered in its efforts to generate revenues with services that customers would be willing to buy. Technical problems delayed deployment of new services, and relationships between iTalk's engineers and operations personnel became strained. Facing these and other problems in the summer of 2001, iTalk was not in a strong position to go public or to raise additional funds from private investors. Worse, iTalk had become potential prey for takeover attempts.

ORGANIZATION BACKGROUND

In the summer of 2000, iTalk's stylish new building on Silicon Valley's Route 101 symbolized the incredible power of the Internet to generate economic value through investments in technology. iTalk was founded by Dennis Henderson in January 1999 with $1 million in private funding. Three months later, iTalk had a working prototype ready to show to venture capitalists, and in May 1999, iTalk closed $11 million in its first round of venture funding. By December 1999, product development was complete, and a free voicemail trial was introduced in the San Francisco Bay area. The trial proved to venture capitalists that iTalk's technology worked, allowing Henderson to raise an additional $40 million in a second round of funding. In March 2000, iTalk's service went nationwide and immediately experienced phenomenal Web site traffic. Over 1,000,000 subscribers signed up for free voicemail within a month after its official launch. Media Metrix, a leading service offering online visitor and Web site usage intensity analysis for different industry segments, reported that iTalk had more than 2,000,000 unique visitors to its Web site in April 2000, second only to AT&T among telecom providers.

Although riding the crest of the dot-com wave in 2000, iTalk later foundered in its efforts to generate revenues with services that customers would be willing to buy. Technical problems delayed deployment of new services, and relationships between iTalk's engineers and operations personnel had become strained. Without generating a revenue stream to cover its escalating costs of operation, iTalk was not in a strong position to go public or to raise additional funds from private investors. Worse, iTalk had become potential prey for takeover attempts by larger companies eager to acquire bits and pieces of new technologies. In July 2001, even though iTalk reported $120,000 in revenue through its paid service, it nevertheless ran out of funding. A large media company had begun negotiating to acquire iTalk for about $20-25 million.

The Dot-Com Phenomenon: Internet Startups in Silicon Valley

Silicon Valley is a 30 x 10-mile area in Northern California between the cities of San Francisco and San José. After Netscape Communications went public in 1995, Silicon Valley became the acknowledged center of the emerging Internet economy and a symbol of high-tech entrepreneurship. Silicon Valley was the birthplace of many Internet dot-com companies, including Yahoo.com, Google.com and eBay.com. In 2000, the approximately 4,000 high-tech companies located in Silicon Valley generated approximately $200 billion in revenue, primarily by leveraging investments in information technologies. Several features common to Internet startups enabled the generation of economic value, despite relatively modest investments in human resources and physical assets.

Typically, Internet startup companies reduced the cost of human resources by compensating employees with stock options in addition to salaries. Employees earned stock options based on the length of time they spent with a startup. Typically, one year of employment was needed before employees could receive stock options. Table 1 (see the Appendix) provides an illustration of a typical stock option compensation scheme designed to reward employees for staying with a company.

If and when the startup issued an initial public offering (IPO) of its stock, employees could redeem their stock options. Assuming a company went IPO after 48 months and its stocks were traded at $100/share, an employee's pure profit would be (100-1) x 10,000 = $990,000. Indeed, startup companies that went IPO, in Internet startup jargon, reportedly generated an average of 65 new paper millionaires every day in Silicon Valley during the height of the dot-com boom (Sohl, 2003). The stock option incentives attracted many software engineers to startup companies, because their potential IPO success could be the source of nearly instant wealth. However, startups also were notorious for demanding long working hours. Most of the companies even offered free drink and food in company facilities so that employees could work longer hours without interruption. Not surprisingly, many employees experienced high stress while working for startup companies.

Because many startup companies were not carried successfully to the IPO stage, a common practice in Silicon Valley was for software engineers to jump from startup to startup after they were quarter invested. Instead of working in one startup for several years, they changed companies every year. In this way, they could accumulate 2,500 shares of stocks from each of a portfolio of startup companies. If only one of them went IPO successfully, they could gain a share of the proceeds by redeeming their stock options for cash.

To reduce the cost of physical assets, Internet startups sought to generate economic value by using the principles of virtual organization design. For more than a decade, going virtual had been an organizing strategy recommended for companies seeking to create economic value with a limited investment in physical assets (Mowshowitz, 2002). A virtual organization's contributing partners often included manufacturers, logistics companies, and credit card companies. Thus, Internet startups primarily orchestrated the efforts of other organizations while limiting their own investment in physical infrastructure. Establishing such partnerships required skilled negotiations, often with larger companies. Startups also needed to create efficient business processes enabled by information technologies that linked themselves and their partners together.

In order to survive, it was important for Internet startup companies to acquire substantial funding from investors by convincing them of the potential of the startup's business models. This process required entrepreneurs to present their business ideas and to demonstrate their software products to different investors. After funding was granted, startup companies felt pressure to develop and implement their business plans rapidly so that they could go IPO as soon as possible. Although the majority of dot-com founders intended to build a sustainable business and carry it into and beyond the IPO stage, some entrepreneurs built to sell, hoping that their startups would be acquired by more established companies. In such cases, startup founders often benefited financially, while their employees faced the risk that acquiring companies would choose not to honor the stock options previously granted by the startup company.

The success of the dot-com formula was apparent in the eyes of the investment community, which often placed extraordinary value on companies that had demonstrated modest financial performance. Compared to the 1980s, which saw a modest level of IPO activity (about $8 billion in issuing activity per year), the period from 1990 to 1994 yielded issuing volume of $20 billion per year, followed by increases to $35 billion per year from 1995 to 1998 and $65 billion per year from 1999 to 2000 (Ritter & Welch, 2002). The Internet economy also was credited with creating 1.2 million jobs and generating an estimated $301.4 billion in revenue in 1998, including an estimated $160 billion contributed by online intermediaries and Internet commerce companies (Barua et al., 1999). Five years after the introduction of the World Wide Web, the Internet economy rivaled century-old sectors like energy ($223 billion), automobiles ($350 billion), and telecommunications ($270 billion).

However, an economic downturn in 2000 cast doubt on the viability of the Internet economy and its celebrated dot-com companies. According to a report published by webmergers.com (2003), at least 962 Internet companies shut down or declared bankruptcy in a three-year period. All were substantial companies that had raised at least $1 million from venture capitalists or angel investors, and many had gone public. An additional 3,892 Internet startups were acquired by other companies (webmergers.com, 2003). In retrospect, the longer-term financial performance of Internet startups that had gone public was unimpressive at a time when the stock market performed overall exceptionally well.

SETTING THE STAGE

The Voice Messaging Industry in the Late 1990s

Between 1998 and 2001, the convergence of the Internet and telecommunication industries coincided with growing consumer demand for enhanced telecommunications services, including call management, call return, caller ID, call completion, call waiting, call forwarding and voicemail. International Data Corporation (IDC) estimated that the consumer voice messaging service market alone grew from $1.3 billion in 1999 to $2.3 billion in 2003, driven primarily by new subscribers. In 2000, the market penetration for voicemail in the U.S. was only 17% and was projected to grow to 30% by 2003.

Because the number of wireless, Internet, and telephone users had increased dramatically during this period, consumer demand for enhanced telecommunications services outpaced the degree to which the industry could provide streamlined and integrated services. In 1999, many consumers accessing the Internet used dial-up connections from a single home phone number, so they could not receive incoming calls while they were online. Incoming calls went unanswered or were forwarded to voicemail, and callers typically had no way of knowing if the persons they were trying to call were online or not. Consumers with more than one phone number needed to access multiple voice mailboxes to retrieve their messages, which was costly and time consuming.

In the late 1990s, ongoing deregulation of the telecommunications industry had increased competition for Regional Bell Operating Companies (RBOCs) and created opportunities for new entrants, such as Competitive Local Exchange Carriers (CLECs) and long distance carriers, also known as Interexchange Carriers (IXCs), to enter regional markets. Deregulation also had enabled relative newcomers, such as Internet Service Providers (ISPs) and enhanced telecommunications service providers, to enter the market. Major carriers wished to add new features and functionality to their offerings but found that adding enhanced services was both time-consuming and costly, requiring new infrastructure, operations support systems, and expertise in computer-telephony integration (CTI) technologies. For these reasons, many carriers found it more attractive to purchase enhanced services from new industry entrants and to re-brand them as their own.

CASE DESCRIPTION

The conditions described previously created a potential opportunity for the services that iTalk provided: a nationwide voicemail service that answered home, small office, and wireless phones. Calls to the answered phone numbers were forwarded (on *busy* and *no answer*) to iTalk. If the caller left a message, the subscriber was notified via e-mail and could access the voicemail via phone, the Web, or e-mail. iTalk also developed an automated registration and provisioning process for the customer, eliminating the delays often associated with provisioning services from local telephone companies.

In June 1998, Dennis Henderson, iTalk's founder and board chairman, came up with the idea for iTalk. One of Henderson's friends in Silicon Valley had started Hotmail.com, which provided free e-mail, and Henderson got the idea to offer a similar free voicemail service. He described iTalk's advertiser-supported free voicemail system:

We didn't find anybody who was doing what we were planning on doing. We were not sure whether that was good or bad. We thought, gee, the fact that nobody was doing it, maybe there is a reason why it is not being done. We tried to talk to these advertisers to see if they would be in fact interested in advertising on telephone system. Most of them didn't understand what we were talking about, because they never heard of such a thing.

By April 1999, Henderson had developed a working prototype to show to venture capitalists. By March 2000, he had received a total of $53 million in venture capital and launched free voicemail services nationwide.

iTalk's initial service featured only virtual voicemail, in which iTalk did not directly answer the subscriber's home phone. Instead, callers wishing to leave voicemail or check their mailboxes had to dial an assigned 1-800 telephone number, which connected them with iTalk's e-mail server. Virtual voicemail was relatively easy to create, because it did not require cooperation from telephone operating companies. However, the ease of creating virtual voicemail allowed other Internet startup companies to compete with iTalk by offering the same service. Moreover, the use of a unique 1-800 number proved to be inconvenient for subscribers and their callers, who needed to remember two numbers — one for the home and the other for voicemail.

These factors led iTalk's software engineers to explore ways to allow its service to answer customers' home telephones, a feature that became the biggest single reason for iTalk's competitive advantage. As Phil Pierce, one of the core developers of iTalk's proprietary software, explained:

There were a lot of companies all suddenly popped around. So it's like, ok, how we are going to be different? But we have this home answering thing. So let's push that. So we kind of like pushed aside the virtual side and concentrated on home answering.

Initially, iTalk received revenues from advertisers and allowed customers to enjoy free voicemail services. During the signup process, iTalk collected detailed information about their customers, including age, gender, telephone number, and zip code. Although this information was not released to advertising companies, the data enabled iTalk to deliver customized, targeted advertisements via the telephone, Web, and e-mail. The short ads were played just prior to customers' hearing their voicemail messages.

Figure 1 shows the configuration of iTalk's technology infrastructure. Three separate databases contained voicemail messages, customer information, and advertisements to be played. Three different servers were dedicated to interfacing with standard telephone switches, e-mail, and the Internet, respectively. The phone server allowed iTalk to answer customers' home telephone numbers and capture voicemail messages.

Figure 1. Configuration of iTalk's technology infrastructure

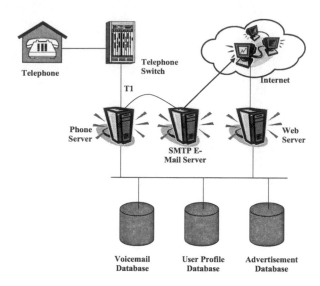

Those messages then could be distributed to customers, along with the proper advertisement, via telephone, e-mail, or the World Wide Web.

Linkages with Business Partners

In order to answer customers' home telephones, iTalk had to rely on telephone companies' provision of the call-forwarding feature of each customer's phone line at the switch level, because only telephone companies had the right to set those configurations. As it expanded throughout the U.S., iTalk needed to negotiate with each of the seven RBOCs to provision services for iTalk customers. Dennis Henderson considered those negotiations to be difficult because of the telephone companies' monopoly statuses. Although required by new laws to cooperate with companies like iTalk, the RBOCs usually were slow to meet their legal responsibilities. He said:

In telephony businesses, there is a lot of regulation. It's not just the legal laws. It is the practical reality of working with telephone companies in terms of they are very slow. They don't always follow law because they don't want to. It's not profitable.

The crucial interface between the RBOCs and iTalk made partners of two very different types of organizations. Telephone companies, their technologies, and business processes existed decades before the Internet even was imagined. By contrast, iTalk was created in 1998, about 100 years after the founding of the U.S. telephone industry. Geographically, iTalk was located in the midst of Silicon Valley, surrounded by scores of startup companies, whose success depended on Internet technology. By contrast, the telephone companies were distributed throughout the U.S., serving local communities.

Whereas iTalk worked at Internet speed, the RBOCs were notoriously slow. Telephone companies enjoyed a dominant power position due to their age, size, and protected industry status. A partnership with the RBOCs was a critical requirement for iTalk's business, but it was insignificant and undesirable from the telephone companies' perspectives, due to the potentially competing nature of the service. John Waldron, iTalk's vice president for operations, put it this way:

To a large degree, it is telephone companies that decide who they want to deal with and when they want to deal with them. It can be a problem if they don't want to deal with you. They are very large, and the decision making structure may be very fragmented, so things move a lot slower.

Given this situation, iTalk adopted a strategy of hiring people from the telephone companies. This strategy imported valuable expertise from the telephone industry, thereby increasing iTalk's ability to negotiate business agreements with the RBOCs. John Waldron himself had more than 20 years of experience in the telephone business and was directly involved in negotiations with the telephone companies. His intimate and extensive knowledge of industry regulations reduced the RBOCs' ability to discourage iTalk with misleading arguments. Waldron explained:

One of the RBOCs told me that they had a whole bunch of people asking for the same kind of things that we're asking for, but we are the only one that got it. And I said "why is that?" And he said "because the rest of them we went 'boo' to, and they ran away. Well, you guys, when we booed, you didn't run." Well, I didn't run because it was not scary to me, because I knew they were wrong, I knew they were bluffing. I knew they could not do that, I knew the laws behind it, I knew what button to push, I knew what phrases to quote, I knew what people to call.

As a result of importing such expertise, iTalk established relationships with all seven RBOCs, allowing it to deliver voicemail services nationwide. This accomplishment helped to set higher entrance barriers for its competitors. Larry Flanagan, iTalk's Web group manager, viewed this accomplishment as a key competitive advantage.

You need to deal with each RBOC around the country, all seven of them, the big ones, individually. And it took about a year and a half to do that. Well, the next competitor that comes after us, it will still take them about a year to do it. So we have a year head start over our competitors for providing that.

The linkages with the telephone companies were not restricted to the negotiation table but also included business processes for provisioning service. In order to obtain free voicemail service, an iTalk customer needed to dial from the specific telephone that he or she wanted iTalk to answer. Those telephone numbers were captured automatically from the phone line and stored in the customer database during the signup process. The same numbers were exported from the database directly into the electronic orders sent to the telephone companies. Thus, iTalk's orders were accurate, assuming that the customer signed up from the desired phone line. Figure 2 provides schematic details of the provisioning process.

Figure 2. iTalk's system for provisioning customer services

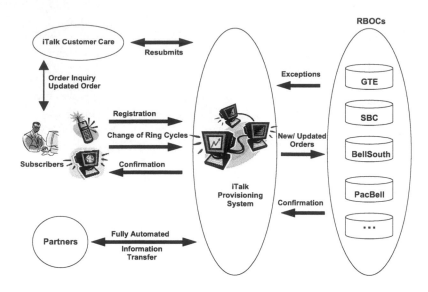

Unfortunately, when iTalk's orders were received by the telephone companies, they encountered provisioning processes that included manual data entry into 30-year-old legacy systems. Because iTalk was powerless to change the RBOC's business processes, its own provisioning was compromised in many cases. Ingrid Baum, who dealt with many customer complaints about provisioning as a member of iTalk's Operations Department, characterized the RBOCs in this way:

I am sure this is what they do: they print our spreadsheet order out, then they write the order, like type it into their machine, then they hand write down all of the order numbers and then they take the handwritten stuff, and then they fill in the spreadsheet, then they e-mail it back to me.

This produced provisioning error rates as high as 26% for iTalk's customers. Provisioning errors meant that service activation was delayed, calls often were forwarded to the wrong number, and the ring cycle was set incorrectly. Many of iTalk's customers attributed these errors to iTalk, and many cancelled their service. Non-subscribers, who were mistakenly provisioned iTalk's voicemail service, were also enraged, jeopardizing the company's reputation. June Schneider, who worked with Baum in operations, explained that provisioning errors drew the ire of people who were not even iTalk customers.

Non-customers would call us and said "you put the service on without my permission. I am going to call an attorney on this." We cannot find the number anywhere. The order typist did it wrong. It's the telephone companies' error; we aren't making it.

Although the proliferation of provisioning errors caused difficulties for iTalk, it did serve as a barrier to entry by making iTalk's business model appear less attractive to competitors. Dan Farmer, iTalk's manager of the Operations Department, explained that working with RBOCs was one thing that most potential competitors wished to avoid.

You know, the reason why we are really going to be successful at iTalk is because what we are doing isn't sexy, and it's hard work. Most people, I think, are going to give up. I mean, this is not fun stuff. This is hard, this is actual work, it's very tedious, and it's boring. But it cannot be avoided because it's in the nature of the phone company. They are a big, bureaucratic, bloated organization that was designed to allow a high level of incompetence in any given job.

Technology's Role

External linkages were enabled by computer software that was designed by iTalk's founding software engineers, which leveraged both the Internet and telephony technologies. iTalk had to receive audio messages, digitize them, store them in databases, and retrieve them when authorized customers used the service, either through telephone or through the Web. All of those steps were handled by iTalk's proprietary software installed on its phone servers and Web servers. As the primary telephony interface of iTalk, each phone server was equipped with dialogic boards that connected directly with telephone lines, while the phone server software controlled the incoming and outgoing voicemail messages. By contrast, telephone switches were highly standardized, allowing each telephone switch to operate with every other, thereby forming the backbone of the telephone network. However, a telephone switch was incompatible with computer communications. Dan Farmer characterized the computer as an alien intruding into the public switched telephone network.

In order to provide voicemail service that leveraged both Internet and telephone technologies, iTalk had to match the reliability of the telephony world. Although the telephone provisioning error rate was very high, the operating reliability of the phone network was much higher than the Internet. Standards for telephone reliability were spoken of in terms of the five 9s; that is, 99.999% reliable. By contrast, Internet reliability was lower, and users were accustomed to temporary losses of service. Dennis Henderson viewed the reliability challenge as one of the key factors that differentiated iTalk from its competitors:

It has been very challenging for us to build our phone servers in the ways that are teleco levels of reliability, not just Web level of reliability. If power goes down, our phone servers still have to work. Anytime someone presses 1 to play a message, no matter how busy the Internet backbone is, it always starts to play the message.

If problems did arise at the interface between telephony and Internet technologies, no one expected the telephone companies to modify their software to accommodate the requirements of iTalk's phone servers. As Dan Farmer sarcastically remarked:

We have situations when all of a sudden our dialogic board that is on the phone servers will knock everybody off the line. Everybody in the Engineering department is insisting it's got to be a problem with our telephone company. They have to figure out how to handle it because telephone companies are never going to change their switch software to go, "Oh nice little iTalk voicemail machines, we don't want to create a condition that causes you to faint."

The need to handle odd conditions in switched telephone networks required iTalk's software engineers to be knowledgeable about computer telephony integration (CTI). Unfortunately, the core group of engineers had joined iTalk from the shrink-wrap computer gaming industry and were relatively inexperienced in CTI. Dave Mobley, one of the original core software engineers in iTalk's phone group, explained the differences between video games and telephony:

When you work on video games, there is a kind of accepted level of instability. Because it's just a game. If it crashes every once in a while, it's not a huge deal, the person just restarts. Obviously, you don't want that to happen, but there is going to be some bugs that get through, and you are not going to fix. Whereas here, you kind of need a higher level of stability.

Because of the underlying lower reliability of the Internet, iTalk's phone servers experienced frequent crashes, which impeded iTalk's service. Customers' messages often were lost or delayed, leading to barrages of complaints to the company.

Growing Pains

After iTalk succeeded in its fund-raising efforts, it expanded its workforce from 50 to 170 within three months. Rapid growth imposed challenges on iTalk's management team. When iTalk was small, communication never had been an issue. There was only one meeting, which everyone attended every day at 11:00 a.m., and everyone knew exactly what was happening in the company. After growing to 170 people, it became nearly impossible to hold meetings that would involve everyone. Although there were still a lot of meetings, employees increasingly complained that only managers got to attend meetings and that there was no feedback regarding what happened in them. As a result, employees lacked a clear vision about iTalk's strategic plan and the future direction of the company.

Many of iTalk's original employees expressed affection for the early days at iTalk. John Waldron reminisced:

At one point, we had 50 people working for iTalk in a space that is supposed to hold maybe 12. So we had people lined up on tables, we had six people in an office that was intended for one, we had cubicles with more than one person sitting in them. We were crammed into this little tiny space. I really miss the closeness. I used to know everyone in the building.

Although iTalk was still quite young, its folklore had already developed. The founding engineers were respected as "magicians" who could make something out of nothing. Ingrid Baum expressed her admiration for Dennis Henderson and the core group of software engineers:

I could never do it. I think they are fabulous, I mean, they take nothing, and they make something. They really do. They produce something, they create this machine. My god, when we first started these guys were working everyday, all day, all night, I mean, they just lived there. And I think they did a tremendous job. And they still are.

Stories about the early days were filled with details about heroic efforts and long working days. One engineer said that he had once worked 42 hours straight in order to bring the service alive during the July 4th weekend. Another reported working 16 hours a day, seven days a week for more than two months. Sometimes due dates for projects were insane, calling for completion within two or three days.

Interdepartmental Relationships

As iTalk grew, the core engineering group remained mostly intact, while new hires came mainly from the telephone companies. The Engineering Department was responsible for the design and implementation of voicemail features that were requested by Marketing and Business Development, and the Operations Department became responsible for installing the software, configuring the network, monitoring and maintaining the system, and responding to customers. Members of the engineering and operations departments differed dramatically in their backgrounds. The engineers were primarily from the shrink-wrap software industry, whereas the operations staff was mainly from the telephone industry.

Most of the engineers were younger people who joined the company early and never had worked in a large corporate environment. Some hated the bureaucratic environment of the big companies. Jerry Gordon, who specialized in marketing and business development, captured this mindset well.

For me, going to work for a giant corporation would just be like a slow death. I think the bureaucracies in those places would just drive me nuts. I wouldn't feel like I was really contributing to anything. I'd feel like I'd be coming in at 9 leaving at 5. There's no purpose to it.

Such views were not held by the majority of people with telephony backgrounds, who referred to large corporations as well-structured, mature environments with sophisticated procedures. They felt that their experience in large corporations could contribute to iTalk's success. For example, Sarah Connor, an operations employee, remarked:

I have been in telecommunications for about 30 years; most of my career is spent in large corporations. Because I had spent so many years in a large corporation where there was a lot of structure, I thought that I had learned quite a bit that would really help, and would be transferable to a small company.

Engineering and operations also had conflicting views on issues, such as internal procedures and documentation. On the one hand, most of the software engineers felt that documentation was an impediment to creative work and irrelevant in an environment where requirements changed so rapidly. On the other hand, employees from large corporate backgrounds viewed lack of documentation as a serious problem that could waste resources. This latter view was expressed by Marty Coleman, who had spent most of his career in telephone companies prior to joining iTalk's Quality Assurance (QA) Group:

I would never work for startup again. It's insane. It's a waste of manpower and resources. Either nothing is documented, or, if something is documented, it is not constantly updated. When you document, when you work with other departments and groups, you are passing your knowledge and wisdom off to a large group of people. By doing that, business-wise, you are not relying on one person.

The difference between engineering and operations was mirrored by the difference between the product orientation of the shrink-wrap software industry and the service orientation of the telephone industry. According to John Waldron:

In a product company, if you are shipping a package of software, and you say "operations," you think of the guy who decides what the fitness of the paper is, and which shrink-wrap you are going to do. In service businesses, the operation is the big factor. If you cannot deliver the service consistently, you are in big trouble.

Although they respected the development engineers, people working in operations viewed their own roles as indispensable to company success and thought that they deserved more respect. Erik Morgan worked in the Technical Support Group of iTalk's Operations Department. He expressed the view that:

The only people in this company that really know the true flavor of the product are in operations. Engineers and everybody else, they don't deal with it on the day-to-day basis. They are not bombarded with complaints from customers. I think 99% of this company is based on Operations just to make it work. I don't think anybody is working as hard as operations.

The differences between engineering and operations made communication and coordination between departments more difficult. These problems were aggravated by the technologies used by engineers to develop software products and the production system that the Operations Department supported and maintained. Operations people viewed the cause of service interruptions as software problems, not hardware failures. As Bud Potter, another member of the Technical Support Group, explained:

When it comes to whether our outages are due to hardware or software, 95% of them are self-inflicted software problems. Self-inflicted meaning, our engineers gave us code, the code doesn't work. Or we did an upgrade and that stuff doesn't work.

In operations' view, iTalk's computers did not function properly, because software was released from engineering prematurely. For Erik Morgan, this meant constant pressure to keep the system up and running:

You are literally babysitting all the computers. You have problems every hour; you are rebooting a server every hour. It's just like a bounce cycle that never stops. It's like there are so many people at war with it that they constantly need to be up and down with the machine, their lives are tied with the machine. So it's not 9 to 5, it's not 8 to 10, it's 7 by 24, 365 days a year.

A major factor that contributed to the software bugs was time pressure. Some of the early software projects lacked both elegance and documentation, making them failure-prone and hard to fix. Don Braxton, one of the original members of the engineering group, explained why some of their work might experience problems:

We did whatever we could to get things done, even though the code may be ugly or something. We didn't have time to think. We always thought after we get this done, out of the door, and working a little, we could rewrite the product, the whole program. Make it better, stable and everything, and we never had that chance.

The problem of software stability was aggravated by problems of scalability. Although iTalk's software had worked perfectly under lower load conditions, service reliability issues surfaced after acquiring 1,000,000 subscribers, because a higher load threshold was passed. As Dave Mobley explained:

Our code doesn't have a lot of mileage and experiences done on it. Also, as we're doing this, our customer base keeps growing. So our load keeps going up. Every time our load builds up a little, you know, we uncover some problems that we never knew about.

Minimal testing was common practice for many companies using the Internet as a delivery platform, because it was relatively easy for them to correct mistakes by releasing updates and patches through their Web sites.

It was also the case that a truly valid test of iTalk's software was only possible under live conditions. This was because it was impossible to simulate the crucial interface between iTalk's Internet technologies and the switched networks of the telephony world. As a result, Quality Assurance (QA) was not able to test the software properly before it went into service. Thus, customers were the first people to encounter software bugs and complain to operations. As the number of updates and bug fixes coming from engineering increased, QA also had little time even to attempt testing. Marty Coleman described how the pressure to release software fast undermined effective testing:

A different version of phone server code, at least once a day, is given to the QA engineers. They were often told by the VP of engineering "you have an hour to test this build." So you want to know why there is no relevance to the QA engineer group here? Because you cannot test anything in an hour, in a day, in a week, thoroughly, accurately, precisely, and correctly. With that said, we now know why QA engineering is a joke — time constraints.

The Shifting Business Model

As iTalk grew, it shifted its business model to attract revenue from premium services that were billed to customers rather than deriving revenue from advertising. The shift was necessary, because advertising revenues were not sufficient to ensure profitability. Shifting the business model required significant reconfiguration of iTalk's technical infrastructure and the creation of reliable new features that targeted customers would pay for. Dennis Henderson wanted these changes to happen rapidly, which placed even greater pressure on the strained relationships between the engineering and operations departments. According to Dexter Rollins, who worked in the Technical Support Group of operations:

Marketing is going "oh, get more features, get more features," but then operations are going "no, no, no, we are not that stable." From beginning till now, we have not really tried to stabilize the system; it is always the new features — features, features, features! I don't think that is the appropriate way to do it. Because eventually, the more features you add on an un-stabilized product, these will come back and bite you.

With the new emphasis on product development, some engineers felt that their heroic efforts to develop software for the former business model were wasted. They had emotional attachments to the software they had developed and did not appreciate seeing it die. Anita Scott, who had worked in the phone group of the Engineering Department, commented:

I looked at the code I wrote last summer, and they are not used right now, like the whole idea of what we were doing has been thrown out. This one piece of code that I wrote, they just decided to use a different technology, so look back, about a period of like three months is kind of wasted.

SUMMARY

In July 2001, Dennis Henderson reflected on the mounting problems of running iTalk. He knew that there were several avenues that required attention. After grabbing a late supper at iTalk's cafeteria, he returned to the board room alone and jotted some questions on the white board:

- **How can we improve the external relationships?** iTalk's virtual organization model depended heavily on external partners. Yet, it seemed to Henderson that the big telephone companies were far less dependent on iTalk than iTalk was dependent on them. How could the problems of provisioning and service reliability be solved? Were there other partners that iTalk could turn to?
- **How can we resolve conflicts between operations and engineering?** Matters had gotten worse, and it seemed to Henderson as though two different cultures had developed at iTalk. How could these groups be made to work together more effectively?
- **How can the software be made more reliable?** Henderson increasingly was concerned with the frequent server crashes that interrupted customer service. What could be done to make the service more stable while still adding attractive new features to serve customers better?
- **How can we change the business model to make more revenue?** When iTalk's service was free, customers appeared willing to use it and even tolerate some service interruptions and, of course, the targeted advertising. But what kind of

services would customers really be willing to pay for? What other sources of revenue could be found to cover the growing costs of running the company?

As he sat down and pondered these questions, Henderson was well aware that iTalk already was targeted for acquisition by a much larger and more established company. Although he had envisioned iTalk as being built to last instead of built to sell, he wondered if the end of iTalk as a separate business was imminent. The price offered to iTalk was expected to be in the range of $20-25 million, less than half the amount invested in the company to date. Nevertheless, he mused, selling the company at this point might be the best course of action. But if that happened, what would happen to the people he had brought into iTalk? Surely, their stock options would be worth little after a takeover. Would this be a fair reward for their personal investments of time and energy to get iTalk up and running?

Henderson sighed, got up, and went to the white board again. This time he wrote a single word — People?

REFERENCES

Barua, A., Pinnell, J., Shutter, J., & Whinston, A. (1999). *The Internet economy: An exploratory study* (Research Report). TX: University of Texas at Austin, Center for Research in Electronic Commerce.

Mowshowitz, A. (2002). *Virtual organization*. Westport, CT: Praeger.

Ritter, J. R., & Welch, I. (2002). A review of IPO activity, pricing, and allocations. *The Journal of Finance, 57*(4), 1795-1828.

Sohl, J. (2003). The US angel and venture capital market: Recent trends and developments. *Journal of Private Equity, 6*(2), 7-17.

webmergers.com. (2003). *Report: Internet companies three years after the height of the bubble*. Retrieved June 2, 2003, from http://www.webmergers.com/data/article.php?id=67

APPENDIX

Table 1. Typical plan for compensating employees with stock options

Working Period	Stock Options Gained/Month	Total Stock Options Gained at End of the Period	Value to Employees at $1/Share
1st – 11th month	0 share	0 share	$0
12th month	2,500 shares	2,500 shares	$2,500
13th – 48th months	208 shares (1/48th of 10,000)	10,000 shares	$10,000

Daniel Robey is professor and John B. Zellars chair of information systems at Georgia State University. He earned his doctorate in administrative science from Kent State University. Professor Robey is editor-in-chief of Information and Organization *and serves on the editorial boards of* Organization Science, Academy of Management Review, *and* Information Technology & People. *He is the author of numerous articles in such journals as* Management Science, Organization Science, Information Systems Research, MIS Quarterly, Journal of Management Information Systems, Academy of Management Review, *and* Academy of Management Journal. *His research includes empirical examinations of the effects of a wide range of technologies on organizations. It also includes the development of theoretical approaches to explaining the consequences of information technology in organizations.*

Leigh Jin is an assistant professor in the Information Systems Department at San Francisco State University. She earned her doctorate in computer information systems from Georgia State University in 2002. She received her MBA and BBA in management information systems from Beijing University of Aeronautics and Astronautics. Her recent research interests include open source software adoption and use, e-business, and virtual organization/community.

This case was previously published in the *International Journal of Cases on Electronic Commerce*, 1(3), pp. 21-36, © 2005.

Chapter II

The Value of Coin Networks:
The Case of Automotive Network Exchange®

Andrew Borchers, Lawrence Technological University, USA

Mark Demski, Lawrence Technological University, USA

EXECUTIVE SUMMARY

As a response to strong competitive pressures, the U.S. automotive industry has actively employed Electronic Data Interchange in communications between suppliers and carmakers for many years. This case reviews the recent development of ANX®, a COIN (Community of Interest Network) intended to provide industry-wide connectivity between carmakers, dealers and Tier suppliers. The authors identify technical and business challenges to the success of ANX®.

BACKGROUND

During the past 20 years the U.S. automotive industry has gone through significant change, heightened competition and increasing globalization. The industry can be characterized as a small number of manufacturers (Ford, GM, DaimlerChrysler, and Japanese and European transplants) that obtain automotive components from several thousand part suppliers. These manufactures then sell their products through a network of thousands of independent dealers. Through the 1970s, 1980s and 1990s the industry has gone through wrenching changes as it faced the challenge of globalization and significant over-capacity. Two of the major strategic efforts made by U.S. manufacturers include the increased use of parts suppliers (so called "outsourcing") and an increase in the use of electronic data interchange (EDI) to facilitate communication between trading partners.

The suppliers that provide parts to the auto industry are categorized in a "Tier" structure. Those that deliver parts directly to a manufacturer are categorized as Tier 1 suppliers. Tier 1 suppliers, in turn, receive parts from a network of Tier 2 suppliers. Based on automotive industry estimates, there are approximately five thousand Tier 2 suppliers supplying a few hundred Tier 1 suppliers. The Tier 2 suppliers receive additional parts and service from Tier 3 suppliers bringing the total population to the tens of thousands of firms worldwide.

The supplier industry has evolved over the years. Earlier in the century, manufacturers largely took on the responsibility of creating their own components. Over time, manufacturers have migrated toward using outside suppliers. In the past the supplier community created individual components, typically for a single manufacturer. Now, due to industry consolidation, suppliers find themselves doing business with more than one manufacturer and supporting operations on a global basis. Further, manufacturers expect suppliers to engineer and manufacture entire sub-assemblies delivered "just in time" and sequenced for immediate assembly, rather than shipping individual parts. Suppliers face strong price competition from their peers and on-going expectations from manufacturers to lower their cost and improve their quality.

SETTING THE STAGE

Beginning in the 1970s and 1980s, manufacturers introduced the concept of Electronic Data Interchange (EDI). Each of the carmakers created a proprietary network and required their major suppliers to connect to this network. Since suppliers typically focused on a single manufacturer, they could standardize on whatever single platform was used by this manufacturer.

With suppliers changing to supply multiple carmakers, they had to maintain duplicate data connections to network with the various manufacturers or Tier 1 suppliers they did business with. For example, a single supplier may have a dedicated point-to-point data connection to Manufacturer A, a high-speed modem connection to a Tier 2 supplier, and a Frame Relay data connection to Manufacturer B. In addition, different applications (e.g., CAD/CAM or mainframe inventory systems) mean "a supplier may have a requirement for multiple connections to the same manufacturer based on different applications" (Kirchoff, 1997). Suppliers, or in some cases the carmaker they are supporting, are responsible for installing and maintaining these data connections and the hardware (such as modems and routers) necessary for the connections.

The auto industry established the Automotive Industry Action Group (AIAG) to create standards for the exchange of information between industry partners. Historically, AIAG's focus has been on application level standards. Known as "transaction sets," AIAG standards for various business documents, such as purchase orders or advanced shipping notices, simplified EDI for the industry. However, AIAG's focus at application level standards did not address lower level connection issues.

As one would suspect the more data connections a firm maintains the higher the cost to the supply chain. These costs ultimately add to the cost of parts. Members of AIAG analyzed several business processes that contribute to the manufacturing of automobiles. AIAG determined that the most addressable business process was electronic information exchange and the cost of the data connections throughout the automotive chain. Members of AIAG costed out the various data connections that a typical supplier needed to maintain. By reducing the number of data connections, AIAG estimated that the cost of manufacturing an automobile could be reduced by $71 a unit. (AIAG, 1998).

This potential savings comes on top of a documented history of cost reductions in the automotive industry through the use of EDI. Mukhopadhyay (1995) conducted an ex post facto study at Chrysler over a nine-year period. In this research the author identified savings of $60 per vehicle in manufacturing and logistics costs from the use of EDI. Further, Chrysler saved an additional $40 per vehicle from electronic document transmission and preparation.

The benefits of EDI, unfortunately, have mostly been to the carmakers, not to suppliers. In a survey of 250 Tier suppliers, AIAG found that 95% of the respondents could not identify a single advantage to their firm from the use of EDI (Wallace, 1998). Suppliers comply with carmakers EDI requirements as a condition to remain in business. Most, however, have not determined how to make EDI work to their benefit.

CASE DESCRIPTION

With application level standards firmly in place, the question of a common communications network and protocol came to the forefront in the mid-1990s In 1994, the AIAG trading partners defined and published a document entitled "Trading Partner Data Telecommunications Protocol Position." This publication recommends the data network protocol TCP/IP (Transmission Control Protocol/Internet Protocol) as the standard for transport of automotive trading partner electronic information. A year later Chrysler, Ford, and General Motors endorsed TCP/IP as the standard protocol suite for inter-enterprise data communications among automotive trading partners (www.aiag.org, 1998).

The concept of using TCP/IP for inter-enterprise communications came after a number of firms found TCP/IP useful within their organization, but had trouble using it between enterprises (www.aiag.org, 1998). In response to this experience, the AIAG Implementation Task Force developed ANX® as a single, secure network.

ANX® Design Alternatives

There were several options considered for the design of ANX® (AIAG, 1998). One was the implementation of a private network, which would entail installing private point-to-point data circuits between all of the trading partners in questions. Point-to-point networks, however, are not very scaleable, especially if one requires a fully meshed network where each node is connected to every other node. For example, a 5-node network would need 10 circuits to interconnect all nodes. In general, the number of connections is a function of the number of nodes (N) and can be calculated as N*(N-1))/

2. To interconnect 2,000 Tier 1 suppliers, that would require 1,999,000 private line connections. The cost of such a network as a whole would be very high.

Another option was to use the public Internet. A number of firms have created "extranets" to connect themselves with trading partners. Such networks utilize the public Internet with encryption to ensure privacy. The key problem, however, with using the public Internet in the automotive industry is response time and reliability. Indeed, with just-in-time inventory systems, carmakers often communicate part requirements with only a few hours of lead time. A missing or late transmission could shut down an assembly plant, with attendant costs of $1,000 per minute or more.

The AIAG considered the public Internet, realizing that it offered great flexibility. AIAG felt the quality of service, however, would be "completely unpredictable because no service provider can offer guarantees beyond its own network borders" (AIAG, 1998). They also noted "a vast majority offer no meaningful guarantees even within their own networks." Indeed, most of the connections between Internet Service Providers (ISPs) are at public peering points such as Metropolitan Exchange Areas — MAE East and MAE West. These points have a reputation of congestion due to the explosive growth of the Internet. Indeed, carmakers did not want to see production impacted by unusual Internet loads, such as come with the outbreaks of viruses.

Virtual Private Networks (VPN) were another option. Such networks are often implemented using Frame Relay or ATM. The AIAG felt that this solution improved security and reliability, but believed "this model would require the industry to choose a single provider, giving one company monopoly powers, with little incentive to enhance capabilities or keep prices competitive over time." (AIAG, 1998).

The fourth option is what the AIAG referred to as the Automotive Network eXchange (ANX®) model. Under the ANX® model, multiple service providers would be certified to transport data traffic, subject to strict technical requirements on accessibility and packet loss. AIAG requires these providers to interconnect with each other. AIAG felt this would foster competition, yet maintain the flexibility of the public Internet. The AIAG made the decision to move forward with the ANX® model in 1995.

Certification standards cover eight areas, including: (1) network service features, (2) interoperability, (3) performance, (4) reliability, (5) business continuity and disaster recovery, (6) security, (7) customer care, and (8) trouble handling. The ANX® Overseer must certify that a potential provider is 100% compliant in order for them to become certified (www.aiag.org, 1998).

The ANX® model is an excellent example of a Community Of Interest Networks (COIN). COINs do not typically have a single carrier providing data transport circuits. Instead, the trading partners in a COIN usually have the option to choose from a select list of carriers that meet the requirements that the user community set forth. The administrator of a COIN is typically an industry interest group, such as the AIAG, made up of delegates from companies in that particular industry or community. This administrator is the catalyst to bring users onto the COIN network and to ensure that the best business interests of the user population are addressed.

Comparing the difference between the emerging COIN and private or virtual private networks is important. By selecting a COIN, AIAG focused on maximizing connectivity between partners while ensuring high quality connections and ensuring communication firms cannot monopolize network connections. In return for this benefit, member firms turn over control of their networks to an industry group. Alternatively, if individual carmakers employ a VPN, they can exercise much greater control of the network and ensure that the network is optimized for their needs. However, a firm that employs a VPN must periodically put the VPN out to bid if they want to ensure competitive pricing.

Network Operation

To better understand how ANX® works, consider the network diagram included in Appendix 1 and the relationships between the entities:

- **Automotive Network eXchange Overseer (ANXO):** This company has direct operations and management responsibilities over the ANX® service. They are under contract to the AIAG.
- **Trading Partner (TP):** These firms are the actual end users of the ANX® network. They include automotive component suppliers (such as Lear, Dana, Goodyear, Delphi) and manufacturers (DaimlerChrysler, Ford, General Motors).
- **Certified Service Provider (CSP):** These firms comprise the telecommunications and IS companies that meet certain network performance and trouble reporting criteria set forth by the AIAG and ANXO. Examples include EDS, MCI and Ameritech.
- **Certified Exchange Point Operator (CEPO):** This company manages the Asynchronous Transfer Mode (ATM) switch that will interconnect all of the Certified Service Providers.
- **IPSEC Function:** IPSEC is a highly secure data encryption algorithm that will keep the information in IP Packets scrambled as they leave the TP's location. At each TP's site a separate piece of hardware is needed to encrypt the data.
- **Certificate Authority Service Provider (CASP):** This company provides "electronic certificates," adding another level of security. Certificates, based on an IP addresses, guarantee that a TP is actually who they say they are. An example of this occurs whenever one downloads any software from Microsoft's Web site. A certificate appears confirming that one is downloading from an actual Microsoft Web site.

At a high level, the ANX network works in this fashion. Trading Partners connect through a Certified Service Provider to an exchange point. The exchange point will switch the data traffic appropriately to the destination Trading Partner. Note that in Appendix 1, the line drawn to the public Internet and other ISPs is not possible due to the inability of keeping public Internet traffic from "leaking" onto the ANX® network and visa-versa. As a result, the ANX® network is a separate network using IP and is not connected to the public Internet. ANX® TP's can only electronically communicate to other ANX® TP's via the ANX®. If an ANX® TP wants to electronically communicate with a non-ANX® TP (for example, a Japanese or European carmaker), then the data will not traverse the ANX® network at all. Instead, the TP will have to maintain a separate connection to the non-ANX® TP.

Implementation

The first stage was to pilot the proposed architecture and learn from a working model. AIAG launched the pilot in September of 1997. It consisted of three CSPs and approximately 30 TPs. These firms uncovered several issues during the pilot phase. Among the issues were trouble resolution procedures and the actual cost of ANX® when rolled out for production. There were several vendors providing IPSEC encryption devices, but participants identified serious interoperability problems. There was no Certificate Authority Service Provider during the pilot. Finally, in September of 1998, AIAG launched ANX® in production, over three years after its official inception.

As of this writing, there are reportedly technical interoperability issues between the IPSEC vendors and the Certificate Authority company. Costs are a serious concern. One effort to reduce costs, a proposal to permit dial-up access to ANX®, has not been accepted due to packet-loss requirements. Carmaker support for ANX® remains strong, but Tier suppliers and communication vendors are less certain of ANX®'s success.

CURRENT CHALLENGES/PROBLEMS

A successful implementation for ANX® will require resolution of a number of problem areas. Among these are:

- Making "Co-opetition" work
- CSP Costs and Profit Potential
- TP Costs
- Does ANX® have a viable business model?
- Technical direction
- ANX®'s place in a firm's IT architecture

"Co-Opetition"

As indicated earlier, one major concern of the AIAG was that they did not want to give any single network carrier "monopolistic" powers over price in this network. Hence the ANX® model was that of "co-opetition." The carriers, especially in the pilot, needed to simultaneously cooperate and compete with each other. By "cooperation," they had to agree to interconnect with each other in a manner unlike any other interconnection agreement, via the ANX® exchange. During the pilot, "cooperation" was tremendous between all of the carriers because the initial quality of service parameters set forth by the ANXO required significant network upgrades and investments by the carriers. By "compete," each carrier's rate structure will be published publicly, creating immediate competition. Given the high level of standardization, this model will "comoditize" ANX® services. The question is "Will vendors participate in this business arrangement, both in a competing and cooperating fashion?"

CSP Costs

As noted above there are eight areas that a CSP has to comply with to become certified. These metrics include carriers developing a specific trouble ticket format, meeting strict packet-loss metrics, developing a specific billing format, and creating a dedicated network management infrastructure. In short, there is a significant upgrading of infrastructure that the carriers must invest in. How much of that cost is passed on to the trading partner depends on the size of the trading partner population and the level of competition between CSPs. Determining the true size of the ANX® community and factoring in the "commoditizing" of the network is a real concern for CSPs. Given the fact that CSPs will have little to differentiate themselves on in providing highly standardized services, they could be looking at very thin profit margins. Early estimates by carriers showed an increase in the cost of a standard connection that the supplier community is already using of three to four times.

TP Costs

The costs that a Trading Partner incurs are also a point of concern. Prior to implementing ANX®, TPs face a variety of costs for EDI-related services that are not always easily identified and aggregated. Many suppliers, for example, have a collection of point-to-point circuits (and related modems and routers) to specific assembly plants and engineering facilities. The supplier typically pays for these circuits. However, in some cases these are treated as an extension of the assembly plant's infrastructure and the cost of these circuits are paid by the automakers. Communications between suppliers and automakers also take place over X.25 or dial-up services.

With the arrival of ANX® came the potential to reduce this redundant collection of data circuits and replace it with a single connection. There are, however, a number of new cost elements to be considered. Overall, AIAG documentation (1998) shows ANX® to be a cost savings proposition for different sized trading partners. Not all participants have seen the situation this way, however. Indeed, AIAG's case is dependent upon a trading partner being able to centralize their data communications through a single connection. Given the geographic spread of some Tier suppliers' organizations, this may not be feasible. Further, it may lead suppliers to "back haul" data from remote locations to a central point, creating a load on their internal networks.

First, there are costs charged by the ANXO and the Exchange Point Operator to fund the management and oversight of the entire process. It is understood that there are start-up and on-going costs involved with all project initiatives. Indeed, the population that

Administrative cost for a 64 Kbps (DS0) connection

	Payment Destination	Payment Frequency	Payment Amount
ANX ® Registration Fee	AIAG	Annually	$400
ANX ® Subscription Fee	ANXO	Annually	$1,300
ANX ® Assessment Fee	ANXO	One-time	$500

Administrative cost for a 1.544 Mbps (T1) connection

	Payment Destination	Payment Frequency	Payment Amount
ANX ® Registration Fee	AIAG	Annually	$1,600
ANX ® Subscription Fee	ANXO	Annually	$4,800
ANX ® Assessment Fee	ANXO	One-time	$2,300

Administrative cost for a 3.0 Mbps (2 T1s) connection

	Payment Destination	Payment Frequency	Payment Amount
ANX ® Registration Fee	AIAG	Annually	$3,600
ANX ® Subscription Fee	ANXO	Annually	$10,900
ANX ® Assessment Fee	ANXO	One-time	$2,900

is supposed to benefit from ANX®, the trading partners, bears the cost for these administrative costs. These costs include fees to become a certified trading partner, and recurring costs to maintain connections to the ANX® network.

One-time costs, as identified by AIAG (1998), include subscriptions, IPSEC software, training and systems integration. AIAG estimates show these costs to run from $10,000 to $24,000, depending on the size of the TP and the number of connections the TP needs.

In addition to these one-time costs, there are recurring costs. To get an idea of the fees the ANXO and AIAG charge the TP's per connection consider the ANXO Fee Schedule for ANX® Release 1. Note that these costs are above and beyond the cost of a data circuit from a CSP.

Beyond these fees, trading partners have to secure data circuits from a CSP. AIAG estimates put this cost at $2,000 per month for a 56 Kbps connection and $4,000 per month for a T-1 connection. In practice, CSPs provide bundled services that include routers, management, and administrative costs.

Although ANX® may in time help suppliers eliminate a myriad of redundant data circuits, these costs appear high to many Tier suppliers. By way of comparison, typical T-1 connections to the public Internet cost significantly less. Costs of $500 per month for a T-1 circuit and $800 for an ISP connection are not uncommon.

A Sustainable Model?

As is common with new technology, there will be a flurry of "early adopters" that will embrace new technologies. If a product is viable, then a majority of the population will adopt the product after they see they reach a certain comfort level with it and find a reason for it. With high technology, especially data network technology, new technologies are introduced sometimes before the comfort levels are reached with the general population. This may be because the early adopters are merely test-driving products. One of the root problems with Automotive Network eXchange® is the possibility that the architecture and fee structure will render this an expensive and relatively short-lived initiative. The challenge for all participants in ANX®, particularly those that have to invest significant dollars, is simply "Will there be a return for my investment?" and "Will this initiative last?"

So far, the automakers have remained strongly behind the ANX® network. Tier suppliers have little choice but to comply as dictated by the automakers. CSP responses are particularly noteworthy, however. Some firms invested the minimum to be involved with ANX®, creating only a single network entry point into the ANX® network. Other CSPs fully committed themselves by making their entire network support ANX®.

Technical Choices

One of AIAG's major challenges is what technology to use in future generations of ANX®. The choice of TCP/IP in 1994 was a visionary move. The challenge is what sort of network to implement ANX® over. Each of the choices noted above (dedicated circuits, the public Internet, a VPN or a COIN) has strengths and weaknesses.

The original decision to go with a COIN, made in mid-1990s, may not be appropriate today. The telecommunications industry is on the cusp of a tremendous cost and reliability revolution. In the coming years several "second generation" telecommunications companies: Qwest, Level3, and IXC, will appear online. These networks do not use the old circuit-switching technologies currently used by the current long distance carriers (such as AT&T, MCIWorldcom, Sprint). Instead, they are based on the native IP routing protocol. This is the same protocol that the ANX® is standardized on. These new firms are buying high capacity next-generation Lucent fiber optic technologies that have unprecedented reliability and packet-loss statistics. These next generation carriers may meet the ANX® network throughput metrics with little or no costly upgrades.

These next generation firms will likely use private-peering arrangements as traffic volumes render public-peering points inadequate. Robust private-peering agreements already exist today between the large Internet providers. Private-peering agreements will increase to accommodate the huge data pipes being employed to ensure delivery and reliability of IP Packets. Again, the private automotive exchange point model may no longer be necessary to achieve high qualities of service. Indeed, ANX® service levels that seem so high today, may become the norm in the future.

Another technical issue is the relationship of ANX® and the public Internet. Almost all of the suppliers that do business electronically have some sort of connection to the Internet. Indeed, the larger suppliers have high speed dedicated connections to the Internet. Under the ANX® model TPs are unable to consolidate ANX® and Internet traffic. Even if such consolidation was permitted, the routing characteristics of the Internet could lead public Internet traffic to "contaminate" the ANX® network through the exchange point. There is also a less controllable problem with cross-contamination of data packets. This resides at the trading partner's site itself. Manufacturers and larger suppliers have already made their IP-enabled applications accessible via the Internet. Tremendous investment has been made in configuring their servers with adequate firewall and encryption devices to protect their own IS structure. If a trading partner were to open a server to the ANX® network and advertise the same IP addresses that are used on the public Internet domain name services, they would in turn become an exchange point between the ANX® and the public Internet. The only solution so far is to have the trading partner invest in an entire duplicate set of servers with different "ANX®" IP addresses to mirror their Web-enabled applications.

ANX®'s Place in the Firm

The IS professional in the automotive industry is faced with an interesting situation today. An increasing number of applications are becoming IP enabled, and their industry interest group (AIAG) is advertising an IP solution that will be a connectivity cure-all

(Steding, 1998). There is a tremendous amount of analysis that needs to be done to justify migration from one's current IS structure. The questions that need to be asked include:

What IP-enabled applications does the firm have? Do these run on the public Internet or ANX®?

In interviewing members of manufacturing and the supplier community, it is clear that there are several applications that can benefit from an IP scenario. These include e-mail, FTP File transfer of CAD/CAM diagrams, and Web-enabled EDI. There are also plans to put real-time production and scheduling applicications on to ANX®. The question is should firms operate on dual TCP/IP networks-namely, ANX® and the public Internet? Or should they focus on a single TCP/IP network?

What physical locations does the firm need an ANX® connection to?

Larger suppliers have also invested heavily in their own internal WANs and VPNs. Does it make sense to install ANX® connections at all of a firm's locations? Or should all traffic be collected to a central point and then transferred to ANX®? It may not make economic sense to purchase several connections. However, a company's WAN architecture may or may not be adequate to "back-haul" the information to the locations that need it. If inadequate, firms may have to upgrade their internal WANs. Does this defeat the purpose of having an ANX® network?

Does the firm want users to access servers from the Internet also?

Of the existing IP applications one hosts, it will be necessary to purchase new hardware and software to duplicate these applications onto the ANX® network. Another proposed solution to address the Internet/ANX® incompatibility is to allow one leg of the round-trip (typically return trip) made by the ANX® IP packet to traverse the public Internet. Does this defeat the purpose of the current ANX® architecture?

What are the current and potential future costs of connections?

Costing the migration from existing to ANX® connections can be complex. One must determine if he is maintaining several different connections and aggregate the costs of existing connections. Next, one will have to obtain costs of ANX® connections from the list of CSPs and add the administrative costs listed above. Further, there may be other hardware needed and possible additional costs from a system integrator (i.e., Compuware, Anderson Consulting) to do the work if a TP does not have the staff internally. Only then can the decision be quantified from a cost perspective.

What bandwidth is necessary?

Probably the most alarming business issue with the ANX® business model is the disincentive for high bandwidth. As illustrated in the fee structure, there is more overhead associated with higher bandwidth. This does not reward users for putting more and more applications onto the ANX® network.

What are the business/political costs associated with ANX®?

Sometimes the decision to move forward on new initiatives is not made to meet an organization's own business needs and objectives. Suppliers must determine their customers (e.g., carmakers) and executive management's stance on the ANX® issue. Carmakers will likely not mandate ANX® connections across the board for existing supplier relationships. Instead, the carmakers plan to make ANX® a requirement in the specifications for new business opportunities. A supplier's analysis will need to go beyond the dollars and cents of the migration, and factor in the cost of future business opportunities.

SUMMARY

The discussion above has pointed out a number of challenges in implementing ANX®. It is important to remember some of the positives. First, the AIAG determined that the automobile supply chain will achieve cost reductions through standardizing on a single protocol. The protocol selected (TCP/IP) is a wise decision based on the acceptability and availability worldwide. Another major contribution was getting several firewall vendors (including Checkpoint and Vanguard) to inter-operate based on an agreed-upon standard (IPSEC).

The major challenge for the automotive industry is to determine how ANX® can change over time to take advantage of the latest technical advancements and changes in the industry. The AIAG designed the ANX® architecture to address the technological shortcomings of the mid-1990s. It has recently launched this architecture into production, and there have been no changes to take advantage of major improvements in processing and fiber optic speed. When new generation carriers come online with newer, lower cost technology, the cost structure will totally change due to their use of native IP architecture rather than legacy circuit switching technology used today. It may become even more difficult to justify the costs of being on the ANX® network when the data connections used today reduce in cost.

REFERENCES

AIAG Board of Directors. (1998). *The business case for ANX® Services*. Retrieved from http://www.aiag.org

Kirchoff, D. (1997, November). ANX — Making the connection. *Actionline Magazine*.

Mukhopadhyay, T., & Kekre, S. (1995). Business value of information technology: A study of electronic data interchange. *MIS Quarterly, 19*(2), 137-156.

Steding, P. (1998, April). Connectivity cure-all. *Actionline Magazine*.

Wallace, B. (1998). Suppliers slow to profit from EDI. *Network World, 5*(6), 1-8.

www.aiag.org. (1998, November).

ENDNOTE

Automotive Network eXchange® and ANX® are registered in the U.S. Patent and Trademark Office as service marks by the Automotive Industry Action Group (AIAG).

APPENDIX 1:
ANX® NETWORK ARCHITECTURE

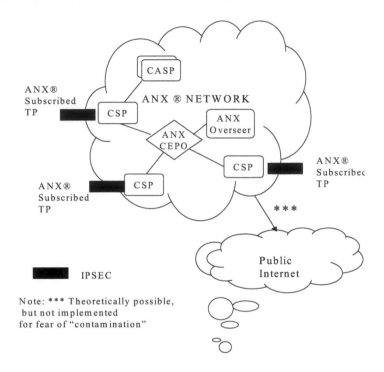

This case was previously published in the *Annals of Cases on Information Technology Applications and Management in Organizations*, Volume 2/2002, pp. 109-123, © 2002.

Chapter III

Up in Smoke:
Rebuilding After an IT Disaster

Steven C. Ross, Western Washington University, USA

Craig K. Tyran, Western Washington University, USA

David J. Auer, Western Washington University, USA

Jon M. Junell, Western Washington University, USA

Terrell G. Williams, Western Washington University, USA

EXECUTIVE SUMMARY

On July 3, 2002, fire destroyed a facility that served as both office and computer server room for a College of Business located in the United States. The fire also caused significant smoke damage to the office building where the computer facility was located. The monetary costs of the disaster were over $4 million. This case, written from the point of view of the chairperson of the College Technology Committee, discusses the issues faced by the college as they resumed operations and planned for rebuilding their information technology operations. The almost-total destruction of the college's server assets offered a unique opportunity to rethink the IT architecture for the college. The reader is challenged to learn from the experiences discussed in the case to develop an IT architecture for the college that will meet operational requirements and take into account the potential threats to the system.

ORGANIZATIONAL BACKGROUND

Western University[1]

Western University (WU) is a public, liberal arts university located on the west coast of the United States. The university's student enrollment is approximately 12,000. WU focuses on undergraduate and master's level programs and is comprised of seven colleges, plus a graduate school. WU receives much of its funding from the state government. The university has earned a strong reputation for educational quality, particularly among public universities. In its 2003 ranking of "America's Best Colleges," *US News & World Report* ranked Western University among the top 10 public master's-granting universities in the United States. The university community takes pride in WU's status. According to Dr. Mary Haskell, President of WU, "Continued ranking as one of the nation's best public comprehensive universities is a tribute to our excellent faculty and staff. We are committed to maintaining and enhancing the academic excellence, personal attention to students and positive environment for teaching and learning that has repeatedly garnered Western this kind of recognition."

College of Business Administration

The College of Business Administration (CBA) is one of the colleges at WU. CBA's programs focus on junior and senior level classes leading to degrees in either Business Administration or Accounting. In addition, CBA has an MBA program. About 10% of the students at WU (1,200 FTE) are registered as majors in the CBA. Each year, CBA graduates roughly 600 persons with bachelor's degrees and about 50 with MBA degrees.

CBA has four academic departments — accounting, decision sciences, finance and marketing, and management — each of which has about 20 full-time or adjunct faculty members and an administrative assistant. Other academic and administrative units for the college are the college office, three research centers, and the CBA Office of Information Systems (OIS) — about 20 persons total.

Organizational Structure of Information Systems at Western University and CBA

The organizational structure of information technology (IT) support services at WU includes both centralized and decentralized units. Figure 1 shows a partial organizational chart for WU that depicts the different groups and personnel relevant to the case.

WU Information Systems

WU's Office of Information and Technology Services (OITS) group is a large centralized organization that provides university-level services to the WU administration, the seven colleges, and students. OITS is headed by the Vice-President for Information Systems Services Ken Burrows and is organized into three areas: User Support (US), Administrative Services (AS), and Information Technology Services (ITS). OITS has offices in Thompson Hall, which houses the WU administration offices, and the Administrative Center located approximately one mile off campus on 23rd Street (and more often called "23rd Street" than by its proper name of "Administrative Center").

Figure 1. Partial organization chart for Western University

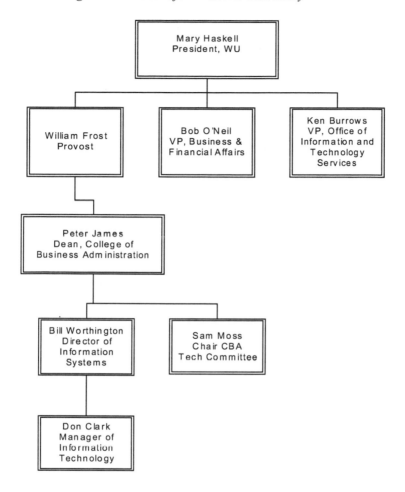

The 23rd Street facility contains a secure, fire-protected, emergency power equipped server room. The OITS main offices and US are located in Thompson Hall, while ITS and AS are located at 23rd Street. The key server computers that are operated by OITS are summarized in Table 1.

CBA Information Systems

While some of WU's colleges rely almost completely on the OITS group for all of their IT service needs (e.g., management of data servers, end-user support, etc.), other colleges — including CBA — have created their own decentralized IT unit to address specific information service needs. CBA's Office of Information Systems (OIS) was created in 1985 to provide systems support to CBA's microcomputer lab and faculty workstations, which were the first on campus. Although OITS now provides support for

Table 1. Server computers at Western University and CBA

Server Name	Operating System	Category (see note)	Comments
WU OITS Academic Support servers. (List does not include Administrative Support servers.) All at 23rd St. Location.			
Kermit	Netware 5.1	Production	Faculty accounts
Bert & Ernie (two machines)	Netware 5.1	Production	Student accounts
Oscar	Win NT 4.0	Production	Course support
Grover	UNIX	Production	Web pages
CBA servers at the time of the fire. All but Iowa were located in MH 310 and destroyed in the fire.			
Wisconsin	Netware 5.1	Production	Faculty and staff file and print services
Missouri	Netware 5.1	Production	Student file and print services
California	Linux	Production	Restricted file services
Jersey	Win NT 4.0	Production	Web server
Maryland	Win 2000	Production	CBA web site and SQL DBMS server
Alabama	Win 2000	Production	Tape backup – Windows servers
Massachusetts	Netware 5.1	Production	Tape backup – Netware servers
Indiana	Linux	Production	Help desk
Dakota	Win 2000	Academic	Web server
Washington	Win 2000	Academic	SQL DBMS server
Carolina	Netware 5.1	Academic	File and print services
Virginia	Win 2000	Academic	Web server, SQL DBMS server
Colorado	Linux	Academic	E-mail services
Iowa	Win 2000	Research & Development	Web server, SQL DBMS server

these types of systems, CBA has elected to continue its OIS function. CBA's OIS staff provides most of the information systems and technology services used by CBA, including its 55-workstation computer lab.

CBA's OIS group is located in Mitchell Hall, which is CBA's office building. Mitchell Hall is an older building that lacks a number of features that are common in newer buildings. For example, at the time of this case, Mitchell Hall had fire alarms that could be triggered manually, but it did not have automatic fire alarms that would detect a fire and then contact the fire department and shut off the ventilation system (shutting down the ventilation system is critical to reduce smoke damage). Office and work space in Mitchell Hall was at a premium. The OIS group's computers and operations were limited to a cramped office that was about 30 feet long and 15 feet wide. This space was home to 13 computer servers, a variety of technical equipment, numerous file cabinets and bookshelves, and the OIS staff.

Table 1 lists CBA servers that existed at the time of the fire. CBA servers were named after states. While some of the servers were new and powerful machines, others were upgraded workstations that were used for low demand special purpose applications. The age of the CBA servers ranged from one year old to six years old.

Due to a tight operating budget, the staffing resources for CBA's OIS group were very limited. The OIS group had two full-time employees who were supplemented by numerous part-time student employees. The director of information systems for OIS was Bill Worthington, who had been in the position for over five years. Worthington had extensive operating experience with a number of technologies and operating system environments.

In early 2001 CBA hired Don Clark to serve as the OIS group's second full-time employee. Clark's title was manager of information technology, and his job was to manage operations for the OIS group. Clark knew the people and information systems of CBA very well, since he had worked under Worthington the previous two years before he graduated from CBA with an MIS degree.

CBA's Technology Committee

CBA has a Technology Committee that provides advice to the dean and OIS concerning a variety of matters involving the planning, management, and application of technology in CBA. At the time of the case, the committee was chaired by an MIS professor named Sam Moss. Other members of the committee included other MIS faculty and representatives from each of the CBA departments — usually the most technology-literate persons in those departments.

SETTING THE STAGE

On the evening of July 3, 2002, a fire destroyed many components of the information systems infrastructure for CBA, including virtually all of the server machines for CBA's computer network. To help provide a context for understanding the issues associated with IT disasters, this section provides an overview of the topic of disaster recovery, followed by a description of the fire disaster incident at CBA and a summary of the key people and places involved with the disaster and subsequent recovery process.

Information Technology Disasters and Disaster Recovery

An information technology (IT) disaster can be considered any type of event that may significantly disrupt IT operations (Lewis, Watson, & Pickren, 2003). Examples of disasters range from events that are more ordinary (e.g., an IT move or upgrade) to those that are more dramatic (e.g., a hurricane). A recent study analyzed all disaster recovery events (a total of 429) handled by a large international provider of disaster recovery services during the period from 1981 to 2000 (Lewis et al., 2003). This study identified the most common types of disaster events, as well as the degree of disruption caused by the events. Seven types of disasters were found to be most common, with the disaster categories of "natural event" and "IT failure" being the most frequent types of disasters encountered. While disasters associated with fires and IT moves/upgrades were not as common, the study found that the potential for extended periods of disruption for these types of disasters were particularly high (Table 2).

While IT disaster recovery has long been an important issue for organizations (e.g., Rhode & Haskett, 1990), the issue of IT disaster recovery gained a higher profile and

Table 2. Most frequent types of IT disasters (adapted from Lewis et al., 2003)

Disaster Category	% of Disaster Events (out of 429 events)	Days of Disruption (minimum – maximum)
Natural Event - e.g., Earthquake, severe weather	28 %	0 – 85
IT Failure - e.g., Hardware, Software	23 %	1 – 61
Power Outage - e.g., Loss of power	16 %	1 – 20
Disruptive Act - e.g., Intentional human acts (e.g., bomb)	7 %	1 – 97
Water Leakage - e.g., Pipe leaks	7 %	0 – 17
Fire - e.g., Electrical or natural fires	5 %	1 – 124
IT Move/Upgrade - e.g., Data center moves, CPU upgrades	3 %	1 – 204

sense of urgency following the terrorist attacks on September 11, 2001 (e.g., Eklund, 2001). Although many organizations may not feel threatened by terrorism, most organizations are vulnerable to some form of disruption to their IT operations. For example, the electrical blackout that hit the northeastern section of the United States in August 2003 exposed many businesses to vulnerabilities in their ability to handle a disaster (Mearian, 2003). Given the importance of IT operations to contemporary organizations, it is critical for organizations to be prepared. Unfortunately, industry surveys indicate that many organizations are not as prepared as they should be (Hoffman, 2003). In particular, small- and medium-sized organizations tend to be less prepared, often due to limited IT budgets (Verton, 2003).

Disaster planning can require significant time and effort since it involves a number of steps. As described by Erbschloe (2003), key steps in disaster planning include the following: organizing a disaster recovery planning team; assessing the key threats to the organization; developing disaster-related policies and procedures; providing education and training to employees; and ongoing planning management.

The Night of July 3, 2002

The fire disaster event at CBA occurred on the night of Wednesday, July 3, 2002. During this evening, the WU campus was quiet and deserted. The following day was Independence Day, which is a national holiday in the United States. The July 4th holiday is traditionally a time when many people go on vacation, so a large number of WU faculty and staff who did not have teaching commitments had already left town. The library and computer labs on campus were closed at 5 p.m. in anticipation of the holiday, and custodial staff took the evening off.

The quiet mood at the WU campus on the evening of July 3rd changed considerably at 10:15 p.m. when a campus security officer spotted large amounts of smoke emerging from Mitchell Hall. The officer immediately reported the incident to the fire department. Within minutes, Mitchell Hall was surrounded by fire trucks. The fire crew discovered that a fire had started on the third floor of Mitchell Hall. Fortunately, the fire fighters were able to contain the fire to an area surrounding its place of origin. Considering that Mitchell Hall did not have an automated smoke alarm system, the structural damage to the building could have been much worse. As noted by Conrad Minsky, a spokesman for the local fire department, "If it hadn't been for [the security officer's report of the smoke], the whole building probably would have burned down." Unfortunately, the room where the fire started was Mitchell Hall 310 (MH 310), the central technical office and server room for CBA — where virtually all server hardware, software, and data for CBA were located.

The dean of CBA, Peter James, and the director of CBA's OIS, Bill Worthington, were called to the scene of the fire late on the night of July 3rd. Their hearts sank when they learned that the fire had started in MH 310 and that it had completely devastated the contents of that room. Due to the fire department's protocols involving forensic procedures at a disaster site, neither James nor Worthington was initially allowed into the area near MH 310. However, based on what they could see — numerous fire fighters in full gear coming in and out of the Mitchell Hall building — they could imagine a grim situation inside the building.

There wasn't much that James or Worthington could do that evening, other than wonder if anything could be saved from the room and what it would take to recover. James found a computer on campus from which he could send an e-mail message to the CBA staff and faculty, most of whom were blissfully unaware of the disruption that was about to happen in their lives. The e-mail sent by Dean James is shown in Figure 2.

After the Smoke Had Cleared

After the fire fighters managed to put out the fire and clear the area of smoke, the fire department's forensic team closely examined the burn site. The extent of the damage to MH 310 and its contents was complete. According to the fire department spokesman Conrad Minsky, "It was cooked beyond belief. It was so hot in there that a lot of the [forensic] evidence burned ... All we saw was just huge clumps of melted computers." The extent of the damage is illustrated in Figure 3, which shows a picture of a MH 310 server rack shortly after the fire was extinguished. Based on the evidence available at the scene, the forensic team concluded that an electrical fire in one of CBA's oldest server computers had started the blaze. Once the fire started in the server, it slowly spread to other servers in a slow chain reaction. Since the servers were not in contained fire-proof server racks, there was nothing to stop the initial server fire from spreading. Ultimately, the entire server room and all of its contents (including technical equipment, furniture, back-up tapes, and papers) were destroyed.

Figure 2. E-mail message from Dean James

From: Peter James
Sent: Thursday, July 04, 2002 1:03 AM
To: Faculty and Staff of CBA
Cc: Mary Haskell (President); William Frost (Provost)
Subject: FIRE IN MITCHELL HALL
Importance: High

You may have heard that there was a fire in Mitchell Hall that started sometime prior to 10 PM Wednesday evening. It is now 12:55 Wednesday evening, and Bill Worthington and I have been there to review the situation.

The fire appears to have started in MH 310, which is the server/ technology room. The room and all of the contents have been destroyed according to the fire personnel, and we were prohibited from entering that part of the building pending an investigation by the fire inspector. There is substantial smoke damage to all of the third floor, and to a considerably lesser extent the fourth floor as well. Vice President O'Neil, who was also present along with his crew and Western's Police Chief, indicated that clean-up will start at the latest Friday morning. There was little to no damage on the lower floors, although considerable water had been pumped into MH 310 as well as the seminar rooms, and possible water damage had not yet become evident.

I am requesting that you ***do not come to the office on Friday*** unless absolutely necessary. Third floor offices will be in the process of being cleaned, and some of them may not be habitable because of the smoke damage. An assessment will be completed over the next few days.

We do not yet know what we will do regarding the loss of the servers and associated data and software.

Peter

Figure 3. Computer server rack in Mitchell Hall 310 following the fire

CASE DESCRIPTION

This section discusses the key activities and issues that confronted the members of the CBA IT staff and administration following the fire. As is the case with many disaster events, it took some time for the people involved to get a complete understanding of the extent of the disaster and its implications. Once the situation was assessed, a number of issues needed to be addressed to deal with the disaster. These issues included data recovery, the search for temporary office space and computing resources, rebuilding Mitchell Hall, and disaster funding.

The Morning After

Sam Moss, MIS Professor and Chair of the CBA Technology Committee, awoke on the morning of July 4th looking forward to a day of relaxation — a bit of work around the house capped off by a burger and a brew while he and his friends watched the civic fireworks show that evening. Little did he realize that the real fireworks had happened the evening before.

Moss always checks his e-mail, even on a holiday. He was, of course, rather disturbed to see the message from Dean James about the fire in Mitchell Hall. Moss's first reaction was to try to visit a Web site that he maintained on his experimental server, Iowa, which was located in his office. Much to his relief, the Web site responded and appeared to be working as usual.

Moss also tried to access the CBA Web site, which he had developed with another colleague (Chanute Olsen), as well as files maintained on the CBA faculty data server. Neither of these last two attempts was successful.

Upon first reading, the dean's message did not convey the full impact of the loss of systems to Moss, who hoped that the servers might be recoverable. Moss knew that the CBA office maintained tape backups of the directories containing the Web site, faculty, and student data. While he expected a few weeks of disruption, he anticipated that CBA's IT operations would be in pretty good shape shortly.

Moss was very familiar with the techniques for remote access to Iowa and other college and university systems (see Microsoft Corporation, 2004a, and Novell Corporation, 2004, for examples of remote access software). Concerned that access to his office, where Iowa was located, might be limited for a few days, Moss immediately copied the backup files of his database projects and those of his students to his home machine. He also copied ASP (active server page) files and any other documents that would be difficult to replace.

Moss contacted Dean James later in the day via e-mail. Moss made his remarks brief: "Iowa appears to be OK, we can use it as needed as a temporary server." … and … "Would you like me to convene the College Technology Committee to provide advice for the rebuilding of our technology support?" James's response was that he appreciated the offer of Iowa, and yes, he wanted advice on "How should we rebuild?"

The following sections summarize what was learned and what was done by key players during the month after the fire. These items are not in chronological order; they are a summary of what Moss learned as he prepared to address the issue of rebuilding the college's IT architecture.

Determining the Extent of Data Loss

The first question, of course, concerned the state of the data back-up tapes of the CBA servers. Unfortunately, the situation regarding the back-ups was discouraging. While nightly data back-ups had been made, the tapes were stored in MH 310. Copies were taken off-site every two weeks. At the off-site location, the most recent copies of data and documents from Wisconsin and Missouri were dated June 16 — the last day of Spring Quarter. All of the more recent backup tapes were located on a shelf near the server machines and had been destroyed in the fire. On Maryland, the static pages and scripts to create dynamic pages (e.g., a list of faculty drawn from the database) of the CBA Web site had been backed up, but the SQL Server database (which included data as well as stored procedures and functions) on that machine had *not* been backed-up. According to Worthington, the college was waiting for special software needed to back- up SQL Server databases.

In addition to the back-up tapes, there were several other sources of backed-up data; however, these sources were rather fragmented. For example, many professors worked on the files for their university Web sites from home and used FTP (file transfer protocol) to copy the updated files to the Web server. Their home copies were, in effect, back-ups. Moss had made some changes to the CBA Web site pages the evening of July 2nd and had copies of those pages on his home system. The Iowa server in Moss's office had been used as the test bed for the database portion of the Web site; therefore it contained the structure of the database, but neither the most recent data nor the stored procedures and functions.

Although no desktop machines, other than those in MH 310, were burned in the fire, they were not readily available immediately after the event. Smoke and soot permeated the building. There was a concern that starting these machines, which were turned off at the time of the fire, might lead to electrical short circuits and potentially toxic odors and fumes. The decision was made to clean these machines before restarting them.

Recovery and Repair of Smoke-Damaged Systems

Within 48 hours of the fire, the university contracted with BELFOR International to help with clean-up. BELFOR specializes in disaster recovery and restoration. Examples of high profile disasters that BELFOR responded to in 2003 included the major wildfires in California and hurricane Isabel in the eastern United States (BELFOR, 2004). With extensive experience with disaster recovery and regional offices in Seattle, Portland, and Spokane, BELFOR was well-positioned to respond to the Mitchell Hall fire incident. BELFOR's areas of expertise included restoration of building structures, office equipment, and paper documents.

Based upon BELFOR's initial analysis, none of the computer systems, nor any of their components, from MH 310 was deemed recoverable. On the other hand, all other computers in the building escaped heat and water damage and could be "cleaned" and continued in service. These machines contained valuable data and were set up with individual users' preferences. Recovering them would be a major benefit.

The cleaning procedure for computers was a labor intensive process that involved immersion in a special water solution followed by drying and dehumidification. Systems units, monitors, and most printers could be cleaned. Keyboards, mice, and speakers were cheaper to replace than to clean. Most of these systems were cleaned in the 30 days following the fire and then stored until the reopening of the building in mid September. The college identified a few critical systems, based on the data they contained, for expedited cleaning and return to service.

Sources of Funding for Disaster Recovery

The cost of cleaning Mitchell Hall and rebuilding the computer system was significant. Estimates rose dramatically during the period that followed the fire. The initial estimate was $750,000. However, ongoing investigation of the damages and building contamination indicated that the total cost would be much higher. By the end of July, the university issued a press release indicating that the costs required to replace equipment and make Mitchell Hall ready for classes would be approximately $4 million. In fact, the final cost was about $4.25 million.

The repair costs were high because of smoke and water damage. The ventilation system in Mitchell Hall had used fiberboard ducts instead of metal ducts, and these had absorbed smoke and particles during the fire. The ducts could not be cleaned and had to be completely replaced on the top three floors. This required new ceiling tile installations to accommodate the new locations of the ventilation ducts. Carpets on these three floors also had to be replaced. MH 310 was completely gutted and rebuilt starting with the metal studding in the walls. The remainder of the third floor had the drywall walls torn out and replaced. In addition, the electrical and network infrastructure was rebuilt, which required completely rerunning the associated wires back to an electrical closet on the third floor. BELFOR handled the cleaning for all computers and other electronic equipment that was of sufficient economic value to justify the work. These cleaning costs totaled more than $100,000. The replacement costs for computer equipment and software that had been lost in the fire was approximately $150,000 — an amount that did not include the labor required to install and test the hardware and software.

The cost of the recovery was paid from two sources. WU had a specialized Electronic Data Processing (EDP) insurance policy to cover hardware losses. This policy provided up to $250,000 (on a "replacement cost" basis) less a $25,000 deductible. The other source of funds was a self-funded reserve set up by the state. The amount of money provided from this fund was determined by the State Legislature. President Haskell invited several legislators to campus and conducted first-hand tours of Mitchell Hall so that they could view the damage. Based on their review of the situation, the State Legislature provided a sufficient amount of funding to WU to rebuild Mitchell Hall and to replace all the hardware and software that had been destroyed. In a press release, President Haskell stated "We are very pleased with the timely and supportive action from legislative leaders and the Governor's office."

Temporary Quarters

Office space on the WU campus is at a premium. Fortunately, there was a small area in WU's administration building, Thompson Hall, which had been recently vacated. This office, with perimeter cubicles surrounding a large open area, became the temporary location of CBA's department offices. Each department chair and each department secretary had a workspace with a desk, phone, and limited storage for files and books. The offices had connectivity to the campus network. An open area was converted to a "bull pen" for faculty use and small group meetings.

Because the incident happened in the summer, it was relatively easy to relocate the classes. During the school year, university classrooms are at 100% utilization during mid-day class hours but in the Summer Quarter, less than 50% utilization is common. The only class sessions lost were for the few classes scheduled to meet on Friday, July 5th. Classes resumed as usual on July 8th, although in different rooms. Those that had been assigned to meet in the Mitchell Hall computer lab were relocated to another lab on campus.

Computer Hardware Resources Available for Immediate Use

For the immediate term, CBA was able to make use of 15 computers that had been received but not yet deployed in Mitchell Hall. These machines had been ear-marked as replacements for older systems and for recently hired faculty due to arrive in the fall. Fortunately, these systems were being held in the 23rd Street facility. They were quickly configured and delivered to the offices in Thompson Hall.

The MIS program at Western had a set of 20 notebook computers that were used for classroom instruction as part of a "mobile computer lab." At the time of the fire, they were stored in a locked cabinet in a second floor room in Mitchell Hall. These computers were examined by the health inspectors and deemed safe to use (i.e., cleaning was not needed). These machines contained wireless network cards and were configured with the Microsoft Office suite. No MIS classes were scheduled to use this lab during the summer, so these computers were signed out to faculty who did not have sufficient capacity in their home systems. The combination of new systems, the notebook computers, and individuals' home systems was sufficient to equip all staff and faculty with adequate workstations.

The loss of servers was much more problematic. The college lost every server it had with the exception of Iowa — the research and development server running Windows 2000 Server and MS SQL Server 2000. Iowa was located in Moss's office on the third floor. Although this location was less than 100 feet from the scene of the fire, Iowa had been relatively protected in Moss's office. Because Iowa had been operational continuously since the start of the fire and had not shown evidence of any problems, Moss and Worthington decided to keep it in operation (not clean it). Iowa was relocated to a well-ventilated area in the Thompson Hall offices.[2]

The university maintains many servers: a series of Novell systems used for student and faculty data, a server running Windows NT devoted to academic use, and a UNIX server used for most university, faculty, and student Web pages. Worthington was able to negotiate for space on one of the Novell servers to hold the data that had been stored on Wisconsin and Missouri.

Recovering Files from Novell OS Servers

Immediately following the fire, Worthington obtained temporary workspace in the 23rd Street building and set out to recover the data from the June 16th tapes. He quickly ran into a severe stumbling block. CBA used digital tape equipment that was newer than the drives attached to Kermit, WU's Novell server. Fortunately, a tape drive in the correct format was attached to Grover, WU's UNIX server. To recover data from Wisconsin and Missouri, it was necessary to copy the tape contents to a disk on Grover; then write the data to tapes in the format of the drives attached to Kermit, and finally restore from those tapes. Most of the data was available within a week, thanks to Worthington's diligent efforts.

Back on the Web

The Windows servers that were used to support CBA's WWW sites and database applications were also lost. As it turned out, the process of recovering and recreating this data took much longer than the Novell files and was not completed until July 31st. Worthington was able to recover portions of the CBA Web site which had been backed up on June 16th. The recovered data consisted of those folders that professors and department chairs had prepared for their Web sites. As the files were recovered, they were copied onto the Iowa server. Unfortunately, the database data for the CBA WWW site, including the views and stored procedures used to execute queries, had not been backed-up to tape and were lost.

Sam Moss and Chanute Olsen set about to reestablish the site as quickly as possible. Once Iowa was operating in its new location in Thompson Hall, they arranged for the DNS entry for the Web site to point to Iowa's IP address. They quickly rebuilt portions of the database and copied as much data as they could from other sources. On his home computer, Moss had current copies of many of the Active Server Page (ASP) files that were used to extract data and format the data for the page displays on CBA's WWW site. Moss immediately loaded these files onto Iowa. Unfortunately, Moss did not have copies of ASP files on the Windows server that had been maintained by others. These files ultimately had to be recreated from scratch.

Much of the information displayed on the college Web site was drawn from a database. Maintenance of the database data was done by the department chairs and secretaries. Once Iowa was fully operational, these persons had to re-enter data about their departments.

Moss and Olsen contacted the professors who had sites on the college Web site. For those who had home back-ups of their data, it was easy to recreate their sites on Iowa. Those who did not have a back-up were forced to wait until files were copied from the back-up tapes. By the first of August, all Web sites had been restored and most of the database entries had been recreated.

CURRENT CHALLENGES/PROBLEMS FACING THE ORGANIZATION

During the period following the fire, Moss thought about the challenges that the college would face once the immediate tasks of restoring data and temporary operations were accomplished. He knew that a major task would be rebuilding the College's IT operations. Moss would need to work with his College Technology Committee through the months of July and August to come up with recommendations. As Moss pondered the situation, he realized that there were many unanswered questions, which he organized into two categories:

- **Planning for IT architecture:** What system applications and services were needed at CBA? What were the hardware/operating system requirements for these applications? How should the applications and services for CBA be allocated across servers? How many servers should CBA acquire? Where should the servers be physically located, and under whose control?

- **Assessment of threats:** What were the disaster-related threats to the CBA information systems? Would it be possible to design a CBA IT architecture that would minimize the impact of disaster and facilitate disaster recovery?

Planning for the Information Technology Architecture

Issues concerning IT architecture were at the top of Moss's list. As discussed by Seger and Stoddard (1993), "IT architecture" describes an organization's IT infrastructure and includes administrative policies and guidelines for hardware, software, communications, and data. While one option for CBA would be to replicate the architecture that had been destroyed, Moss did not believe that this would be the best way to go. The array of servers and systems before the fire had grown on an "as needed" basis over time, rather than guided by any master plan. While the previous system had worked, Moss wondered if there might be a different, more effective IT architecture for CBA. Since CBA would need to rebuild much of its IT operations from the ground up in any event, Moss thought that it would be worthwhile to take this opportunity to re-examine the IT service requirements and the way that the requirements would be implemented. By doing so, it was possible that the college might adopt a new IT architecture.

Moss was aware that an important first step in IT architecture planning is to determine the applications and services that will needed (Applegate, 1995). To kick off the IT architecture planning process, Moss sent an e-mail message to all faculty and staff on the morning of July 8th. The purpose of the message was to solicit input from the CBA user community regarding IT service requirements. The e-mail message is shown in Figure 4.

By learning more about the desired IT applications and services, Moss and his committee would be able to determine the types of application software that could be used to provide the specified services. This information would be useful, as it would have implications for the types of operating systems that would be needed. Due to operational considerations, Moss anticipated that no more than one type of operating system would reside on any given server. After the services were identified and the operating systems requirements were determined, the next step would be to decide on the number and type of server machines. Based on input from Worthington and others, Moss knew that an "N-tier" type of IT architecture would offer useful advantages for CBA. An N-tier design (also called multi-tiered or multi-layered) would involve separating data storage, data retrieval, business logic, presentation (e.g., formatting), and client display functions. One benefit of an N-tier design is that the development staff can focus on discrete portions of the application (Sun Microsystems, 2000). Also, it makes it easy to scale (increase the number of persons served) the application and facilitates future change because the layers communicate via carefully specified interfaces (Petersen, 2004). Although multiple tiers may reside on the same server, it is often more efficient to design each server to host specific parts of the application. For example, database servers perform better with multiple physical disks (see McCown, 2004). The use of more than one server can also complement security offered by network operating systems. For example, CBA provided file storage services for both faculty and students. Although

Figure 4. E-mail message from Sam Moss

From: Sam Moss
Sent: Monday, July 08, 2002 9:43 AM
To: CBA Faculty and Staff
Subject: Thinking about CBA Server needs
Importance: High

Colleagues,

As we rebuild from the fire, we have an opportunity to rethink our server configurations. The first step in such a process is a requirements analysis (we all teach it this way, don't we?). To that end, I would appreciate your input on the following questions:

What are the data and application program storage needs for CBA?
What service requirements do we have?

Don't answer this question in terms of a proposed solution (i.e., "We need two Netware servers and one NT server under CBA control.") but rather in terms of what must be delivered, as independent of technology as possible, to our faculty, students, and external publics.

--Sam Moss
          Chair, CBA Technology Committee

these services could be hosted on a single machine, allocating the services on separate machines, and not allowing any student access to the "faculty" machine, would provide additional security for faculty files (e.g., student records, exams, faculty research).

As Moss received responses to his e-mail message, he organized the responses in a table to help him and his committee analyze the requirements. As indicated in Table 3, Moss had organized his list of application and service requirements for CBA into different categories based on the type of server that might be used to host the services. A variety of different types of servers may be used in an N-tier architecture. Examples of some of the different types of servers and their functions include the following (*TechEncyclopedia*, 2004):

- **Application server:** The computer in a client/server environment that performs the business logic (the data processing). For WWW-based systems, the application server may be used to program database queries (e.g., Active Server Pages, Java Server Pages).
- **Database server:** A computer that stores data and the associated database management system (DBMS). This type of server is not a repository for other types of files or programs.
- **Directory services:** A directory of names, profile information and machine addresses of every user and resource on the network. It is used to manage user accounts and permissions.
- **File server:** A computer that serves as a remote storage drive for files of data and/ or application programs. Unlike an application server, a file server does not execute the programs.
- **Print server:** A computer that controls one or more printers. May be used to store print output from network machines before the outputs are sent to a printer for printing.

Table 3. CBA IT requirements

Type of Server Requirements for Services/Applications	Suitable Software or Operating Systems
File Server 1. Individual directories and files (faculty and student documents) 2. Class directories and files 3. Homework and exercise drop-off 4. College documents and records	Novell NetWare MS Windows Server UNIX and derivatives (e.g., LINUX)
Print Server 5. Printer queues for faculty and staff 6. Printer queue for student lab	Novell NetWare MS Windows Server
Database Server 7. Data to support CBA web site 8. Student Information System Data 9. Faculty Information System Data	MS SQL Server Oracle MySQL or another RDBMS
Application Server 10. Web site active server pages 11. Student and faculty info. systems	MS IIS was previously used for item 10. Item 11 is a future project
Web Site Server 12. Static and dynamic pages for the college and departments 13. Individual faculty pages 14. Student club pages	Static pages could be on any of several systems, currently on MS Windows and UNIX servers Dynamic pages work in conjunction with item 10
Teaching Server 15. Web development class 16. Database management class 17. Enterprise resource mgmt. class 18. Network administration class	Classes 15-17 currently use Windows 2000 Server Class 18 uses Linux, Novell NetWare and Windows 2000 Server
Research and Development 19. Database management 20. Web site development	Current research all conducted on software that requires Windows
Tape Back-up Server Tape backup of all servers	Veritas Backup Exec IBM Tivoli
Directory Services File and Print Services login (for Items 1-6 listed above) College Web server and database server login (Items 7-14, 19 & 20) Academic Course logins (Items 15-18)	Novell eDirectory (formerly Novell Directory Services – NDS) Microsoft Active Directory (AD)

- **Research and development server:** A local name used at CBA to describe a computer that is used for developing and testing software and programs. A single R&D server often hosts multiple tiers because demand (i.e., number of clients served) is light.
- **Tape back-up server:** A computer that contains one or more tape drives. This computer reads data from the other servers and writes it onto the tapes for purposes of backup and archive.
- **Teaching server:** Another local name used at CBA to describe a computer that is used for educational purposes to support classroom instruction in information systems courses (e.g., WWW development, database management, network administration).
- **Web server:** A computer that delivers Web pages to browsers and other files to applications via the HTTP protocol. It includes the hardware, operating system, Web server software, TCP/IP protocols, and site content (e.g., Web pages and other files).

SUMMARY

It was clear to Moss that the College had not been well positioned to handle an IT disaster. Although many books on IT management include references to disaster planning (e.g., Frenzel & Frenzel, 2004), the college had been too busy handling day-to-day IT operations to devote resources to disaster planning. Moss resolved to ensure that the next disaster, if there was one, would not be as traumatic. As part of this process, Moss wanted to identify the threats to CBA's IT system (Microsoft Corporation, 2004b). As suggested by Table 2, Moss knew that IT disaster threats can come from the physical environment (e.g., WU's location makes it susceptible to both earthquake and volcanic activity), from machine failure, and from human causes. With the help of the technology committee, Moss set out to identify the primary threats.

Once Moss and his committee had generated the list of threats to CBA's IT operations, he hoped to design the IT architecture in a way to minimize the impact of any future disasters and facilitate future disaster recovery. Toigo (2000) has pointed out that decentralized IT architectures, such as a N-tier design, involve special considerations for disaster planning. To help minimize the impact of disaster, Toigo recommends that a decentralized system include high-availability hardware components, partitioned design, and technology resource replication. For example, having similar machines in multiple locations may provide an immediate source of back-up computing resources (as illustrated by the use of Iowa and Kermit after the CBA fire). Moss planned to arrange meetings with both OIS and OITS personnel to determine how CBA and WU could shape its IT architecture plan to minimize the impact of any future disaster.

REFERENCES

Applegate, L. M. (1995). *Teaching note: Designing and managing the information age IT architecture.* Boston: Harvard Business School Press.

BELFOR. (2004). *BELFOR USA news.* Retrieved May 30, 2004, from http://www.belfor.com/flash/index.cfm?interests_id=41&modul=news&website_log_id=27802

Burd, S. D. (2003). *Systems architecture* (4th ed.). Boston: Course Technology.

Eklund, B. (2001, December, 20-25). Business unusual. *netWorker.*

Erbschloe, M. (2003). *Guide to disaster recovery.* Boston: Course Technology.

Frenzel, C. W., & Frenzel, J. C. (2004). *Management of information technology.* Boston: Course Technology.

Hoffman, M. (2003, September 1). Dancing in the dark. *Darwin.*

Lewis, W., Watson, R. T., & Pikren, A. (2003). An empirical assessment of IT disaster risk. *Communications of the ACM, 46*(9ve), 201-206.

McCown, S. (2004, February 20). Database basics: A few procedural changes can help maintain high performance. *InfoWorld.*

Mearian, L. (2003, August 25). Blackout tests contingencies. *ComputerWorld.*

Microsoft Corporation. (2004a). Windows 2000 Terminal Services. Retrieved May 27, 2004, from http://www.microsoft.com/windows2000/technologies/terminal/default.asp

Microsoft Corporation. (2004b). *Microsoft operations framework: Risk management discipline for operations.* Retrieved June 1, 2004, from http://www.microsoft.com/technet/itsolutions/techguide/mof/mofrisk.mspx

Novell Corporation. (2004). Novell iManager. Retrieved May 27, 2004, from http://www.novell.com/products/consoles/imanager/

Petersen, J. (2004). *Benefits of using the N-tiered approach for Web applications.* Retrieved June 1, 2004, from http://www.macromedia.com/devnet/mx/coldfusion/articles/ntier.html

Rhode, R., & Haskett, J. (1990). Disaster recovery planning for academic computing centers. *Communications of the ACM, 33*(6), 652-657.

Seger, K., & Stoddard, D. P. (1993). *Teaching note: Managing information: The IT architecture.* Boston: Harvard Business School Press.

Sun Microsystems (2000). *Scaling the N-tier architecture.* Retrieved June 4, 2004, from http://wwws.sun.com/software/whitepapers/wp-ntier/wp-ntier.pdf

TechEncyclopedia. (2004). Retrieved June 2, 2004, from http://www.techweb.com/encyclopedia

Toigo, J. W. (2000). *Disaster recovery planning* (2nd ed.). Upper Saddle River, NJ: Prentice-Hall.

Verton, D. (2003, March 10). Tight IT budgets impair planning as war looms. *ComputerWorld.*

ENDNOTES

[1] The facts described in this case accurately reflect an actual disaster incident that occurred at a university located in the United States. For purposes of anonymity, the name of the organization and the names of individuals involved have been disguised.

[2] In the two years since the incident, Iowa has operated continuously with no hardware problems. The monitor was cleaned and the mouse and keyboard were replaced, but the only cleaning the system unit received was a wipe of the outer surface. Iowa initially emitted a pungent smell, but that odor disappeared after two to three weeks.

Steven C. Ross is an associate professor of MIS at Western Washington University (WWU) and was previously on the faculties of Montana State and Marquette Universities. He holds degrees from Oregon State University (BS) and the University of Utah (MS, PhD), and conducts research in the areas of microcomputer software, systems development, and applications of database technology.

Craig K. Tyran is an associate professor of MIS at WWU and was previously on the faculties of Oregon State and Washington State Universities. He holds degrees from Stanford (BS, MS), UCLA (MBA), and the University of Arizona (PhD), and conducts research in the areas of technology support for collaboration and learning.

David J. Auer is the director of information systems and technology services in the College of Business and Economics at WWU. He holds degrees from the University of Washington (BA) and WWU (BA, MA, MS) and has published several textbooks.

Jon M. Junell is the manager of information systems in the College of Business and Economics at WWU. He holds a BA from WWU and has worked in IBM's field services operations.

Terrell G. Williams is a professor of marketing at WWU and was previously on the faculties of California State University, Northridge, University of Colorado, Denver, and Utah State University. He holds degrees from the Universities of Wyoming (BS, MS) and Arizona (PhD) and has conducts research in the areas of consumer behavior, marketing strategy, and marketing education.

This case was previously published in the *Journal of Cases on Information Technology*, 7(2), pp. 31-49, © 2005.

Chapter IV

Policy Processes for Technological Change

Richard Smith, Simon Fraser University, Canada

Brian Lewis, Simon Fraser University, Canada

Christine Massey, Simon Fraser University, Canada

EXECUTIVE SUMMARY

Universities, among the oldest social institutions, are facing enormous pressures to change. There have always been debates about the university, its purpose, its pedagogical program, and its relationship to other social and political structures. Today, these debates have been given renewed vigor and urgency by the availability of advanced information and communication technologies for teaching and learning. These include computers and computer networks, along with the software and telecommunications networks that link them together. When these technologies are used to connect learners at a distance, they are called "telelearning technologies." When referring to their use more generally, to include local as well as remote teaching innovations, they are sometimes called "technology-mediated learning" (TML). Despite much media attention and recent academic criticism, pressures on universities are facilitated, but not caused, by telelearning technologies. Change in universities is not simply a result of forces acting upon universities, but is the result of a complex interaction of internal and external drivers. The use of telelearning technologies intersects with a host of social, political, and economic factors currently influencing university reform. Technology, in this context, has become the catalyst for change, reacting with other elements in a system to spark a reaction and a change in form and structure. This chapter examines policy processes for the introduction of technology-mediated learning at universities and colleges. It is based on the results of a two-year research project to investigate policy issues that arise with the implementation of telelearning technology in universities and colleges. The focus was on Canadian institutions of higher learning, but the issues raised are common to higher educational institutions in other countries. The study scanned a large number of institutions, reviewed documents, and interviewed key actors including government and institutional administrators, faculty, and students, to discover the range of issues raised by the implementation of telelearning technologies. This chapter discusses these issues and findings.

- *What policies or processes are in place to guide change in colleges and universities? Who knows about these policies and participates in them?*
- *What are the forces behind technological change in higher education organizations? Are they external or internal?*
- *Can technology be used as a tool for achieving meaningful and positive change or is it an end to itself?*
- *In what ways can technology be used to increase access to education?*

DOING THE RIGHT THING AND DOING THINGS RIGHT

Organizations implementing telelearning technologies often find themselves facing a variety of new issues not encountered when delivering courses in traditional formats. For example, telelearning technologies can provide access to courses for a broad range of new users. What kind of new or different support services will these new students require? On the flip side of the access issue, students are often concerned about who will have access to files that have stored their electronic discussions, how their identities are safeguarded, and how long these files will be stored. These concerns regarding the implementation of telelearning technologies can be broadly classified as concerns on how to implement these technologies, or "doing things right."

These micro issues of implementation, however, quickly raise questions about "doing the right things," the larger, often politically charged questions that form the policy environment for telelearning technologies. These issues are about why telelearning technologies are used and often evoke preconceived notions of economy, society, and education. These issues are concerned with power relations and the very nature of educational institutions. Examples of these issues would be the purpose of education, the role of professors/trainers, and the goals of business-education partnerships — not only "how" a subject is taught, but what, when, why, by whom, and for what purpose. These broad policy debates, while easily becoming polarized, can help to define an institution's goals so that choices about implementing telelearning technologies become clearer.

Clearly, the two aspects of telelearning policy, "doing things right" and "doing the right things," are linked and both must be dealt with in organizational policies and practices. The importance of sound policy processes that can deal effectively with both aspects cannot be overstated.

One could argue that universities already have well-established mechanisms in place to make these kinds of decisions. After all, universities have long traditions of collegial decision making. But it is a peculiar feature of decisions about technology that these well-worn processes are seldom respected, as the wisdom of how and why to use technology is expected to be apparent to all.

The issues raised by telelearning technologies suggest a need for a systematic approach that honors collegiality while ensuring that the difficult questions can be dealt with in ways that do not overwhelm the process but serve to facilitate choices about implementation. One danger is that policy processes focus solely on "doing things right," trying to avoid controversy with broader political questions. The decisions that result from such processes risk being dismissed by those affected as ill considered and will not be supported. Another danger is that "doing the right thing" questions can overwhelm all discussion, with no progress made on making any decisions for the institution. In the end, decisions are often made anyway, but without consultation, behind the scenes, and as surreptitiously as possible, to avoid getting caught up in an endless and unproductive process.

POLICY PROCESSES

Drivers for Policy Processes

Telelearning technologies serve to amplify a variety of pressures acting on universities and colleges today. For example, the post-secondary sector is experiencing greater competitive pressures than ever before. Institutions can no longer count on their geographical "turf" as being safe from poaching by other institutions. New public and private institutions are emerging to offer popular programs. Telelearning technologies serve to magnify these competitive pressures as online courses attract students from all over the world and as entirely "virtual" institutions are created with no campus infrastructure and no tenure.

At the same time, the demand for post-secondary education is increasing. This demand is coming increasingly from adult workers who are returning to school to upgrade their skills and seek higher professional degrees. These students are seeking more flexible schedules, up-to-date curriculum, and high levels of support services. The more traditional student cohort is seeking similar flexibility as more of these students have part-time jobs and are taking longer to complete their degrees. In this case, telelearning can be an opportunity for universities and colleges to expand their student base and to create new revenue streams through the remote delivery of courses.

The temptation for university administrators in the face of these threats and opportunities is to try to respond quickly, that is, without consultation with their existing constituencies in faculty and students. Consultation, as seen later in the chapter, takes many forms but it is first and foremost an attempt toward inclusiveness in the decision-making process. It is more important than ever that universities establish policy processes that can help them establish priorities and directions to guide planning and to enable rapid responses to threats and opportunities.

Strategic Planning

Strategic planning is a business concept that has migrated recently to universities and colleges as they seek processes to direct their future development. The process can be initiated for a number of reasons. Many institutions feel the need to identify a "niche" for themselves in an expanding marketplace by identifying specific areas where the institution will focus its efforts. In other cases, a strategic plan is useful for convincing others — the Senate, faculty members, and students — that change is necessary (Tamburri, 1999, p. 10).

But the translation of strategic planning from business practice to one appropriate to post-secondary institutions is not automatic. Strategic planning cannot be applied in universities and colleges in the same way in which it is applied in the private sector. Organizational goals in higher education are often vague and, even when well defined, contested. The division of responsibility for priority setting between disciplinary units and the organization as a whole is unclear. But vagueness can be a virtue within post-secondary institutions. Individual units are continually scanning their own discipline's environment and are making informed judgements about their specialized unit. These judgements may conflict with judgements made for the organization as a whole. In the end, contradicting strategies may coexist in the university at the organizational level and at the level of the individual unit (Norris & Poulton, 1991).

Cynthia Hardy argues that many university strategic plans display a fatal lack of emphasis on implementation. She shows how an "executive management" model of strategic planning cannot be imported into universities since it assumes a unitary organization with a common goal. In fact, universities are pluralist organizations where different groups often have competing visions. This means that difficult decisions, such as the reallocation of funds or the elimination of programs, never occur or are made in ways that treat everyone equally since the plan avoids conflict by ignoring how power is distributed and how decisions are really made within the institution (Hardy, 1992).

In light of these concerns, Olcott (1996) suggests a variation on strategic planning specifically designed for aligning institutional academic policy with distance education practice. The need for alignment will become more important as distance education continues to move progressively from the periphery to the core of institutional functions. Olcott argues for a reciprocal adaptation of both distance education units and institutional policy and practice; distance education systems must adapt to create an environment that values mainstream academic norms, and institutional practices must recognize the advantages of the distance education approach. This rapprochement can be achieved by avoiding traditional areas of discontent and agreeing on a commitment to educational values such as quality, access, and responsiveness. He suggests a range of areas where policies can be reformed: recognition of distance education teaching for tenure and promotion purposes, academic residency requirements, and intellectual property.

Strategic Planning for Technology

While strategic planning has begun to play an important role in university and college planning processes, what is different today is the addition of a new function for information technology-teaching and learning.

All too often, computing plans are focused on technology itself, rather than on how technology enables faculty and students to achieve some of the key instructional or research goals of the institution. (Hawkins 1989, cited in Nedwek, 1999)

Still, while many universities may be aware of the need for planning, fewer have successfully extended this process to information technology. The 1998 Campus Computing Project report is instructive. Just under half of U.S. colleges reported having a strategic plan for information technology:

[Fully] 60 percent do not have a financial plan for information technology and less than a third have a plan for using the Internet in their distance learning initiatives. (Green, 1998)

While information technology planning for educational technology is still not widely observed, there are some lessons that we can draw from information technology planning generally. A study of 150 technology officers in universities in the U.S. found that approximately 10% of respondents participated in no technology planning at all, saying that it was a frustrating, time-consuming endeavor that distracts instead of contributes to their day-to-day tasks. Nonetheless, this study found that a majority of technology officers devoted a considerable amount of time to strategic planning. The successful processes were able to distinguish between the two functions of technology planning: socioeconomic goals and strategic goals (Ringle & Updegrove, 1998). Socio-economic goals for technology planning were issues concerned with process. In this case, the goals for a planning exercise were to:

1.   Align technology with other institutional priorities
2.   Disseminate knowledge about technology needs and constraints
3.   Build alliances with key decision makers
4.   Lobby for and obtain financial and other resources
5.   Address existing technology needs
6.   Keep an eye on the leading edge

These process goals were the most important function of technology planning for these technology officers. The second function of technology planning — the strategic — is concerned with technical issues. Given the speed at which the technology is changing, few technology officers were confident in being able to predict their institution's needs two or three years down the road. For this reason, technology planning needed to focus primarily on the process issues and not get bogged down in technical details (Ringle & Updegrove, 1998).

Ringle and Updegrove's (1998) findings correspond to this chapter's findings about the important role of policy processes for institutional telelearning policy. Technology needs will change quickly and unpredictably. It is crucial, however, that a forum exists for addressing the role and function of technology in the institution. This same study found that the least successful technology plans were those that were marginalized and set apart from overall institutional strategic planning (Ringle & Updegrove, 1998).

John Daniel addresses the development of technology strategies for teaching and learning extensively in his book *MegaUniversities and Knowledge Media: Technology Strategies for Higher Education.* He makes the point that change works best if it is supported by peer groups and training and if research results are used to demonstrate the reasons for change. It is unrealistic to expect single technology decisions for entire universities. However, the organization as a whole can support technology in strategic ways while allowing units to determine the best way in which to carry out this priority for their students and discipline (Daniel, 1996).

Alberta's Learning Enhancement Envelope program, which provides funding to that province's post-secondary institutions for technology-enhanced learning, makes an institutional technology plan a requirement of funding. As a result, institutions in this province are developing a body of knowledge about technology planning for teaching and learning.

In Canada, the Standing Committee on Educational Technology of BC has developed a guide to educational technology planning. Their plan describes an inclusive process with advisory and communication processes to assist in getting "buy-in" from different internal groups. They avoid the common pitfalls of strategic planning by focusing on implementation. A regular process of revision ensures that any plan is not set in stone for a period of longer than two years, allowing for negotiation and adaptation to new circumstances. The plan is meant to be flexible and adaptable to the specific cultural and institutional circumstances of different colleges and universities (Bruce et al., 1999).

Fair Process

Another danger of strategic planning within universities is that they fall prey to internal lobbying and opposition. As a result, controversial proposals are eliminated before they reach fruition (Tamburri, 1999, p. 11). This is not to say that it is necessary to create division in order to create change. Kim and Mauborgne (1997) note that it is more important that decision-making processes be fairly carried out than that they accommodate everyone's interests. Fair process was the key factor in the cases they studied on the diffusion of new ideas and change in organizations. Kim and Mauborgne identify three key elements to fair process: first, it engages people's input in decisions that directly affect them; second, it explains why decisions are made the way they are; and third, it makes clear what will be expected of organizational members after the changes are made (1997).

Clearly, fair process can only do so much. Policy processes must negotiate between a set of prior normative issues and a set of practical issues associated with achieving a particular outcome or decision. A successful policy process for change achieves a balance between these two elements, satisfying employee needs for procedural justice with the organization's need to reach decisions and to move forward.

Organizational Change

Much of the discussion so far has concerned organizational change in our universities and colleges. According to Hanna, for change to occur in established organizations, three conditions must be met: (1) enormous external pressures; (2) people within the organization who are strongly dissatisfied with the status quo; and (3) a coherent alternative embodied in a plan, a model, or a vision (Hanna, 1998, p. 66).

It is this third condition that presents perhaps the greatest challenge to higher education institutions as they chart their course in this emerging environment. Sound policy processes are a crucial part of the development of this alternative plan since "the collegial tradition of academic governance makes it unlikely that a technology strategy developed without extensive faculty input would have any impact" (Daniel, 1996, p. 137).

It has been suggested that the challenge to using educational technology effectively in universities and colleges is threefold (Morrison, 1999):

1.   Technical: adequate support and training
2.   Pedagogical: helping faculty reorient their teaching to best exploit the technology
3.   Institutional: reorienting the institution to the effective deployment of educational technology

The first two issues can be addressed with changes in policy and funding. The final step, however, requires something more difficult — leadership and vision.

Organizational change in universities and colleges, therefore, requires a delicate balance of collegial and collaborative policy processes that are championed by a leader with a vision for the institution. Such grandiose organizational change projects are clearly not suited to all institutions — most would surely fail. All, however, are capable of beginning to address the place of educational technology in their teaching and learning.

Part of the process is simply to allow innovation to make its way through the institution more effectively. Universities and colleges have been described as organized along a "loose-tight" principle. That is, as long as an organizational member's behavior is generally aligned with organizational values, individual creativity and innovation are supported. If the individual's behavior moves outside the realm of these core values, the organization "tightens" as a response to guide behavior back to the core values (Olcott, 1996).

Part of the challenge for post-secondary institutions in finding their way with telelearning is to create an environment in which it is not only safe to experiment on the periphery, but also where it is safe to fail in the center; where it is "safe to take the risks needed to improve learning and teaching in times of constant, accelerating change" (Gilbert, 1998). The alternative is to have innovation continually at the margins without ever affecting the core. Kay McClenney observed this trend about innovation in U.S. colleges. She notes that despite mounting pressures for change, most innovative practices are kept at the margins of institutions, thus relieving pressure on the college to truly transform the institution (Gianini, 1998).

PUTTING IT ALL TOGETHER: TEACHING AND LEARNING ROUNDTABLES

One of the most useful models for introducing pedagogical and technological change is the Teaching, Learning and Technology Roundtable (TLTR) program coordinated by the Teaching, Learning and Technology Group (TLT Group), an affiliate of the American Association for Higher Education (http://www.tltgroup.org). The TLTR program provides a set of tools for institutions to help shape goals, facilitate discussion, and organize the implementation of strategies, outside of the bureaucratic structure. A set of structured activities helps evaluate institutional values and pedagogical principles over the use of technology. For example, participants are asked what it is they most value about their institution and would hate most to lose. Only then is technology examined to see how it might support stated values and principles.

A TLTR-style committee should approach its membership strategically. In general, it should be broadly representative of key units in the institution. It is important to have the support of senior-ranking individuals, but they need not be members. The most useful members will be at the operational levels — those who either work with technology or would be expected to.

One of the strengths of the TLTR model is that it places telelearning issues firmly within the context of "doing the right thing" questions. Many technology-planning exercises have failed because they were marginalized from broader questions of institutional mission and purpose. A TLTR group ensures that there is a forum within the institution for those with concerns about the use of technology to have their issues dealt with.

Another strength of the TLTR model is that it is a broad set of organizing and operating principles that can be adapted to local circumstances. Steven Gilbert suggests that TLTR groups not assume a formal policy function in their institution. Some TLTR groups, however, have found that over time, their credibility within the institution is such that they are asked to take on a policy role. Each group will establish itself differently. The TLT Group sponsors national events to help local TLTRs get started and to allow established Roundtables to share experiences and strategies. In this way, local strengths are supplemented by a national network of support.

Finally, the greatest contribution that TLTRs seem to make to institutional telelearning policy is in their communication function. Staff and faculty from dispersed units in the institution discover shared experiences and learn about useful innovations. Support units discover ways in which they can more effectively support telelearning activities. For decentralized institutions, like many universities and colleges, this is an important achievement.

The University of Ottawa and Carleton University jointly sponsored a TLTR workshop in 1998 to begin these processes at their institutions. They have developed an excellent resource on TLTR at (http://www.edteched.uottawa.ca/ottacarl/TLTR/). Although it is still early in the process, several other Canadian institutions have expressed interest in the model and are considering adopting it or trying parts of it in their own sites.

SUMMARY

This chapter has shown that policy processes are critical to the development of sound policies and strategies for telelearning technologies. Based on this research, the following institutional policy processes should be considered:

1. Use a transparent process of deliberation and implementation.
2. Make decisions based on research. Since academic culture values research, the basis for technology decisions needs to be clearly communicated and documented.
3. Enable faculty to feel in control of the technologies and that they fulfill an academic purpose.

The issues associated with online learning are quickly and easily polarized, linked as they are to fundamental ideas about the purpose of education, the role of professors, and the sharing or wielding of power.

Based on this research on policy processes, there are two key areas that need further study. First, there is a need for more research on the impact of policy processes. In the area of telelearning technology, studies are being done to evaluate the technology in terms of cost-benefit, learning outcomes, and pedagogical approaches. More research is also needed on the most effective way to enable universities and colleges to make decisions in this area.

Second, on a broader scale, there is a need for research on the management of change in universities that recognizes and works to uphold those values that make universities unique public institutions — including an unfamiliarity with and even an abhorrence of "management" itself. It is also important that whatever guidelines are developed, these must be sensitive to the variations and differences between universities.

As higher education administrators, teachers, and students seek to maneuver their way through the challenges ahead, they will need to find ways to negotiate change, identify priorities, and find solutions that work. In this context, policy processes become critical. This study has shown that the selection and application of appropriate policy processes for the introduction, application, and use of technology-mediated learning plays a key role in managing technological change in an institution.

DISCUSSION QUESTIONS

1. Which of the policy processes discussed here seem to fit with your organization? What steps would you take to see these processes put in place?
2. Who is involved in the technology planning process in your organization? Could more people be involved in the process?
3. Is the process of technology planning regarded as legitimate by the members of your organization? What role do students play? What about teachers? Others?
4. What are the drivers of change in your organization?
5. Should organizational change be included as part of an information technology strategic plan?

REFERENCES

Bruce, R., Bizzocchi, J., Kershaw, A., Macauley, A., & Schneider, H. (1999, May). *Educational technology planning: A framework*. Victoria, British Columbia: Centre for Curriculum, Transfer and Technology. Retrieved from http://www.ctt.bc.ca/edtech/framework.html

Daniel, J. S. (1996). *Mega-universities and knowledge media: Technology strategies for higher education*. London: Kogan Page.

Gianini, P. (1998, October). Moving from innovation to transformation in the community college. *Leadership Abstracts, World Wide Web Edition, 11*(9).

Gilbert, S. W. (1998). AAHESGIT Listserv, Issue 195. Retrieved from http://www.aahe.org/technology/aahesgit.htm

Green, K. C. (1998, November). *The Campus Computing Project: The 1998 national survey of information technology in higher education.* Encino, CA. Retrieved from http://www.campuscomputing.net/

Hanna, D. E. (1998, March). Higher education in an era of digital competition: Emerging organizational models. *Journal of Asynchronous Learning Networks, 2*(1), 66-95.

Hardy, C. (1992). Managing the relationship: University relations with business and government. In J. Cutt, & R. Dobell (Eds.), *Public purse, public purpose: Autonomy and accountability in the groves of academe* (pp. 193-218). Halifax & Ottawa: Institute for Research on Public Policy and the Canadian Comprehensive Auditing Foundation.

Kim, W. C., & Mauborgne, R. (1997). Fair process: Managing in the knowledge economy. *Harvard Business Review, 75*(4), 65-75.

Morrison, J. L. (1999). The role of technology in education today and tomorrow: An interview with Kenneth Green, Part II. *On the Horizon, 7*(1), 2-5.

Nedwek, B. (1999, March 21-24). *Effective IT planning: Core characteristics.* Presentation to the Society for College and University Planning Winter Workshop, Information Technology Planning, Hawaii.

Norris, D. M., & Poulton, N. L. (1991). *A guide for new planners.* Ann Arbor, MI: Society for College and University Planners.

Olcott, D. J., Jr. (1996). Aligning distance education practice and academic policy. *Continuing Higher Education Review, 60*(1), 27-41.

Ringle, M., & Updegrove, D. (1998). Is strategic planning for technology an oxymoron? *Cause/Effect, 21*(1), 18-23. Retrieved April 20, 1999, from http://www.educause.edu/ir/library/html/cem9814.html

Tamburri, R. (1999). Survival of the fittest. *University Affairs,* 8-12.

This case was previously published in L. A. Petrides (Ed.), *Cases on Information Technology in Higher Education: Implications for Policy and Practice*, pp. 34-42, © 2000.

Chapter V

Implementing Software Metrics at a Telecommunications Company:
A Case Study

David I. Heimann, University of Massachusetts Boston, USA

EXECUTIVE SUMMARY

Establishing and using fundamental measures of software development progress is an essential part of being able to predictably produce high-quality customer-satisfying software within a reasonable agreed-to timeframe. However, in many organizations such measurement is done incompletely or not at all. While various forces encourage measurement programs, others form barriers to such implementation. This case study explores a formal Metrics Program established to track and analyze the development of a new version for the major voicemail product of a mid-sized telecommunications company. The study addresses the evolution of the company's organizational structure and culture that led to the adoption of the program, the components and structure of the program, the implementation of the program, its effects on product quality and timeliness, and what happened thereafter. The study also raises questions with respect to facilitating an organizational atmosphere where a Metrics Program can flourish.

BACKGROUND

This case study investigates a formal Metrics Program established to track and analyze the development of the major voicemail product for a mid-sized telecommunications-related company, which we shall call "Telogics." It is described from the point of view of the metrics lead for the program.

The study describes how, in an overall downward trend for process and quality assurance, a window of opportunity can open to carry out significant product-enhancing work in these areas. It also raises the issue of how to preserve the gains and historical memory when the window closes and the downward trend guarantees that it will not reopen in the near future.

Aside from the use of metrics to improve the development of a major software product, three other major themes appear in this case:

1. The slow but steady absorption after the merger of two companies of one legacy company by the other one, and the organization evolution brought about but this absorption.
2. The shift by Telogics away from a mature product with a large but static customer base to a multiple line of older and newer products with many new customers, and prospective markets, with an accompanying refocus of Telogics from engineering-driven to market-driven). This maturation of product line with an accompanying flattening of revenues and shift toward new products is occurring within much of the telecommunications industry, not just Telogics.
3. The implementation of a major quality process in a major division of Telogics as the company itself was moving away from a process/quality orientation due to its changing products and organization, and the economic adversity being visited on it and many other companies in the telecommunications industry.

Telogics makes communications systems and software that phone companies use to offer call answering, voice or fax mail, communications surveillance and recording, and other services. Currently (from financial reports in early 2003 for the year 2002), it has around $750 million in revenue, with around $180 million in operating losses and about 4,800 employees. During the period of the case study (approximately the year 2000), it had approximately $1.2 billion in revenue, with around $250 million in operating income and about 6,000 employees. It started in the 1980s as a company producing a large-scale voicemail system.

During the case study period its products were based on two versions of this voicemail system, and they were beginning to branch out into new products such as text-to-speech and intelligent-network applications. It is currently continuing this expansion of its product range, as well as revamping its voicemail-based product after since having eliminated one of the two original versions. As with many telecommunication-based companies, its profits and stock price, very high during the case study period, began declining near the end of that period and has since significantly declined. However, the price is currently stable at the lower level. The corporate organizational structure has evolved over Telogics' history. Because of the key impact this had on the existence, form, and progress of the Metrics Program, the organizational evolution during the 1998-2001 timeframe is tracked in detail in this case study.

History

For most of the time in its earlier years, the style of the company (we shall call the company in its earlier years "Techanswering") was the traditional "cowboy" or "death march" one, very much at the initial process level in the Capability Maturity Model® (Humphrey, 1989). There was hardly any process, a lot of what was there were in people's heads, their notebooks, or their e-mails. Schedules were determined haphazardly, as were requirements. Towards the ends of projects, Herculean efforts were expended to get the job done, with 70- to 80-hour weeks common. However, frequently the product was delivered late, often with significant major defects. However, with large-scale voicemail being early in its product maturity stage and with few competitors, customers were willing to accept this in order to obtain the then-unique product.

However, it gradually became apparent to Techanswering that the climate was beginning to change, with competitors appearing and customers becoming less tolerant. A major wakeup call was the rejection by a large foreign telephone company of a major Techanswering proposal, a significant setback. Techanswering had "quality assurance" groups that were actually testing groups for each of the various development centers within the Research and Development organization; but partly in response to this rejection, it formed a Corporate Quality group to guide the software development process for the entire Research and Development (R&D) organization. One of Corporate Quality's functions was to identify, gather, and report metrics relating to defects and timeliness.

In mid-1998, Techanswering merged with a foreign voicemail system producer, which we shall call "Techvox," to form Telogics. The two companies had complementary voicemail products and markets. Techanswering built a larger, wireline-based, more complex product, which they sold to comparatively few but large customers, mostly in the United States, South America, and Asia. Techvox built a smaller product, mostly wireless based, which they sold to a larger number of smaller customers, mostly in Europe and the Middle East. As a single company the merged entities therefore commanded a significant market share of a wide-ranging market.

Though the merger was more or less one of equals, Techvox was more "equal" than was Techanswering, an important factor later on during the Metrics Program. Techvox was financially more conservative, less of a risk taker, and more business-oriented than technology-oriented. Especially initially, Techvox supplied a counterfoil to Techanswering's let's-go-for-it somewhat disorganized style, and there are some within the legacy of that company who have said that if the merger had not occurred, Techanswering would have become bankrupt or bought out, considering the way they were doing business.

Despite the more careful style of Techvox compared to Techanswering, it was the latter who continued its interest in software process improvement, quality assurance, and metrics. In fact, by the time of the merger, the scope of its Corporate Quality group had been expanded to cover the whole of the company.

In the spring of 1999, the U.S. product organization (hereafter called The Division) of Telogics was renamed after its primary voicemail product to reflect that its focus was on designing, developing, serving, and supporting this product. Sales and marketing were a corporate-wide function with geographically based divisions, with The Division "contracting" with them to produce orders that these marketing divisions secured. The Division was therefore primarily engineering driven.

Figure 1. Telogics organization before case-study period

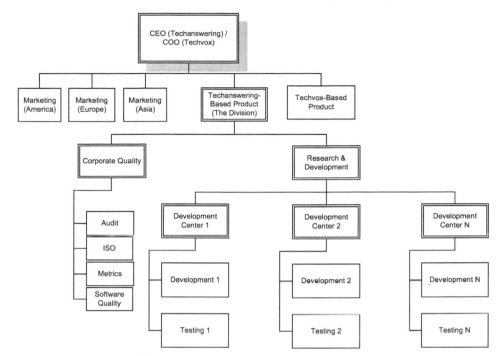

Organization Just Prior to the Case Study Period

At this point, The Division was carrying out the quality assurance function as follows (see Figure 1): Each of the development centers had a testing group, which reported to the manager of that development center. They in turn reported to the Deputy Vice President (DVP) of R&D.

Within Telogics' Corporate Quality organization was the Software Quality (SQ) group. The SQ group was responsible for final product quality. It reported to Corporate Quality, a peer of R&D, thus providing an oversight of product quality independent of the development center managers.

Corporate Quality had another key function as well; the metrics function. This function was to identify, gather, and report metrics for project and corporate-wide activities. This covered metrics for product-related defect occurrences, process-related defect occurrences, and on-time performance.

SETTING THE STAGE

Motivation for the Metrics Program

Early in the year 2000, The Division delivered a new version of its flagship voicemail product, Version B, to its major customers. Unfortunately, the delivery did not go well.

There were a large number of customer-visible defects in the delivered product, many of them major. In addition, frequently the installation at customer sites had problems, some of which affected service. Furthermore, the delivery was several weeks late. Some of the customers warned The Division that this kind of delivery could not happen again if they were to remain customers.

As planning began for the next version, Version C, due out in a little over a year, The Division's R&D organization, as well as the Corporate Quality group, began to develop a plan to make the quality and timeliness of Version C much more predictable than had been the case for Version B. A significant part of that effort was to develop and carry out a metrics gathering and analysis program (see also McAndrews, 1993; Giles, 1996). This was motivated by two major factors:

1.   When a technology reaches the "early majority" or later stage in its maturation, customers regard quality as a major decision factor, whereas earlier in the cycle, new technology features and strong performance are more the major factors (Rogers, 1995; Moore, 1995; Norman, 1998). Telogics' voicemail product was moving into the "late majority" stage, and the major customers, as demonstrated by their reaction to the mediocre quality of Version B, demanded quality in the product.
2.   Defects detected and repaired early in the software development life cycle (SDLC) cost significantly less than if they are detected and repaired later in the cycle. Several sources (Boehm & Papaccio, 1988; Willis, 1998; McConnell, 2001) have propagated a rule of thumb that the cost of a defect increases by an order of magnitude for each step in the SDLC that passes by before it is repaired.

A Major Organizational Change

In early July 2000, just as the Metrics Program was being planned and begun, a major reorganization occurred (see Figure 2) that would eventually significantly affect the program. Instead of one U.S. and one foreign corporate organization, with the U.S. R&D and marketing departments all reporting to the U.S. Operations vice president, each

Figure 2. Telogics organization during case-study period

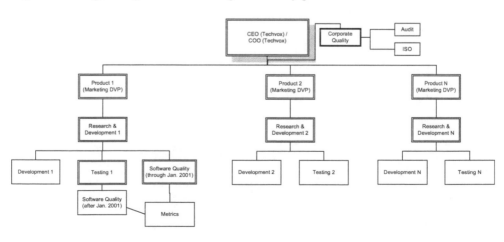

product line became its own division, with its own R&D department reporting to a marketing-oriented VP. Each division would now have its own testing group, reporting to that division's R&D department. Whether or not a division had a quality group was left up to that division. The overall Corporate Quality organization became much reduced, concentrating only on audits and ISO 9000 registration.

The metrics group, including the metrics lead, was merged into the SQ group, which in turn was moved into the division responsible for the major voicemail product. It reported only to R&D for that division; no other corporate divisions retained any connection with the group. Furthermore, the U.S. Operations vice president, a major champion of software quality within Telogics, was stripped of his major corporate responsibilities and became simply an administrator of the U.S. facilities.

A short time later, the CEO, the only former Techanswering executive in a major top corporate post, was moved into a ceremonial position. No Techanswering executives now occupied any top corporate posts. The focus of Telogics became more oriented to its Techvox legacy than its Techanswering legacy and became more marketing driven than engineering driven.

Establishing the Metrics Project

With the July reorganization, an executive who had been a key direct report for the DVP for R&D in The Division became the DVP for R&D in the new voicemail division. He became a strong champion for improving the predictability and hence the quality and timeliness of the Version C product. As long he remained a strong champion, the metrics project, as a part of the predictability effort, moved steadily forward.

The Version C predictability improvement effort started in June and accelerated with the corporate reorganization in July. Four teams were formed to develop plans for the effort: Metrics, Development Lifecycle, Defect Management, and Estimation. The SQ manager, the metrics lead, and an outside consultant developed a metrics plan, and the Metrics Team approved it, forming the foundation for the Metrics Program.

CASE DESCRIPTION

The Metric Profiles

The major basis for the Metrics Program was a set of Metric Profiles derived from templates in Augustine and Schroeder (1999) which formed a set of measurements from which the actual measures used in the effort were chosen. The profiles encompassed four areas: Quality, Functionality, Time, and Cost. The measures were designed to answer key organizational questions, following the Goal-Question-Metric approach (Basili & Rombaugh, 1988) and the metrics planning process of Fenton and Pfleeger (1997). A sample Metric Profile is shown in the Appendix. Note that the measurement state-of-the-art continues to move forward, with approaches being developed such as Measurement Modeling Technology (Lawler & Kitchenham, 2003).

While The Division did not formally seek a CMMI Level 2 status, the Metrics Program held to the spirit of the Measurement and Analysis Key Process Area (KPA) for CMMI Level 2 (CMMI Product Team, 2001).

Which Managers Received Reports and How They Used Them

The metrics-reporting program started in mid-August 2000, beginning with weekly reports. They evolved and expanded, eventually reaching several reports per week, through the delivery of Version C in May 2001.

Initially the report went to the project manager. However as time passed other people were added to the list, so that eventually it also went to the DVP of R&D, the director of testing, the two testing managers on Version C, the Version C development managers, and the two people maintaining test-case, defect-fix, and defect-verification data.

The project manager used the report in his weekly status reviews with the DVP. In addition he used the information as part of his presentation at the R&D Weekly Project Reviews, a management forum to communicate status for projects in R&D. Others in the division also used the report. For example, the director of testing used it in his meetings with the testing managers to make sure that the all their quality-related data were consistent.

Overall, the uses of the metrics by management were as follows:

- *By project management:* To track defect-resolution and testing progress, to become alert to problems early, to manage development and testing resources, and to respond to senior management.
- *By senior management:* To assess the health of the project both absolutely and in comparison to the other projects within the organization, to communicate with customers on product progress, to plan ahead for product delivery and follow-up efforts.

What Metrics Were Reported

1. Pre-Test-Stage Defects

The initial period of metrics reporting, during the summer and early fall, focused on pre-test metrics, i.e., defects, comments, and issues arising from the system analysis, system design, and coding phases of the SDLC. These characteristics were reported according to a Telogics procedure called the "Document and Code Review Process for Version C" (2000). The procedure introduced a consistent yardstick for declaring whether a characteristic was a defect or not. Each characteristic was judged to be one of three possibilities (see Figure 3 for a sample tabulation):

Defect   A defect, with a severity of either major or minor.
Comment   Not a defect.
Issue   Further investigation is needed to determine defect status.
              To be resolved to either a "Defect" or a "Comment."

2. Test-Stage Defects

The most active period of metrics reporting, during the fall and winter, focused on metrics relating to testing and defects found during test. The testing took place in four

Figure 3. Sample table of weekly pre-test defects

Week Ending	Reviews Held	Total Incidents	Comments	Issues	Defects			
					Total	Major	Minor	Severity Omitted
8/12/00	5	41	25	11	5	0	3	2
8/19/00	9	10	5	3	2	1	1	0
8/26/00	3	1	1	0	0	0	0	0
9/2/00	7	45	15	12	18	6	7	5
9/9/00	4	5	4	1	0	0	0	0
9/16/00	2	2	1	0	1	0	1	0

Figure 4. Sample defect graph

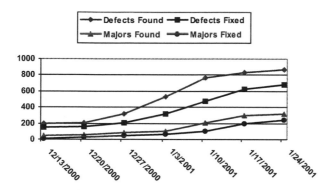

phases-preliminary (Phase 1), integration (Phase 2), full-scale or quality-assurance testing (Phase 3), and system testing (Phase 4). All of the test phases met their scheduled completion dates.

The defects were kept in a commercially available database called the Distributed Defect Tracking System (DDTS). Since DDTS was an Oracle-based system, databases such as Access could link to it, and thus automated metrics tracking systems became possible. The metrics lead used Access queries and programs extensively to obtain the defect data for the required reports (see sample chart in Figure 4), which made it feasible to produce presentations in a day. This allowed for the variety and volume of reports that management required, as well as reports based on real-time data often no more than an hour or two old.

3. Tests Scheduled, Executed, Passed, Failed, and Blocked

The tracking of tests scheduled, executed, passed, failed, and blocked (Figure 5) was maintained by an internally developed test tracking system, which will herein be called TestMate. TestMate was developed using Visual Basic and Access, and included password-protected updating, identification of defects by feature sets within the

Figure 5. Sample test cases table

Test Cases 12-Jan-2001	
Total Cases	2,024
Pending	1,553
Executed	471
Passed	424
Failed	37
Blocked	21

products and by features within the feature sets, identification of DDTS defect numbers for tests finding defects, and identification of testers and developers.

Since TestMate was available Telogics-wide, it allowed anyone to see how testing was proceeding, either overall or for a particular feature. This provided an openness that greatly aided the communication among the various organizations working on Version C and allowed the metrics lead to produce metrics reports in real time.

4. Code Length and Code Changes

To put the number of defects and the testing effort in perspective, one needs to know the size of the software product. This is not as straightforward as would appear (Fenton & Pfleeger, 1997). The usual measure of size, the number of lines of code in the software, may be misleading. Some lines may be far more complex than others, and languages require different amount of lines to represent an action. More sophisticated approaches to code size than lines-of-code exist, such as function points or code complexity. However, these require more sophisticated tools and skills, neither of which Telogics had or was willing to acquire. Therefore, lines-of-code was indeed used to measure code size.

Even using the lines-of-code measure is not straightforward, however (Humphrey, 1995). While Version C had around 2 million lines of code, most of it was code carried over from previous versions. Only several hundred thousand lines were actually created, changed, or deleted from Version B (see Figure 6). The code counter had to include all builds within version C in order to provide a meaningful size metric.

The code size and activity information was very valuable:

- It provided a perspective of the defect exposure and risk, as well as determining whether a surge in defects was due to defect proneness or a high amount of activity.
- One could determine when development activity was basically complete, so that defect tracking and analysis no longer had a moving target and better estimates could be made.

5. Build, Schedule, and Cost Reports

Part of the project planning included a build schedule (i.e., evolution of product functionality), which included all the builds to be created, the contents of each build, and the planned date for each build. As work proceeded and each build was created, the actual date was compiled and tracked against the planned date.

Figure 6. Sample lines of code activity by build

Build	Added LOC	Changed LOC	Deleted LOC	Total
1	32,428	3,851	1,889	38,168
2	1,746	11,385	472	13,603
3	167	0	0	167
4	283	84	0	375
5	106	37	36	181
6	481	96	20	577
7	50	77	9	136
8	2,329	797	382	3,508
9	1,496	1	0	1,497
10	7,913	10,195	1,473	19,581
…	…	…	…	…
Total	87,462	40,111	25,772	153,395

Another metric tracked was schedule. This was done through tracking major milestones for development, testing, and release. The milestone accomplishment dates were provided by the project manager who reported release timeline and content on a weekly basis.

The one significant schedule event was a slip for the beginning of full testing, due to delays in development for a critical feature, from September to December 2000. However, preliminary testing continued through this time on the significant portion of the project already developed, so by the time full testing formally began, much of it was already completed. The date for release testing therefore was met, as was the date for clean pass delivery and customer acceptance testing.

Cost was included at the start of the metrics effort. However, as the project proceeded, it became clear that cost was not a high priority for management, as compared to quality, functionality, and schedule. Hence early in the metrics effort, cost was dropped as a metric.

Evolution of the Metrics Reports

The reports evolved as time went by, with increasing acceptance and understanding, broader and higher distribution and exposure, ongoing requests for more information to be included, and a request from upper management for analyses in addition to reporting. By late in the project cycle, the reports covered a broad spectrum, including the following:

- Cumulative defects from review summaries and reviews held, by week (total and majors) and by lifecycle stage.
- DDTS defects found or fixed by week, by when found, and by product, projected date when defect criteria will be met, and defects found by function (development, testing, etc.).
- Estimated defects found (total and majors), estimated defects fixed, estimated defects remaining, projected defect count, estimated versus actual.
- Defects fixed but not verified, projected date when verification backlog will have been processed.

- Test cases to be run, executed, passed, failed, blocked (especially for one critical product), projected date when all tests will have been run, project date when tests-passed criteria will be met.
- Resources available and used (testers and fixers and verifiers).
- KLOCs by build (created, changed, deleted, total), etc.
- Analyses-valuation of current data, alerts as to upcoming problems, de-alerts as to problems addressed, conjectured reasons for discrepancies and problems, projected defects outstanding, tests passed, fixes verified date projections, and likelihood of target dates being met.

This evolution was a natural process, moving ahead as the metrics lead became more familiar with the analysis process and the overall project, as the project and management users became more familiar with the reports and how they could be used in day-to-day administration and project guidance and planning, and as data providers became more familiar with the data to produce, and as automated tools were developed and used. Especially key was the request by management at a relatively early stage in the reporting evolution to have the metrics lead include analyses and recommendations.

Managerial and Other Impacts of the Metrics Program

Version C delivered on time, whereas Version B was late, and Version C had much fewer customer-visible defects, especially major ones, than was the case with Version B. Precise numbers are not available, but a good relative approximation is:

	Version C	Version B
Open customer — visible major defects	2	20
Open customer — visible minor defects	30	150
Delivery vs. schedule	On Time	Two weeks late

Version C also had many fewer installation problems.

This outcome likely helped to restore Telogics' reputation with the three major customers whose relationship had been strained by the experience of Version B. This was a major positive impact for senior management, who had become worried about these customers. The following aspects of the Metrics Program were instrumental in having this happen:

- *A much broader and deeper spectrum of metrics were reported.*

Before the Metrics Program only a few metrics were being reported such as defects, test cases, and broad schedule adherence. As a result of the Metrics Program, the metrics became broader and deeper. Defects before testing were reported, as well as those during and after testing. Defects were reported by feature set as well as overall and were also broken out by new defects versus forwarded open defects from previous versions. Code size was tracked over the full build cycle, with specific additions and changes noted. Both test cases and schedule adherence were broken down by feature set.

All of this richer information allowed more comprehensive and specific analyses, often pinpointing alerts in specific areas even when the defect level was OK on an overall basis. The implications for management were that they could identify defects at earlier stages in the SDLC when they were much less expensive to repair, as well as being much more able to control the stability of their code size and build patterns.

- *Important issues were detected before reaching crisis levels.*

As the Metrics Program and the software development continued and evolved, the metrics lead and others were able to detect and alert anomalies in important areas before they became far off course and reached crisis level. Examples of these were:

- One of the feature sets in Version C was a new function, in fact a key deliverable for this version. Several times during the project, an unusually large number of defects began occurring in this function. Within days after such a trend began, the metrics lead detected it and notified the project manager and the DVP, resulting in corrective action being taken. Upon delivery, this function — a subject of worry during the project planning and development — operated smoothly for the customer, a key to their satisfaction.
- Open major defects, a major phase exit criterion, were tracked and supported. On several phases the metrics lead's projections of open majors at the intended milestone date showed that the number of open majors would exceed the exit criterion. Management was alerted and told where in the product the large numbers of open majors were so that testing and repair efforts were focused on those areas. As a rule the numbers of open majors then decreased, resulting in the milestone criteria being met. Version C met the criterion of zero (or very few) open majors at delivery, whereas Version B had not.
- As Telogics moved through the final testing phase before delivery, they had resolved what was originally a large backlog of open defects down to a reasonable level. However, the metrics lead noticed that many resolved defects had not yet been verified and thus could not be considered completed. In fact, the backlog of resolved-but-not-verified defects began to increase to substantial proportions. Project and senior management were notified, who began assigning development staff to assist in verifying these defects. After several weeks of these efforts, the substantial backlog had been fully addressed.

Both project and senior management could respond much faster and effectively to these developments, a significant boon for both levels of management.

- *Management was able to give emphasis to areas of the project where it did the most good.*

With more detailed reports and the inclusion of analyses, both project and senior management had information by which they could emphasize areas of the project where they were most effective. A critical example was a particular new feature that was key to Version C but had risk in developing; the Metrics Program provided specific defect, code build, schedule, and testing information on this feature. This allowed management to use its scarce time more effectively in guiding the Version C development.

CURRENT CHALLENGES/PROBLEMS FACING THE ORGANIZATION

A Fatal Organizational Shift

The Metrics Program succeeded in what it set out to do, and the Version C product was much improved from its predecessor. However, despite the Metrics Program's success, larger corporate and organizational issues adversely changed the atmosphere in which the Metrics Program, and indeed the overall corporate quality function, operated. This changed atmosphere first undermined the organization, SQ, in which the Metrics Program resided, then did the same with the Metrics Program itself. The loss of support became apparent during the last several months of the project. The Metrics Program ended with the delivery of Version C, with no continuation to future projects.

In retrospect this was a culmination of an evolution that began taking place well before the July 2000 reorganization and may have begun with the merger of the two legacy companies in 1998 — the struggle between the Techanswering part and the Techvox part of Telogics for dominance. While the merger was technically one of equals, it soon became clear that the Techvox part of Telogics was politically the stronger entity.

This seriously affected the long-term prospects for metrics, process improvement, and software quality assurance (SQA) as a whole. The former Techanswering had been very much engineering oriented and relatively friendly to SQA, while the former Techvox was more finance oriented and relatively unfriendly to SQA. So long as economic times were good and the Techanswering influence on Telogics was still significant, the environment was friendly to SQA. However, as the Techvox influence became more and more pervasive, and as economic times began to decline, this environment became much less so.

Before the July 2000 reorganization, the R&D and service portions of Telogics were organized as a U.S.-based division and a foreign-based division. This allowed the Techanswering atmosphere to thrive in the engineering-oriented U.S.-based division. This division had a strong Corporate Quality organization, independent from the R&D organization, which had direct ties to a similar organization in the old Techanswering.

The July 2000 reorganization changed Telogics' structure from a horizontal one to a vertical one. Each major product had its own organization, with a sales and marketing group, an R&D group, and other functions. It became up to each organization as to whether or what quality activities they would carry out. Significantly, these organizations were headed by marketing executives who now reported directly to top management, bypassing engineering entirely.

Paradoxically, in the short term the reorganization helped the Metrics Program. Before, metrics had been a part of Corporate Quality, thus a staff activity removed from the R&D effort. However, with the reorganization the DVP of R&D for Telogics' largest product organization wanted a strong metrics activity and established it as part of an independent SQ group reporting directly to him. Metrics therefore became a line activity with direct influence within this organization.

However, the general atmosphere continued to become more metrics-unfriendly. A few months after the reorganization, the overall corporate leadership of Telogics changed. This removed all American influence from the top corporate levels, from which the support for quality, process improvement, and SQA as a whole had come, and eventually affected the R&D DVP's support.

In the early fall of 2000, the DVP hired a new director of testing for the division. Despite the DVP's earlier emphasis on quality and predictability, the man he hired did not favor this approach; he emphasized testing to ensure quality, with little or no overall process improvement or measurement to ensure product quality before the start of testing.

This represented a major turning point in the division climate. A major unanswered organizational question is why the R&D DVP, who had been such a strong champion for predictability through measurement and process improvement, hired a key subordinate with such an opposite outlook. This move did not appear to be a hiring error, but rather the beginning of a significant change in the senior manager's own outlook, as he over a short time became far less sympathetic to what he had championed earlier. Whether he himself had a change of heart or whether the hire and the change of policy were forced on him by upper management is still uncertain, but in retrospect the fate of future metrics efforts and process improvement was sealed at this point.

The new testing director did not want to have a separate software-quality group reporting directly to the DVP. He influenced the DVP to abolish the SQ group and absorb its members into the testing organization. The Metrics Program was now under the management of a person who opposed such a program.

Early in the spring of 2001, the main testing of Version C was completed and final testing began. At that stage the project manager, likely emboldened by the revised organizational situation, attempted to exclude the metrics lead from the reporting process. The DVP intervened and demanded the reporting continue. However, shortly thereafter even that support diminished and disappeared. In the spring of 2001, Telogics began facing a business slowdown, part of the slowdown in the overall telecommunications sector. In a layoff shortly thereafter, nearly all personnel, including the metrics lead, associated with metrics and software quality were released.

The slowdown hit Telogics hard. In the summer of 2001, the stock price fell sharply and has not recovered to date. Earnings dropped significantly and further layoffs occurred. The entire Telogics division in which the Metrics Program took place was phased out in favor of the division making the other version of the voicemail product, a division and a product with origins in the old Techvox.

Questions for the Organization

Why, with the evidence of the Metrics Program's success, did the division choose to eliminate it? This is a difficult question to answer explicitly; the managers directly involved might not consciously know, and even if they did, would probably consider it sensitive and proprietary even within Telogics. However, a number of hypotheses can be raised:

- Financially tough times were beginning, as the stock price drop eventually was to show. Telogics did what a lot of companies do in such circumstances: cut costs and programs, especially newer, more innovative ones.
- The division was one that was not going to be a part of the company's dominant future. It produced one of two versions of Telogics' legacy product. Only one

version was likely to survive, and since the other version derived from the dominant wing of the company, it was becoming ever clearer that the division was going to eventually lose out. The DVP may have given up hope for the long-term future of the division, so was no longer oriented towards long-term thinking.

- The DVP may have been the sort of person who goes with the dominant paradigm, whatever it is at the time. At the beginning of the metrics effort, that paradigm was still quality and predictability, while at the end it had become strictly maintenance and legacy.
- The testing director was not on board when the Metrics Program was created, so he did not have any ownership in it.

The question for Telogics is how to assure quality and predictability of its products in the adverse times it is facing and to preserve the legacy of its Metrics Program. In these times the temptation is strong to cut all possible costs, spending only what is absolutely necessary to create and market its products and services. Moreover, many of the quality and metrics champions have either voluntarily or involuntarily left Telogics. However, even in adverse times customer satisfaction is very important. In fact, customer satisfaction is even more important in adverse times, since there are fewer customers and smaller orders, and a loss of a customer has much more impact and is harder to replace. The legacy of the Metrics Program may still be available for Telogics, both in terms of its physical artifacts and processes, and in terms of corporate memory. Since to ignore quality in these times is to invite a "death spiral," this legacy is a valuable asset to Telogics, already paid for. However, this legacy is perishable, and the window of opportunity may be closing. The major challenges for Telogics are therefore:

a. How does Telogics best use this hard-earned legacy to assure quality as it continues to do business in these critical times?
b. How can Telogics best preserve the tools, artifacts, and data from the Metrics Program so that these assets can be easily recovered in the future?

Should Telogics consider a future metrics or other software process improvement program, they will need to address a number of questions arising from the Metrics Program of this case study. These questions apply not just to Telogics but also to other companies considering such efforts.

1. What degree of independence must a Metrics Program have, and how does one maintain that independence in the face of the natural inclination of a project manager and a testing director to want to control the formulation, compilation, and reporting of data? Note: there may be parallels here with the relationship of the Bureau of Labor Statistics to the rest of the Department of Labor.
2. How does one effectively obtain defects before the testing phase in the software development life cycle (SDLC)? How does one define a defect in the requirements, analysis, or design phases? How does one define severity, and how does one define when such a defect is corrected?
3. How does one effectively and feasibly carry out requirements traceability (the linking of software defects to the design elements, specifications, or requirements they affect)? This was one of the most difficult elements of the Metrics Program.

4.   Quality, process improvement, and metrics efforts often have variable support. There are periods where strong support exists, especially when times are good. However, when times become difficult, too often the support wanes, and such programs become one of the first cutbacks.

   a.   How do quality champions create an environment where quality and metrics efforts are faithfully followed in good times or bad?
   b.   In order to make continuous overall progress in the variable environment, a company must generate strong advances during the favorable times, but then preserve them during the unfavorable times until support again emerges. Note that since during unfavorable times the   personnel carrying out these efforts may likely leave voluntarily or be laid off, the preservation effort must survive such occurrences. What is the best way to make this happen?

5.   What is the best way to effectively market a Metrics Program in an organization, both to get it established in the first place and to maintain its political support once it is in operation?
6.   In the time span during which the Metrics Program operated, the Internet (and intranets and extranets) has become an increasingly important manner of communication. Any future Metrics Program should make full and effective use of the Internet. What is the best manner to use the Internet to collect metrics data, to calculate the metrics, to formulate effective reports, and to disseminate (active) and allow easy access (passive) to them?
7.   How does one develop a win/win situation between the metrics lead and the project manager, that is, how can one construct the metrics effort so that its success is beneficial personally to the project managers and their careers? Similarly, how does one develop a win/win situation between the metrics lead and the development manager, as well as the metrics lead and the testing manager?
8.   In relation to the previous question, a possible mechanism is to establish, through software process improvement organizations or industrial consortiums, a knowledge base of successful and unsuccessful metrics efforts (articles such as Goethart & Hayes, 2001, represent a step in this direction). In this way if loss of support causes the loss of corporate memory and infrastructure for metrics, it is preserved within the community organizations for other enterprises to use, or even the original company when support once again arises. What is the best way to bring this about?

Covey, Merrill, and Merrill (1994) describe four quadrants of activities, as to whether or not the activities are urgent and whether or not they are important. They emphasize "Quadrant II activities" — those that are important but not urgent-stating that these tasks require continual oversight, lest they remain unaddressed because they are not urgent, but hurt the individual or organization because they are important. Software process improvement (SPI), in particular software measurement and analysis, is for most organizations a Quadrant II activity. Unlike testing, a close relative, it is not required in order to ship a software product. However, those organizations that neglect SPI over the long haul suffer declines in the business worthiness of their software products, as product quality declines because of a lack of quality assurance and as product quality becomes more important as a product matures (Rothman, 1998). The experience of this case illustrates the challenge in establishing and maintaining an SPI as a Quadrant II activity, but also SPI's corresponding significant value to product quality, customer satisfaction, and customer retention when that challenge is met successfully.

REFERENCES

Augustine, A., & Schroeder, D. (1999, June). An effective metrics process model. *Crosstalk*, 4-7.

Basili, V., & Rombaugh, D. (1988). The TAME Project: Towards improvement-oriented software environments. *IEEE Transactions on Software Engineering, 14*(6), 758-773.

Boehm, B. W., & Papaccio, P. N. (1998). Understanding and controlling software costs. *IEEE Transactions in Software Engineering, 14*(10), 1462-1477.

CMMI Product Team. (2001, December). *Capability Maturity Model® Integration (CMMI), Version 1.1; CMMI SM for Systems Engineering, Software Engineering, and Integrated Product and Process Development (CMMI-SE/SW/IPPD, V1.1) staged representation* (Technical Report CMU/SEI-2002-TR-004).

Covey, S., Merrill, R., & Merrill, R. (1994). *First things first: To live, to love, to learn, to leave a legacy.* New York: Simon & Schuster.

Fenton, N., & Pfleeger, S. L. (1997). *Software metrics — A rigorous & practical approach.* London: International Thomson Computer Press.

Giles, A. (1996, April). Measurement — The road less traveled. *Crosstalk*.

Goethart, W., & Hayes, W. (2001, November). *Experiences in implementing measurement programs* (Technical Note CMU/SEI-2001-TN-026). Software Engineering Institute.

Humphrey, W. (1989). *Managing the software process.* Reading, MA: Addison-Wesley.

Humphrey, W. (1995). *A discipline for software engineering.* Reading, MA: Addison-Wesley.

Lawler, J., & Kitchenham, B. (2003, May/June). Measurement modeling technology. *IEEE Software*.

McAndrews, J. (1993, July). *Establishing a software measurement process* (Technical Report CMU/SEI-93-TR-16). Software Engineering Institute.

McConnell, S. (2001, May/June). An ounce of prevention. *IEEE Software*, 5-7.

Moore, G. A. (1995). *Inside the tornado: Marketing strategies from Silicon Valley's cutting edge.* New York: Harper Business.

Norman, D. (1998). *The invisible computer.* Cambridge, MA: MIT Press.

Rogers, E. M. (1995). *Diffusion of innovations* (4th ed.). New York: The Free Press.

Rothman, J. (1998, February). Defining and managing project focus. *American Programmer*.

Telogics Corp. (2000). *Document and code review process for Version C* (Internal Communication).

Willis, R. R. et al. (1998, May). *Hughes Aircraft's widespread deployment of a continuously improving software process* (Technical Report, CMU/SEI-98-TR-006). Software Engineering Institute.

APPENDIX

Sample Metric Profile

TELOGICS METRIC PROFILE		
Metric Name: Defect Find Rate	**Importance:** High	**Owner:**
Details		
Metric Description: There are pre-test defects detected during requirements, design, and coding, documented in the online notes file. There are test defects, identified by developers during the integration phase and by test personnel during the test and validation phases. There are also defects found in deployment. Test and deployment defects are documented in the defect-tracking system, with severity ratings from 1 (show stopper) to 5 (not a defect).		
Method of Capture: From planning to the start of integration testing, defects are maintained in the online notes file, though in the future defects could be entered into the defect-tracking system. From the start of integration testing through deployment, defects are maintained in the defect-tracking system. There is a procedure to capture data in the defect-tracking system to an Excel spreadsheet.		
Who Is Responsible for Capturing: Software Quality		
When Captured (frequency): Biweekly until start of validation phase, then weekly until deployment.		
When Metric Capturing Ends: One year after deployment.		
When Reported: As required.		
How Reported: The cumulative number of defects found and the cumulative major defects found are reported as two line graphs. The X-axis is the time in weeks from the start of the project, and the Y-axis is the number of defects.		
Who Is Responsible for the Metric Report: Software Quality		
To Whom it Is Reported: Project management		
Who Analyzes the Data: Product and Project management, Quality (Testing, Software Quality, Process Quality), Development managers		
Additional Comments		
Link to Project Goals: Provides a way to better quantify the effectiveness of non-test methods in finding defects early in the development cycle. May provide timely insight into the status of the testing phase and may provide indicators as to when the software product is ready for release. A high incidence of errors after deployment indicates ineffectiveness of test and non-test methods and increases project cost.		
Issues: Currently, early-stage defects reported in the online note file are reported inconsistently, both in terms of whether a defect is reported at all, and the language and format consistency in how they are reported. In order to obtain meaningful metrics, a process needs to operate on when and how early-stage defects are reported. Note that for later-stage defects, there is such a process using the defect-tracking system, which allows for meaningful defect metric in these stages.		

David I. Heimann is a professor in the Management Science and Information Systems Department of the College of Management at the University of Massachusetts Boston (USA). His PhD is in computer science from Purdue University. He has held positions in government and industry, performing activities in reliability modeling, simulation, database management systems, probabilistic modeling, software analysis, and software process improvement. Among his publications are articles in various Proceedings of the Annual Reliability and Maintainability Symposium *and a book-length article on system availability in* Advances in Computers. *In 1984, he received the Pyke Johnson Award for a paper of outstanding merit from the Transportation Research Board.*

This case was previously published in the *Annals of Cases on Information Technology*, Volume 6/2004, pp. 603-621, © 2004.

Chapter VI

Managerial Issues for Telecommuting

Anthony R. Hendrickson, Iowa State University, USA

Troy J. Strader, Iowa State University, USA

EXECUTIVE SUMMARY

In this paper the issues faced by firms in today's telecommunications environment are compared and contrasted with an actual telecommuting case study of Trade Reporting and Data Exchange, Inc. (T.R.A.D.E.), a software engineering company located in San Mateo, CA. Initial results indicate that telecommuting was successful for T.R.A.D.E because the required technology was widely available (the candidate initiated the idea and had the necessary industry and company experience) the organization could provide the flexible work arrangement while retaining a valuable employee, the employees were able to live in a geographic area of their choice, overall costs could be shared by the company and employees, the job category was an ideal fit, and existing procedures were in place for communicating and managing the geographically detached worker. As telecommunications technology evolves this arrangement will continue to challenge the firm and their employees.

BACKGROUND

Company Background

Trade Reporting and Data Exchange, Inc. (T.R.A.D.E.) is a software engineering company located in San Mateo, CA. The firm was created in 1992 to provide international corporations access to a wide variety of international trade information stored on the T.R.A.D.E'.s massive databases. T.R.A.D.E. gathers data from a variety of sources including U.S. Customs documents, Chinese Customs Administration, Dun and Bradstreet, and numerous other government trade documents. They provide general information and specific trading activities on hundreds of thousands of international buyers, distributors, and suppliers.

This data is consolidated and organized for optimal access by T.R.A.D.E. The company disseminates this information via monthly and quarterly updated CD-ROMs sent directly to the firm's subscribers. Hardcopy reports are also available. In addition, the firm offers customized reporting and monitoring services available on an ad-hoc or event triggered notification basis. The CD-ROM includes sophisticated software querying tools and wizards designed to provide subscribers a user-friendly interface to the information.

Although relatively young, the firm has grown significantly and now employs a professional staff of 50. T.R.A.D.E., Inc. is truly an international organization with offices located in Hong Kong, Taiwan, the United Kingdom, and the corporate headquarters in San Mateo. Additionally, T.R.A.D.E. utilizes a number of licensed distributors in a number of other countries. T.R.A.D.E., Inc. was recognized as one of the ten fastest growing, private companies, in Silicon Valley in 1996. While the firm is hesitant to reveal exact figures, revenues were in the multi-million dollar range in 1996 and represents T.R.A.D.E'.s significant share in the business intelligence market.

Telecommuter Background

The specific employee we will chronicle is Dave Tucker. Tucker has worked for T.R.A.D.E. for nearly five years as a software engineer, after working for a major hardware manufacturer in Silicon Valley for nearly seven years. His responsibilities at T.R.A.D.E. include developing advanced database queries, creating programs to interface source databases into T.R.A.D.E.'s massive database system, and creating software applications for the CD-ROMs which subscribers receive.

Tucker relocated to Silicon Valley nearly 11 years ago, after he graduated with a computer science degree from a large mid-western university. Tucker and his wife both grew up in central Iowa. Now with two young children (ages three and five), Tucker and his wife desired to return to central Iowa to be closer to family and friends and to provide their young children with childhood experiences similar to their own. In the spring of 1996, Tucker proposed a telecommuting arrangement to T.R.A.D.E.

T.R.A.D.E. decided to pilot test the arrangement. Tucker identified a college professor in Ames, Iowa who was interested in exchanging homes, for the summer, with someone near Silicon Valley. T.R.A.D.E. agreed to allow Tucker to telecommute for the summer and both agreed to evaluate the arrangement at the end of the summer trial. After the summer trial Tucker and T.R.A.D.E. agreed that the telecommuting alternative had substantial merit and decided to enter the arrangement on a permanent basis. What follows is an enumeration of the issues related to telecommuting and a chronicle of how Tucker and T.R.A.D.E. addressed each issue. These issues include telecommuter selection, organizational issues, social aspects, costs, and productivity measurement.

SETTING THE STAGE

Telecommuting presents organizations and their employees with a number of opportunities and a number of potential problems. Generally speaking, the benefits of telecommuting include more hours worked per day, more work accomplished per hour, and less employee stress due to commuting and/or work/personal conflicts (such as a sick child) (Curran & Williams, 1997; Race, 1993; Schellenbarger, 1993; Townsend, DeMarie, & Hendrickson, 1998). The organization must cope with employee work schedules being more flexible, in that many telecommuters may or may not keep traditional work hours. Additionally, immediate access to employee skills and services is somewhat diminished.

As shown in Figure 1, successful telecommuting requires special attention be given to: enabling technology, selecting the right job category and the right individual for the arrangement, organizational incentives, social impacts of dispersed workers, management of overall costs, and productivity measurement of geographically dispersed work teams. Each of these issues will be examined and compared to the specific T.R.A.D.E. example. While the specific facts encountered by Tucker represent his personal experiences with telecommuting, the issues examined apply to any telecommuting arrangement. Individuals will undoubtedly have varying degrees of problems and success with any given issue depending upon their personal background, skill level and motivations, but the categories of concern should be universal to all telecommuters and their organizations.

Figure 1. Telecommuting issues

Enabling Technology

A number of hardware, software and telecommunication technologies are needed to support telecommuting in the software engineering industry given its requirements for distributed software development and intraorganizational communication. The specifications of the components located with Tucker, as well as some components that Tucker utilizes that are located at T.R.A.D.E.'s headquarters, are described throughout this section. Each of the items was integrated into a complete information infrastructure that supported their telecommuting needs. The components are grouped into four categories including telecommuter hardware and software located at Tucker's home office that support Web application development and intraorganizational communication, Web and database servers located at T.R.A.D.E., and the telecommunications components used for communication between Tucker and T.R.A.D.E. These components are described in Table 1.

The router is a unique item because it will handle both static and dynamic IP addresses. It was used to set up a remote LAN so that Tucker could have static IP addresses so that a firewall could be configured in California to allow his IP addresses

Table 1. Telecommuting information infrastructure components

Telecommuting Information Infrastructure Component Category	
1. Telecommuter Hardware	*Desktop PC* • Gateway 2000 P6-200, 96MB RAM, 5GB hard drive • modem with voice capabilities, Soundblaster card • SCSI DAT tape backup, 8X CD drive • UPS power supply • running Windows NT 4.0 workstation, 10MB Ethernet *Laptop PC* • Gateway 2000 Solo 2200 P5-166, 32MB RAM, 2GB hard drive, 12X CD drive • running Windows 95, 10MB Ethernet PCMCIA *ISDN router* • Webram Entre, 2 POTS lines, 4 Ethernet ports *Test Server* • Gateway 2000 P5-133, 32MB RAM, 2GB hard drive • running Windows NT 4.0 Server and Web services
2. Telecommuter Software	*Web Browser* • Internet Explorer 4.0 *Web Application Development Software* • Microsoft Visual Interdev, Frontpage 97, Visual C++, and Visual Basic *Programming Languages* • C++, HTML, VBScript, JavaScript, and Active Server Pages
3. Corporate Servers	*Database Server* • 512MB memory, 80+GB disk drive • running Oracle 8 on a HP-UX 10.0 server *Web Server* • IIS 4.0 running on an ALR Revolution Quad 4 (4 P6-200 CPUs), 256MB memory, 16MB disk drive
4. Telecommunications	• ISDN line with two voice lines (one of which was used for fax) • Hotfax MessageCenter was used to provide voice mail, fax, and dial in capabilities • US Robotics Sportser Voice 33.6K internal modem • NetINS RemoteLAN ISDN service (provided a remote LAN with five static IP addresses) • NetMeeting was used for some voice over the Internet • No video conferencing products were used.

through. The router also handles dynamic IP addresses given by ISPs each time a connection is made while allowing routing from multiple machines on the local LAN through the dynamic IP. The remaining items are widely available from a number of hardware and software providers.

Most of Tucker's work involved software engineering, system testing, and general troubleshooting tasks related to the development and support of software applications used in the firm's products and services. Thus, a variety of telecommunications technology was utilized to facilitate Tucker's collaborative efforts with others at T.R.A.D.E. Specifically, Hotfax Message Center was used for electronic mail, fax, dial-in capabilities and some limited voice transmissions. Most of the voice interaction was accomplished over an ISDN connection which provided two voice channels, one of which was occasionally used for fax. In some instances it was necessary to utilize both ISDN channels at once in order to provide adequate bandwidth for some collaborative troubleshooting. In those instances, NetMeeting was also used for voice over the Internet via a dial-up connection.

T.R.A.D.E. and Tucker found NetMeeting to be very useful in these circumstances. The telecommuting arrangement did not have any video conferencing applications. Neither T.R.A.D.E. nor Tucker found this problematic since most of the interactions Tucker had were with associates whom he had previously met and had some ongoing working relationship. This is an important point since these voice connections do not provide visual support for collaborative interaction. Body language, eye contact, and physical mannerisms are filtered from the interaction with only voice connectivity (Townsend, DeMarie, & Hendrickson, 1998).

Upon reflection, Tucker expressed the need to integrate more collaborative tools and video conferencing into his telecommuting toolset. Several products were under consideration. The Internet products, such as CU-See-ME are inexpensive and provide some basic level of video interaction. Additionally, these products are relatively easy to install and are cost effective to install in a wide group of users. However, bandwidth limitations can make this solution inadequate.

Intel's Proshare units were an option that appeared very appealing. This product not only integrates the hardware (video camera) and software into one package, but the product provides a very sophisticated collaboration tool. Virtually any personal computer application can be shared between users with this product. The major drawback of this product, however, is price. Each unit is more than $1,000, and the system is proprietary. Thus each person wishing to interact with the telecommuter must also have this system. Due to the fact that Tucker supported a wide range of developers, with little or no ability to anticipate who might need to interact with Tucker on any given day, T.R.A.D.E. felt the investment in this technology to provide video connection with one telecommuter was too expensive. However, since the success of this arrangement was encouraging for future telecommuting arrangements, the firm did not rule out the possibility of this technology for developers in the future.

Similarly, collaborative technologies such as Lotus Notes™ was considered. This package is popular in the support of geographically dispersed collaborative teams. Notes provides collaborative tools beyond simple e-mail. Applications and documents can be shared and maintained by groups instead of individuals. Many organizations are using Notes to connect their entire professional staffs (Townsend, DeMarie, & Hendrickson, 1998). Again, the firm saw some benefit to this software, but the expense in terms of economic resources, training, and time could not be justified to support a limited number of telecommuters.

Obviously, there are an infinite number of combinations of hardware, software and telecommunication components available, but this case does identify one set of information technologies that enabled successful telecommuting. It is apparent that information technology and telecommunications services are widely available which allow telecommuting to be a feasible work arrangement for software engineering companies. But the technology is a necessary, but insufficient, condition for success. The remaining issues we discuss move beyond information technology issues to the more complex managerial and organizational behavior issues.

CASE DESCRIPTION

Telecommuter Selection

The ultimate success of telecommuting may well depend upon the initial selection of appropriate candidates for this alternative. Generally, individuals who have significant experience in a job, individuals who are above average performers, and persons with a broad range of job assignments will adjust better to a telecommuting arrangement than their less experienced, lower performing counterparts (Boyett & Conn, 1992; Fish et. al., 1992; Kraut, 1988; Mokhtarian & Salomon, 1994; Townsend, DeMarie, & Hendrickson, 1998). More specifically, successful telecommuting candidates will have worked for the organization long enough to acquire some affinity for the corporate culture and nature of expectations in this specific work environment. Additionally, workers who have demonstrated superior technical skills (especially with remote working tools, such as a computer, modems, fax, and videoconferencing) will be more successful (Fish et. al., 1992; Kraut, 1988; Mokhtarian & Salomon, 1994). Finally, some consideration for the individual's suitability for telecommuting may depend upon his/her personality profile and acceptance of a more ambiguous work setting.

In our example case, the firm did not initiate the telecommuting arrangement but Tucker's skills and experience are insightful. Tucker had significant experience in both the software development industry and with T.R.A.D.E. specifically. He had spent 11 years as a software engineer, four of those years with T.R.A.D.E. Thus, he had a firm understanding of his job tasks and the specific policies and goals of T.R.A.D.E. Additionally, Tucker was an extremely high achiever. When Tucker first proposed the telecommuting arrangement, T.R.A.D.E. management stated that they would have never considered such an arrangement had Tucker not been an exemplary employee whose services they wished to maintain, even if it meant altering the firm's standard operating practices. Finally, with respect to technical skills, Tucker pointed out that he felt the relationship functioned well due to his experience with the technology and his ability to provide his own technical support.

Organizational Incentives

For the organization the benefits of telecommuting are numerous. Incorporating remote workers into the organization can provide the catalyst for organization flexibility, allowing the firm to expand and contract to adjust to market conditions. As demand increases the organization can expand without the corresponding increases in physical facilities. Thus, telecommuting offers the organization the opportunity to expand with limited cost increases, or in some cases to reduce overhead by eliminating office facilities (furniture, electricity, and environmental conditioning) and space requirements (Mokhtarian, 1988; Mokhtarian 1991a; Nijkamp & Salomon, 1989; U.S. Department of Transportation, 1993).

As highly skilled human capital becomes a scarce commodity in the brain power industries of the 21st century, telecommuting offers the organization a means of attracting and retaining these limited resources (Thurow, 1996). Telecommuting may provide access to skilled employees who do not desire to reside in a particular geographic area, or who are limited due to other personal commitments (such as elderly or childcare responsibilities). Telecommuting also allows organizations to utilize workers in disadvantaged rural or urban areas, as well as incorporate the skills of workers who have left the workforce temporarily due to illness or to spend time with small children (Curran & Williams, 1997).

Although the advantages of telecommuting are numerous, there are some drawbacks to this alternative. Generally, managers are not trained to deal with remote workers, especially in terms of how to manage their activities and how to motivate them. Many managers are uncomfortable with the perceived loss of control created by telecommuting. However, most of the problems are more perceived than manifest. Good management skills in terms of delegation and performance measurement can alleviate most of the disadvantages (Katz & Kahn, 1979; Katz & Tushman, 1978).

Many of these issues were considered by T.R.A.D.E. in dealing with Tucker's request to telecommute. The firm was aware that the task of replacing Tucker may be daunting. Not only would they need to identify an individual with similar skills, but in all probability, they would face a significant task in attracting the candidate away from their present employer. Additionally, a large amount of time and resources would be spent bringing the new employee up to speed in terms of corporate culture and client expectations. Thus, given Tucker's desire to relocate, yet remain as an employee, the situation presented an opportunity for T.R.A.D.E. to retain the skills they require and provide a valued employee the flexibility he sought. Although T.R.A.D.E. would recapture Tucker's physical space, they didn't realize any significant cost savings related to physical facilities since this telecommuting arrangement was not an organization-wide arrangement. Managerial problems were not considered as crucial, primarily due to Tucker's solid history with the firm and the firm's confidence in his personal integrity and ability.

Social Aspects

The personal social benefits are the main attractiveness to most employees. Telecommuters are often unconstrained by traditional office schedules and work environments. They typically have the freedom to set more flexible work schedules to coincide with personal activities such as child, elderly, or disabled care responsibilities. The inherent lack of commuting lowers the stress, time, and financial cost of this activity. Remote workers recapture valuable time spent in daily commutes, along with reducing or eliminating the need for transportation (bus, auto, fuel, etc.) to commute (Mokhtarian,

1988; Mokhtarian, 1990; Mokhtarian 1991a; Nijkamp & Salomon, 1989; U.S. Department of Transportation, 1993).

The benefits are not without real, and potentially substantial, disadvantages (Egido, 1990). Because telecommuters typically work out of their home environments, many develop feelings of isolation. They often feel out of touch with their co-workers and feel that they miss out on informal social interactions that occur naturally in the work environment (Curran & Williams, 1997; Kraut, Egido, & Galegher, 1990; Kraut & Streeter, 1994). The lack of interaction often extends beyond co-worker interaction. Many telecommuters feel disassociated with other business professionals. Relationships that occurred and developed naturally in traditional business settings now must be recreated, often with substantial effort, in the telecommuting environment.

Working at home can cause other problems. Some telecommuters experience a sort of cabin fever due to the lack of external interaction on a routine basis. Additionally, working at home can cause strain on marital and family relationships. The homogeneity of the work and home environment offers limited stimulus variety, and can ultimately be counterproductive. The yearning for peace and quiet in order to accomplish work tasks may soon be perceived as too peaceful and quiet (Curran & Williams, 1997; Race, 1993; Shellenbarger, 1993).

Some telecommuters develop a sense of disconnect between themselves and their organizations. They may feel their skills are being underutilized and that they are out of the loop in terms of office politics and organizational gossip. While the elimination or minimization of office politics is usually cited as a productive benefit of remote work by employers and employees alike, the reality is that humans are social creatures who require some minimum level of social interaction. Sometimes, negative interaction is better than no interaction. The loss of social interaction can be mitigated if special attention is given to substituting the negative interactions with positive reinforcement of the employee's role in the organization (Hiltz, 1993; Katz & Kahn, 1979; Katz & Tushman, 1978).

Tucker experienced more of this than he initially thought possible. While T.R.A.D.E. viewed Tuckers absence as somewhat negative, Tucker initially thought the telecommuting arrangement was ideal. Tucker's superiors (vice president of engineering, president, and CEO) all expressed reservations with the arrangement due to Tucker's more limited interaction in informal decision making. When in the office, his superiors readily would seek out his input and advice on a myriad of business issues. Now that Tucker was not readily available (in person, although he was always accessible at his remote location), his superiors felt they were not utilizing all of his skills.

Tucker often saw this as time consuming and detrimental to his specific task agenda. However, as time passed, Tucker began to feel his political power and informal influence began to wane. He often found himself less informed on critical corporate issues and felt he was not contributing as much to the organization as he had previously. Tucker admits that he now feels somewhat less connected to T.R.A.D.E. His personal identity now revolves less around his role at T.R.A.D.E. and more around his generic software engineering skills. In an effort to recreate some of the social interactions, Tucker has joined a number of professional and business organizations, such as the Lions Club, and the local Chamber of Commerce. The success of this substitution is yet unknown, but Tucker is skeptical concerning his ability to fully recreate the same interpersonal/ professional sociology he once enjoyed.

Management of Costs

The tangible costs and benefits of telecommuting are fairly straightforward and easy to address. For the organization, the costs include installation and monthly charges for long distance services to facilitate remote access to the employee which may include voice, data, and video communication. If periodic travel is required, then the cost of this travel will also need to be factored into the equation. Travel costs will not only include transportation costs but hotel and living expenses incurred during stays at the organization's site. These costs will be offset by savings from reduced costs of space (office and parking), furniture, and facilities (heating, air conditioning, and other space related benefits) (Mokhtarian, 1988; Mokhtarian 1991a; Nijkamp & Salomon, 1989; U.S. Department of Energy, 1994; U.S. Department of Transportation, 1993).

For the individual, the tangible benefits include reduced costs of transportation and the potential for reduced expenditures from unreimbursed activities such as some business lunches. Depending upon the telecommuting arrangement and personal responsibilities, there may be some potential elderly or childcare savings due to the remote worker's presence in the home environment.

In the arrangement between Tucker and T.R.A.D.E., Tucker originally proposed that he would pay the expenses of the arrangement out of his pocket. Thus, he pays for the Integrated Services Digital Network (ISDN) telecommunications lines into his home, the special software that enables him to emulate a users personal computer desktop environment on his personal computer in his home, and the associated travel expenses for the two to three day trips he makes to San Mateo every six weeks. An interesting aspect of this arrangement is that Tucker's superiors at T.R.A.D.E. expressed dissatisfaction with this unilateral arrangement. Since Tucker was incurring the costs, his T.R.A.D.E. superiors did not feel comfortable asking him to return to San Mateo for special projects outside of the normal six-week visits. With a year of experience with this arrangement, the costs are fairly well established. Thus, in the future the firm and Tucker will likely share the costs in some manner.

Productivity Measurement

But, how do we know what you are doing? is a sentiment often expressed by managers to proposals of telecommuting from employees. Metrics of productivity and performance are critical regardless of the work environment. However, less than optimal performance evaluation can be masked in traditional work environments, but vague and ambiguous management approaches fail miserably in telecommuting environments. As discussed in the telecommuter selection section, telecommuting is not for everyone and every job function. Some jobs simply require too much interaction to allow for efficient and effective skill utilization from a remote site.

This problem notwithstanding, many job functions and individuals can readily adapt to a telecommuting arrangement if good management techniques are practiced. Overall in a telecommuting environment, the emphasis shifts from control of processes to control of results. Work tasks must be planned well, delegated specifically, and be accompanied with timetables and deadlines. It is also important to provide a mechanism to periodically assess progress and issue performance feedback (Curran & Williams, 1997). Obviously, these practices are easier to accomplish with job functions that lend themselves to segmented, specific tasks, but these techniques represent superior management practices regardless of the work environment. In traditional work environments, many of these activities can be accomplished informally. With the employee and supervisor both present, progress assessment and performance feedback is relatively easy and often very informal. Planning is often overlooked since managers can easily modify employee assignments during task execution.

Tucker and T.R.A.D.E. found this area to be the least problematic to operationalize. This may be attributable to Tucker's job function. As Curran and Williams (1997) point out, the five job categories most suitable to remote working are professionals, professional support, field workers, IT specialists, and clerical support. Depending on the specific definition, Tuckers job could be considered in any one of three of the five: professional, professional support, and IT specialist. Already a well managed organization, T.R.A.D.E. incorporated most of these management techniques in its operations prior to Tucker's proposal for telecommuting.

CURRENT CHALLENGES/PROBLEMS FACING THE ORGANIZATION

Telecommuting offers substantial benefits for organizations today and into the 21st century. As the demand and competition for high-quality knowledge workers increases, this alternative will allow firms to attract and retain employees with scarce resources. At the same time, workers are demanding more flexibility as alternatives to traditional work relationships. The case study presented throughout this paper is an example of successful telecommuting. It was successful because each of the six sets of issues: enabling technology, telecommuter selection, organizational advantages/disadvantages, social aspects, costs, and productivity measurement, were addressed by both Tucker and T.R.A.D.E.

The technology needed to enable distributed software development and intraorganizational communication is widely available. Hardware, software and telecommunications capabilities continue to improve while their cost continues to decline. At the same time, telecommunication capabilities continue to evolve and these capabilities alter the way in which employees can experience the world. This continues to challenge T.R.A.D.E. and Tucker alike. However, it is apparent that technology is a necessary but insufficient condition for telecommuting success. Most issues do not involve technology implementation, but instead involve more complex managerial and organizational behavior issues.

Tucker initiated the arrangement which showed that he saw telecommuting as beneficial to him, while he also was an ideal candidate for telecommuting because he had significant experience in software development, significant experience with his company, was an exemplary employee, and had experience in doing his own technical support. The organization saw that the advantages of telecommuting for Tucker outweighed the disadvantages. Any inconveniences the arrangement may cause are outweighed by the costs that would be incurred to replace Tucker's skills with a new employee from a tight job market.

Tucker and T.R.A.D.E. saw some disadvantages arise from the lack of direct interaction with co-workers, but the ability to live in a geographic area that better matched his families needs outweighed these problems. Cost sharing arrangements can be adjusted as the telecommuting arrangement between Tucker and T.R.A.D.E. evolves. And finally, there was a fit between Tucker's job category, one of professional, professional support, and/or IT specialist that enabled his productivity to be easily measured through existing procedures. As these factors evolve, T.R.A.D.E. and Tucker must determine how the telecommuting arrangement will evolve.

While originally somewhat skeptical, the firm's experience with Tucker's telecommuting arrangement has been a positive one. T.R.A.D.E. is now more enthusiastic about entertaining similar work arrangements for other employees and contractors. The disadvantages are predictable and can be minimized. The benefits for the firm in retaining and attracting highly skilled personnel in an extremely competitive job market continues to grow. To date no additional arrangements have been undertaken, but several are being considered.

The overall conclusion of this case is that successful telecommuting can take place in today's IT environment when experienced, self-motivated, employees are involved. There are incentives for the organization to provide this flexibility, and there are procedures in place for communicating and managing remotely instead of in a traditional environment.

In the final analysis, telecommuting alters the existing social pattern of work as we have known it for the last one hundred plus years, i.e., workers leave their homes each day to toil at job sites some distance from their homes, and return each evening to rest and renew the process again. This excerpt from the U.S. Department of Energys Office of Policy, Planning, and Program Evaluation (1994) best sums up the future of telecommuting.

The social effects of telecommuting will occur and be judged in the larger context of advances in telecommunications as well as other changes that effectively restructure workplace interactions. It may be difficult to distinguish the effects that are specific to telecommuting. Currently, telecommuting is not widespread, and its social effects are relatively localized. Individual telecommuters, their work organizations, their families, and to a lesser extent, their communities currently are affected. If telecommuting becomes a widespread working strategy, the effects both positive and negative will become more varied and their scale will become larger, affecting individuals, the workplace, families, communities, and the Nation.

The following issues may become particularly important: separation between home and work; increasing fractionalization of the workforce and society at large; and changes in neighborhood and community interactions. Telecommuting from home blurs existing boundaries between home and work. Employees' abilities to separate (or escape) from their work and employers abilities to invade their employees' homes may change the nature of home and work lives. Within the workplace, three major classes of personnel may develop: those who voluntarily telecommute and derive the benefits of autonomy, flexibility, and freedom from interruptions; those for whom telecommuting is required and who may not experience the advantages of the privileged class of telecommuters or have access to the benefit packages associated with long-term employment in an office setting; and those for whom telecommuting is not an option. Widespread telecommuting may enhance the safety and spirit of neighborhood and community life, though these effects may not be distributed consistently throughout a region or the country because of varying composition of communities may be associated with different levels of telecommuting. Finally, widespread telecommuting may change patterns of residential and commercial development, in part because of proximity of the workplace to the labor force would diminish in importance.

REFERENCES

Abel, M. J. (1990). Experiences in an exploratory distributed organization. In J. Galegher, R. Kraut, & C. Egido (Eds.), *Intellectual teamwork: Social and technological foundations of cooperative work* (pp. 489-510). Hillsdale, NJ: Lawrence Erlbaum Associates.

Boyett, J. H., & Conn, H. P. (1992). *Workplace 2000: The revolution reshaping American business*. New York: Plume Publishing.

Curran, K., & Williams, G. (1997). *Manual of remote working*. Hampshire, UK: Gower Publishing Limited.

Egido, C. (1990). Teleconferencing as a technology to support cooperative work: Its possibilities and limitations. In J. Galegher, R. Kraut, & C. Egido (Eds.), *Intellectual teamwork: Social and technological foundations of cooperative work* (pp. 351-371). Hillsdale, NY: Lawrence Erlbaum Associates.

Finholt, T., & Huff, C. (1994). *Social issues in computing: Putting computing in its place*. New York: McGraw-Hill.

Fish, R., Kraut, R., Root, R., & Rice, R. (1992). Evaluating video as a technology for informal communication. *Communications of the ACM, 36*(1), 48-61.

Hiltz, S. (1993). Correlates of learning in a virtual classroom. *International Journal of Man-Machine Studies, 39,* 71-98.

Katz, D., & Kahn, R.(1979). *The social psychology of organizations* (2nd ed.). New York: John Wiley & Sons.

Katz, D., & Tushman, M. (1978). Communication patterns, project performance, and task characteristics: An empirical evaluation in an R&D setting. *Organizational Behavior and Human Performance, 23,* 139-162.

Kraut, R. (1988). Homework: What is it and who does it? In K. Christensen (Ed.), *The new era of home-based work.* Boulder, CO: Westview Press.

Kraut, R., Egido, C., & Galegher, J. (1990). Patterns of contact and communication in scientific research collaboration. In J. Galegher, R. Kraut, & C. Egido (Eds.), *Intellectual teamwork: Social and technological foundations of cooperative work* (pp. 489-510). Hillsdale, NJ: Lawrence Erlbaum Associates.

Kraut, R., & Streeter, L. 1994). Coordination in large scale software development. *Communications of the ACM, 38*(3), 69-81.

Mokhtarian, P. (1988). An empirical evaluation of the travel impacts of teleconferencing. *Transportation Research, 22*(4), 283-289.

Mokhtarian, P. (1990). A typology of relationships between telecommunications and transportation. *Transportation Research A, 24A*(3), 231-242.

Mokhtarian, P. (1991a). Defining telecommuting. *Transportation Research Record, 1305,*273-281.

Mokhtarian, P. (1991b). Telecommuting and travel: State of the practice, state of the art. *Transportation, 18*(4), 319-342.

Mokhtarian, P., & Salomon, I. (1994). Modeling the choice of telecommuting: Setting the context. *Environment and Planning A, 26*(4), 749-766.

Nijkamp, P., & Salomon, I. (1989). Future spatial impacts of telecommunications. *Transportation Planning and Technology, 13*(4), 275-287.

Nilles, J. (1991). Telecommuting and urban sprawl: Mitigator or inciter? *Transportation, 18*(4), 411-432.

Race, T. (1993, August 6). Testing the telecommute. *New York Times*, p. F11.

Shellenbarger, S. (1993, December 14). Some thrive, but many wilt working at home. *Wall Street Journal*, p. B1.

Thurow, L. (1996). *The future of capitalism*. New York: William Morrow and Company.

Townsend, A. M., DeMarie, S. M., & Hendrickson, A. R. (1998). Virtual teams: Technology and the workplace of the future. *Academy of Management Executive, 12*(3).

U.S. Department of Energy, Office of Policy, Planning, and Program Evaluation. (1994). *Energy, emissions, and social consequences of telecommuting*. Washington, DC: U.S. Government Printing Office.

U.S. Department of Transportation, Office of the Secretary. (1993). *Transportation implications of telecommuting*. Washington, DC: U.S. Government Printing Office.

Whitaker, S., & Geelhoed, E. (1993). Shared workspaces. How do they work and when are they useful? *International Journal of Man-Machine Studies, 39*, 813-842.

Anthony R. Hendrickson is an assistant professor of MIS in the Department of Logistics, Operations and MIS, Iowa State University. He received a PhD in business administration (computer information systems and quantitative analysis) from the University of Arkansas (1994). His research interests include virtual organizations, psychometric measurement, and object-orientation.

Troy J. Strader is an assistant professor of MIS in the Department of Logistics, Operations and MIS, Iowa State University. He received a PhD in business administration (information systems) from the University of Illinois at Urbana-Champaign (1997). His research interests include electronic commerce, strategic impacts of information systems, and information economics.

This case was previously published in the *Annals of Cases on Information Technology Applications and Management in Organizations*, Volume 1/1999, pp. 38-47, © 1999.

Chapter VII

The Elusive Last Mile to the Internet

V. Sridhar, Indian Institute of Management, Lucknow, India

Piyush Jain, Indian Institute of Management, Lucknow, India

EXECUTIVE SUMMARY

As organizations continue to Web-enable mission-critical network-centric information systems, reliable and high bandwidth connectivity to the Internet is essential. While glut of Internet bandwidth is being witnessed in the developed countries, Internet services are still in their infancy in developing countries. Getting high-speed access to the Internet, especially in the remote areas of developing countries, poses challenges due to poor telecommunications infrastructure. With limited bandwidth available from service providers, network managers have to find ways other than simple raw bandwidth increments to meet the increasing demand from the users, especially for improved Web performance. While metrics for measuring Internet performance are being refined by researchers, defining service level agreements with the Internet service providers based on these metrics is challenging, especially in an evolving market. This case highlights the experiences and lessons learned from such an exercise at the Indian Institute of Management located in Lucknow, India.

ORGANIZATION BACKGROUND

The Indian Institute of Management Lucknow (IIML) is one of the six national-level management institutes set up by the Government of India in Lucknow, India, in 1984. The Institute's mission is to help improve the management of the corporate and the non-corporate sectors and also the public systems in the country, through pursuit of excellence in management education, research, consultancy, and training. In order to fulfill its objectives, the Institute undertakes a diverse range of academic and professional activities.

IIML has a large information technology (IT) infrastructure and has an annual budget of about 16.7 million Indian Rupees (INR) (equivalent to about $334,000) allocated in recent years for the development of information technology resources. The details of the annual budget allocated for the computer services of IIML for 2002-2003 are given in Appendix A. Students, faculty, and staff of IIML use Internet resources quite extensively for teaching, research, consulting, and administrative activities.

The Institute has about 1,000 computers, connected to the campus intranet spread over a 50-acre campus, situated well outside the city limits. The Computer Center (CC) at the Institute is responsible for all IT services in the campus. CC employs one manager and six system analysts who are involved in the maintenance of IT services of the Institute.

Students, faculty, and staff of IIML use Internet resources quite extensively for teaching, research, consulting, and administrative activities. The IIML Web site (http://www.iiml.ac.in) is viewed by prospective students, alumni, researchers, and scholars at various institutions and corporations around the world. The Web site also provides information to about 120,000 potential candidates who apply every year for the MBA program offered by the Institute. Apart from World Wide Web (WWW), e-mail is another Internet application widely used by faculty, students, and staff of the Institute. E-mail communication by students and faculty spans across geographical boundaries and is one of the mission-critical Internet services of the Institute.

SETTING THE STAGE

There were 142.3 million Internet users in the U.S. in 2001, representing 49.95% of the population (ITU, 2002). By contrast, in developing countries such as India, there were 7 million Internet users who represent 0.69% of the population. However, the Internet Service Provider (ISP) industry has seen a phenomenal growth of subscriber base at a Compounded Annual Growth Rate (CAGR) of 41.13% over the period 1998-2002, thanks to deregulation and competition introduced by the Indian government in 1998. The ISP market is still in its infancy in India. While the U.S. and other developed countries boast of glut in Internet bandwidth, the domestic bandwidth in India crawls at 2.5 Gbps and the international bandwidth at 3 Gbps (Voice & Data, 2002). Educational Institutes in the U.S. and other developed countries typically have very high Internet access bandwidth in the order of 1.5 to 45 Mbps (Utsumi & Varis, 2003). Because of a low level of Internet infrastructure development in the area, IIML was only able to get a meager 64 Kbps Internet connectivity from the erstwhile government-owned monopoly operator, Bharat Sanchar Nigam Limited (BSNL), in 1997.

A typical Internet connection to a campus network, such as that of IIML, includes an access loop normally referred to as "last mile" to the Internet point of presence (POP). The "last mile" connectivity is typically provided by the basic telecom operator (BTO), (similar to the local exchange companies in the U.S.). The Internet consists of many individually managed networks that are interconnected to each other. The ISPs have "peering" arrangements among themselves for exchange of Internet traffic. In India, the

Figure 1. Topology of Internet connectivity

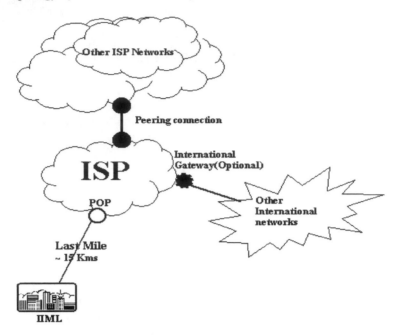

erstwhile monopoly BSNL has the largest Internet backbone network in the country. The new entrants typically have interconnect agreements with BSNL. The ISPs have been allowed to deploy their own international gateway (satellite or submarine optic fiber based) to connect to international operators to route the Internet traffic out of the country. A schematic diagram of the topology of the Internet is presented in Figure 1.

There are various options for a high-bandwidth last-mile Internet connectivity. A comprehensive list of last-mile options for campus network connectivity is presented in Appendix B. Many of the options listed are not suitable for IIML, as the campus is situated 15 Kms from the nearest telephone exchange of the BTO. Hence IIML decided to opt for a terrestrial radio link connectivity of 64 Kbps to connect to the ISP's POP. Terrestrial radio link has the advantages of quick deployment, high availability, and is especially suitable for remote areas where laying cables is difficult and expensive (Stallings, 2001).

Meanwhile IIML deployed a 100-Mbps-Fast Ethernet-based intranet backbone in the campus, laying about 1,500 meters of optic fiber. About 10 servers were installed in the computer center and about 300 more client machines were added to the intranet. Most of the applications and information were Web enabled. With these developments, 64 Kbps was proving to be too little for the growing bandwidth needs of the users of IIML. The users complained of poor response times, high latency, and unreliability of the Internet services.

Recently, the IIML library has started providing electronic access to as many as 695 periodicals and journals available from the publishers. IIML has recently been requested by the government to join a consortium of technology and management institutes in the country to subscribe to electronic resources such as ABI/INFORM, IEEE/IEE journals, and the ACM Digital Library. IIML has been asked by the government of India to discontinue the print subscription of the electronic resources made available through the consortia to reduce overall cost. Access to electronic resources, especially the digital libraries, requires highly available, good-quality Internet connectivity.

With the growing dependence on the Internet for mission-critical services such as e-mail and WWW access, there was a need to assess the existing Internet usage patterns and make improvements in the Internet infrastructure of IIML.

CASE DESCRIPTION

The challenge before Mr. Dilip Mohapatra, Manager of the computer center at IIML, was to determine the optimal way to meet the bandwidth and response time requirements of the many Internet applications in the campus.

Estimating Demand for Internet Bandwidth

The first step in this exercise was to estimate the amount of Internet access bandwidth needed to meet the requirements. The rule of thumb of 22 Kbps per simultaneous user has had widespread acceptance when estimating the appropriate Internet bandwidth (Lewis & Williams, 2002). Clearly the existing Internet bandwidth of 64 Kbps was just not sufficient to meet the requirements of more than 100 concurrent users at the IIML campus. The two widely used Internet applications were e-mail and Web access. E-mail being an asynchronous application does not have strict response time requirements. It is the reliability of the Internet access link, which is more important for periodical delivery and receipt of e-mail messages. It was quite fortunate that the solitary 64 Kbps wireless link stood the test of storm and rain. The link went down only for about six days in the past six years since its inception, giving an availability of more than 99.7%.

However, Web access is much more complex. The Web page load time — from the click on a URL to the point at which the page is completely displayed on the destination PC — is a metric used to measure Web performance (Sevcik & Bartlett, 2001). A response time in excess of four seconds can cause the user's attention to wander (Rubenstein, Hersh, & Ledgard, 1984), and most users give up on Web access after about eight and a half seconds (Cache Flow Inc., 1999). This user perceived latency in accessing Web pages is affected by various Web components: (a) clients, which initiate the requests; (b) servers, which respond back; (c) the proxies, which are intermediaries between clients and servers; and finally (d) the network, which connects the clients/proxies and the servers. A limited research exists where end-to-end performance of the Web has been analyzed (Cohen, Krishnamurthy, & Rexford, 1998). Most often these individual components involved in a Web transaction are improved to realize an overall improvement in Web response time. Sevcik and Bartlett (2001) collected and analyzed the total page load time of the best performance sites benchmarked by Keynote Systems (http://www.keynote.com). They have concluded that beyond 512 Kbps, there is no performance advantage in buying more access bandwidth. Sevcik and Bartlett (2001) observe that the only sure way to improve Internet performance is to move the server closer to the user.

Based on this, Mr. Mohapatra concluded that 64 Kbps is just not sufficient for the user community of IIML and that it has to be increased.

Supply of Internet Bandwidth

Though about 436 companies were given ISP licenses in India in 1998, after the service was opened for private participation, only about 30% of these licensees are operational and only a handful of them are actively providing the services today (ICRA, 2000). Most of the ISPs commenced their operations in more lucrative areas such as metro areas in the country. In the city of Lucknow, in whose outskirts IIML is situated, there are four private ISPs providing services. Out of these, two are local ISPs and do not have much infrastructure to provide high-bandwidth connectivity. HCL Infonet and Satyam Infiway are the only two national ISPs that provide service to business customers in the region and showed interest in IIML's project. The incumbent BSNL, however, showed interest in providing additional bandwidth.

IIML had to decide among the following options:

1.   Continue with the incumbent and increase the access bandwidth from the current 64 Kbps.
2.   Discontinue the incumbent services and procure an increased bandwidth from one of the private ISPs.
3.   Keep the incumbent services at the existing bandwidth and go for additional bandwidth from a private ISP.

Since the ISP industry was evolving in India and the private ISPs were still developing their expertise in providing services, it was risky to go with the second option. Support services from the government-owned incumbent were not found to be satisfactory and hence it was decided to go with the third option. The second Internet access link was planned to complement the existing one so that the Internet bandwidth enhancement proceeds without any interruption to existing services. HCL Infonet was chosen as the service provider based on the presence in the location, and price of the service. It was decided to order 512 Kbps bandwidth from the ISP, keeping in mind the number and requirements of the user community. Since the existing modem and antenna gear were not suitable for bandwidth exceeding 128 Kbps, new sets of modems and antenna were ordered. A separate tower was erected at the IIML campus to house the radio antenna. An annual maintenance contract (AMC) was also signed with HCL Infonet for the upkeep and maintenance of the last mile, and the contract was signed in May 2001. The capital and recurring expenditure of the new Internet service is given in Table 1. The new link went operational on May 15, 2001.

As can be seen from Table 1, Internet service does not come cheap in developing countries. While academic institutions in the U.S. can get 1.544 Mbps (called T-1 service) Internet connectivity at about $7,000 per year (Wiscnet, 2003), which works out to be $4,500 per Mbps, IIML pays more than $40,000 per Mbps connectivity. Cost of the Internet service works out to be about 18% of the budgeted annual recurring expenditure for IIML.

Table 1. Capital and recurring expenditure of the internet service

No.	Items	Price	
		USD	INR
Capital Expenditure			

1	Local Loop Equipments
	(i) 2 numbers of S-Band P-Com M-400 Modems
	     capable of bandwidth up to 8 Mbps
	(ii) 2 numbers of 27 db, 1.2 meter Antenna
	     Connectors and other accessories
		23,000	1,150,000

2	CISCO 2600 Router	4,000	200,000
3	Charges for site survey; radio frequency clearing charges to be paid to the government; charges for installation, testing, and commissioning	2,000	100,000
4	Cable laying and installation of the antenna mast	4,200	210,000
Recurring Expenditure			
1	Internet port charges for 512 Kbps	21,000	1,050,000
2	Annual maintenance charges of the local loop equipment	800	40,000
	Total	**55,000**	**2,750,000**

Quality of Internet Services

Quality of the last-mile link that connects the customer premise with the POP of the ISP is also an important determinant of the Internet performance. The newly installed radio link between IIML and the ISP POP went down barely 60 days after the installation due to a radio modem problem. HCL Infonet, being a new entrant in ISP services, did not have in-house expertise to rectify the problem immediately. It took almost nine days before the problem was solved and the link brought back online. Since redundancy was maintained through the 64 Kbps link, all mission-critical services such as e-mail were routed through the other link, maintaining the connectivity of the IIML campus with the rest of the world. In one instance even the most reliable 64 Kbps went down because of a storm. It took the government operator two full days to bring the link back online. This illustrates the disparity in the quality of service provisioning between a developed country where a matured market exists and a developing country with an evolving nascent market.

Defining quality-of-service (QoS) guarantees over the public network has been very difficult. Most of the ISPs can guarantee only the traffic that travels on its own managed network. Once the traffic hits a peering interconnection point (refer to Figure 1), there is no control over the traffic by the service provider and hence the ambiguity on service guarantees remains. Further, most of the ISPs do not have their own network infrastructure. They lease circuits from other operators such as BTOs to interconnect their POPs in different parts of the country. Hence network unavailability may also arise due to the failure of the leased line about which ISPs do not have direct control. Recently, IIML experienced about a 10-hour failure on January 4, 2003, due to a problem in the circuit leased by HCL Infonet from BSNL.

Even though the Telecom Regulatory Authority of India (TRAI) has announced guidelines for QoS of dial-up access network to the Internet (TRAI, 2001), comprehensive

QoS guidelines and their enforcement is still lacking in India. Hence a service level agreement has to be established and agreed upon between the ISPs and users.

Since "internet" means a "network of networks," the SLA for the Internet can be very complex. While the general format of SLAs differ among the service providers, most of the agreements include language that covers some combination of availability, packet loss, and latency (Passmore, 1999). *Availability* relates to the uptime of the service. Availability guarantee is the most basic service guarantee that the service providers offer. Availability SLAs range from 100% to 95% monthly uptime for the circuit, and they are backed up with an offer of a refund or free service based on the amount of downtime (Passmore, 1999). A circuit that failed at the critical time for a business enterprise causing huge financial damage, but worked fine 99.99% of the total time, would be within such SLA guarantees. Though for educational institutes it may be less critical, network managers should negotiate with service providers for some payback to cover critical downtime losses.

The messages divided into packets and routed through the Internet, from the originating client to the destination server experience network delay, and causes *latency*. The loss of any packet needs retransmission and hence increases the response time (Stallings, 1998). This leads to two more parameters — packet loss and latency — for measuring the Internet performance. Monitoring latency and packet loss is difficult, and the monitoring tools can be very expensive, ranging up to $100,000, requiring software agents and probes to be set up in the entire network (Passmore, 1999). Though most of the ISPs provide the above guarantees for their portion of the network, the SLA guarantees do not hold after the peering point. Hence SLAs for Internet performance are difficult to enforce. A sample of SLA guarantees are given in Appendix C for reference (Wexler, 1998).

Wexler (1998) argues that the service providers differ in their commitments to packet latency and loss. It is also pointed out that providing one without another is not particularly meaningful. While every user wants 100% availability and bandwidth, it is not possible for the ISPs to offer such service levels on the public Internet. Even in developed countries, SLAs offered by ISPs are not complete (Passmore, 1999). In India, where the ISP service is in the initial stages, comprehensive service level agreements are difficult to develop and enforce. The broad terms and conditions of the Internet service agreement, which IIML was able to sign with the ISPs, are given in Table 2.

Policy-Based Network Management

Although bandwidth enhancement has been touted as the easy solution for improving network and application sluggishness, most often it results in a cyclical exercise in user aggravation, producing excessive and unplanned increase in bandwidth (Young, 2002). According to Young (2002), over-provisioning of bandwidth leads to "expectation creep" leading to users demanding more responsiveness and availability from the network without understanding the impact of their demands on the network ecosystem. Policy-Based Network Management (PBNM) bridges the gap between bandwidth requirement and availability. The first step in PBNM is to establish policies, or sets of rules governing priorities and permissions for users and applications throughout the network (Young, 2002). The policies so developed should encourage positive use while simultaneously discouraging negative use of the Internet bandwidth (Tubell,

Table 2. Service level agreements between IIML and the ISP

Service Level Attribute	Guarantee Levels	Penalty Clauses
Network Availability	99% uptime per month, less than 1 hour consecutive downtime (up to a maximum of 7 hours non-consecutive downtime per month) Maximum of 7 outages per month	0.5% of the Total Annual contract Charges (TAC) per additional hour 0.75% of TAC per failure
Maximum Latency	Maximum ping response time of 900 milliseconds for contacting a list of high-performing Internet sites provided by IIML	0.2% of the TAC per hour from the time of logging latency complaint until rectification

Other Service Parameters:

1. Any disconnection, discontinuity, hardware, and link problem must be identified and a solution proposed with in one hour (referred to as the "resolution time"), from the time it is reported in 365 days and (24×7) service mode. If the resolution time exceeds one hour, penalty is calculated at the rate of 0.2% of TAC per additional hour.
2. Vendor should configure related services/software such as Internet Operating System on the routers, Domain Name System, and load distribution, and make them fully operational.
3. Vendor should provide fortnightly report on bandwidth utilization and traffic profile of IIML.
4. Vendor should submit reports for up/downtime before claiming service payment.
5. Vendor should also provide and configure the appropriate bandwidth management software at IIML site to monitor the performance of the Internet link.

2002). The second step is to put in place policy-based software to help ensure compliance with the policy. Thirdly, the network manager must invest the time and effort to use the log information collected by the policy software on Internet usage pattern to determine priorities, schedules, and solutions for the optimal use of bandwidth (Tubell, 2002).

PBNM is closely related to information security management, which protects sensitive organizational information from unauthorized access (Young, 2002). In November 2001, IIML deployed a comprehensive information security infrastructure, to address the security needs of its information resources (Sridhar & Bhasker, 2003). A comprehensive security policy document was prepared to cover areas such as identification and authorization control, incident handling, Internet usage, firewall policies and administration, electronic mail usage, and WWW usage (refer to Sridhar & Bhasker, 2003, for details). To optimally use the available bandwidth among the users and different applications, appropriate policies were developed and included in the security policy document. Young (2002) points out that instant messaging, e-mail with large attachments, and Web browsing add a huge amount of unpredictable data traffic to the network infrastructure and cause latency. To address these issues, the firewall and associated security components were configured to restrict certain services such as Internet telephony and videoconferencing so that the available bandwidth can be used for mission-critical e-mail and Web services. Restrictions were placed on the maximum size of e-mail attachments (to about 2 Mbytes). Filters for blocking e-mail spamming were also configured on the firewall. A policy-based routing was deployed so that all Web traffic was directed through the newly installed 512 Kbps Internet access link, and e-mail and other traffic were routed through the low-bandwidth, yet reliable 64 Kbps link.

Mr. Mohapatra is looking at developing policies for Web usage also. Most of the Web management policies are aimed at preventing/minimizing use of the Internet services for personal use, and blocking access to "undesirable" sites. However, PBNM goes beyond that. As pointed out by Tubell (2002), Web-use policy management strives to ensure that Internet usage confirms both the positive and restrictive aspects of the organization's Internet use policy.

At this stage IIML has deployed the first two phases of PBNM as stated earlier. As an initial step, the reporting module of the firewall has been configured to track the Internet bandwidth usage pattern. The diagram in Appendix D shows the bandwidth usage for inbound and outbound Internet traffic on a typical day. Even though the firewall captures and logs each and every packet going out/coming from the Internet, reporting features in the firewall are very basic. Mr. Mohapatra is looking at various options for the third phase that involves implementing tools for efficient monitoring, and documenting and reporting detailed statistics of the Internet usage.

CURRENT CHALLENGES/PROBLEMS FACING THE ORGANIZATION

Is Bandwidth Ever Sufficient?

Most of the schools in the U.S. and other developed countries have been providing intranet/Internet connectivity to the hostels and student dormitories for quite some time. Even in a developing country such as India, the premier technology and management institutes have also started adopting this model recently. IIML has about 16 student hostels with about 650 rooms. Two of these hostels house executives who participate in the short-term management development programs conducted by the Institute. In January 2002, it was decided to upgrade the existing Fast Ethernet in the campus to a gigabit backbone, capable of providing up to 1 Gbps, and to provide intranet/Internet connectivity to all the student hostels. By June 2002, the gigabit campus network was completed. The new network, consisting of 48 network switches interconnected by 4,000 meters of optic fiber cable and 30,000 meters of copper wiring, provides high-speed connectivity to all the hostel rooms and 60 faculty residences in the campus. The new network was integrated with the old Fast Ethernet network. A schematic diagram of the gigabit network at IIML is given in Appendix E. Soon after this implementation, a number of computers connected to the campus network jumped two-fold and consequently so did the requirement for more Internet bandwidth.

Moreover the existing 64 Kbps link from the government operator, through which all the e-mails were sent/received, went out of operation on October 9, 2002. This necessitated re-routing of all the Internet traffic through the single 512 Kbps link. This in turn deteriorated the Web performance. Since HCL Infonet was not in a position to provide more than 1 Mbps bandwidth, Mr. Mohapatra arranged for immediate upgrade of the link capacity from 512 Kbps to 1 Mbps, to cater to the increased load. However, a long-term solution needs to be formulated by Mr. Mohapatra to cater to the increasing needs of the users. Possible options are listed in Appendix F.

There are some significant developments in the basic telecom industry in India recently. The government opened up the basic telecom service for unlimited private participation in February 2001. License for the service area in which IIML lies was picked up by a private operator called Reliance Inficomm, part of the largest Indian business conglomerate Reliance Group of Industries. Reliance Inficomm is in the process of developing the basic telecom infrastructure in the region and is expected to launch the service in March 2003. Reliance Inficomm has approached IIML and showed interest in providing an fiber-optic-based communication solution including high-speed Internet connectivity. Entry of the private operator has also prompted the government operator to initiate discussions on providing fiber-optic-based last-mile access to the IIML campus. Mr. Mohapatra has to formulate the short-term and long-term solutions for meeting the increasing bandwidth requirements of the IIML user community.

How Should Converged Traffic be Managed?

Most Internet users have encountered the Internet through the World Wide Web and e-mail. However there is a growing demand for a variety of other Internet services as well. For example, Internet Relay Chat (IRC) — offered by service providers such as Yahoo and Microsoft — has become the latest craze of the IIML user community and contributes to a substantial portion of the Internet traffic at certain times of the day. Though in most of the developed countries, Internet telephony was never regulated, developing countries including India banned Internet telephony for the fear of hurting the monopoly government-owned telephone companies. The government of India lifted the ban on Internet telephony in the country in April 2002 (TRAI, 2002). Currently Internet telephony service is blocked by the firewall at IIML to give priority to the other mission-critical traffic. With Internet telephony service providers offering services to the U.S. and Canada at about one-sixth of the price of regular phone-to-phone international calls, many users have started demanding for opening up of these services. With cost of Web cameras on the downhill, desktop-based videoconferencing units have started appearing on the campus, and the users want to take advantage of this service for academic purposes. Institutes all over the world have been trying to grapple with this type of problem (Kirby, 2001).

The problem can be handled by assigning various priority levels for different services at different times of the day. IIML has to formulate time-based priority policies for different kinds of network traffic so that the Internet bandwidth can be effectively and efficiently used.

Will Proxy Caching Improve Web Performance?

Web caching technology has become an attractive solution for improving Web performance, as it provides an effective means for reducing bandwidth demands, improving Web server availability, and reducing network latencies (Barish & Obraczka, 2000). Caching of popular Web objects at appropriate locations has been the topic of research for improving Web performance. A survey on various caching schemes for the Internet is presented in Wang (1999) and Barish and Obraczka (2000). Caching effectively migrates copies of popular Web pages closer to the clients. Caching thus reduces the load on the Internet and improves Web response by serving the requested objects from the local proxy cache servers. Web client users see shorter delays when requesting a Web page and the Internet access link sees less traffic. Moreover, temporary unavailability of the network can be hidden from the users, thus making the network appear to be more reliable (Barish & Obraczka, 2000).

Caches can be installed at three different levels: at client software (browser cache or client cache), at the remote Web server (server cache), and/or in between these two extremes (proxy caches). Caching at the remote Web server is beyond the control of the organization. Client cache is built into Web browsers such as Netscape and Internet Explorer. The third type of cache (proxy cache), which is more effective than the client cache, can be installed by the organization on a machine on the path from multiple clients to multiple servers (Luotonen & Altis, 1994). Proxy cache is normally deployed at the edges of the network (i.e., at campus firewall hosts), so that they can serve a large number of internal users. This type of caching saves disk space by more efficient management of documents requested by multiple users in the organization. The Web requests first go to the proxy server instead of the remote Web server. The requested file is directed to the client if available, otherwise the request is forwarded to the remote Web server, and necessary files are retrieved and directed to the client. Figure 2 shows the location of the proxy cache server in the organization's network.

Figure 2. Positioning of proxy cache server

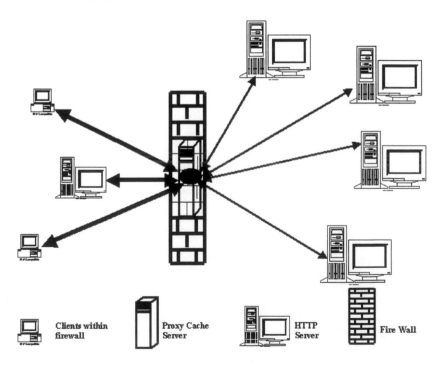

When IIML implemented the security infrastructure, the most popular open-source-based Squid proxy and cache server available as freeware was also installed along with the firewall (refer to http://www.squid-cache.org for details about Squid).

There are several disadvantages to a single caching system such as the one deployed at IIML. The single proxy server becomes the traffic concentration point, and as the number of client requests builds up, the Web performance may deteriorate. The single proxy server also acts as a single point of failure, and hence the reliability of the system is not very high. This has propelled the design of cooperative caching in which a number of cache servers work together to serve a set of clients. In this setting, caching proxies cache the address of the requested objects from other caching proxies (Abrams et al., 1995). There are basically two types of cooperative caching architectures studied by researchers — hierarchical and distributive (Raunak, 1999). In hierarchical caching, caches are kept at different levels of the network, arranged in a tree-like structure. In distributed caching, caches are placed only at the bottom level of the network and there are no intermediate caches. Hence an appropriate number of cache(s), type of cache(s), their architecture and location are critical for improved Web performance. A number of specialized appliance-based cache engines, which perform much better than simple software-based cache servers, are also available in the market today. However these are proprietary and hence are expensive. A comparison between appliance-based cache engine and Squid proxy cache server is given in Appendix G.

It has also been noted by Liu et al. (1998) that simple proxy caching does not always reduce response time. Relevant changes related to proxy caching should also be made to the Hyper Text Transfer Protocol (HTTP) used in the Web. Work on HTTP-Next Generation (NG) is being carried out by the Internet Engineering Task Force HTTP-NG working group (Nielsen, 1998). These developments will lead to better Internet performance.

Performance of the existing Squid cache deployed at IIML in a typical day is shown in Figure 3. The graph illustrates that on average, only 18% of the Web objects requested were served by the proxy cache server. Mr. Mohapatra is concerned about the performance of the existing cache. IIML recently received a grant of about $20,000 from the government for improving the Internet connectivity and security infrastructure. Mr.

Figure 3. Web proxy cache performance

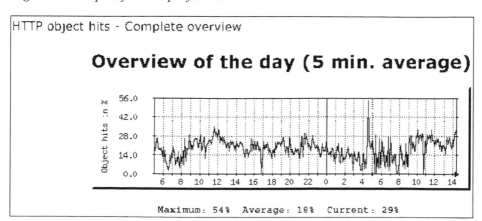

Mohapatra is planning to invest a portion of this grant for upgrading the cache services to improve Web response time.

Bandwidth Monitoring: Whose Job is it Anyway?

Even though an SLA has been signed between the ISP and IIML, Mr. Mohapatra wants more dynamic and granular control over the network traffic. Occasional testing of latency and packet loss does not indicate the health of the Internet connectivity. In response, an increasingly sophisticated set of routers, traffic-shaping appliances, and traffic recognition software has become available to manage the Internet bandwidth utilization (Phillips, 2000).

Continuous monitoring of availability, latency, and packet loss should be done so that the ISPs can be made responsible for any violation in the SLA. However, bandwidth management tools having these features can be very expensive. Can bandwidth monitoring and management be outsourced? Recently most of the ISPs in India have started providing bandwidth management services. However, outsourcing network management requires relinquishing control of customer premise equipment and introduces security risks (Sridhar & Bhasker, 2003). Whether the gear is customer-owned or integrated into the service provided by the operator, tight coordination is required between the ISP and customers to have guaranteed quality of service (Phillips, 2000). Given the still-evolving ISP market in developing countries, is it possible to strictly enforce SLA commitments of the service providers by continuous monitoring of the Internet bandwidth?

REFERENCES

Abrams, A., Standridge, C., Abdulla, G., Williams, S., & Fox, E. (1995). *Caching proxies: Limitations and potentials* (Technical Report TR 95-12). Virginia Polytechnic Institute and State University, Department of Computer Science, USA.

Barish, G., & Obraczka, K. (2000). World Wide Web caching: Trends and techniques. *IEEE Communications Magazine Internet Technology Series, 38*(5), 178-184.

Cache Flow Inc. (1999). Cache Flow White Papers. Retrieved on November 20, 2002, from http://www.cacheflow.com/technology/wp

Cohen, E., Krishnamurthy, B., & Rexford, J. (1998). Improving end-to-end performance of the Web using server volumes and proxy filters. *ACM SIGCOMM Computer Communication Review, 28*(4), 241-253.

ICRA (Investment Information & Credit Rating Agency). (2000). *The Indian Internet business report.* Retrieved on November 20, 2002, from http://www.icraindia.com

ITU (International Telecommunications Union). (2002). *World telecom indicators.* Retrieved on March 25, 2003, from http://www.itu.int

Kirby, R. (2001, April 5). Business case: Rollins College muzzles the Napster mongrel. *Network Magazine.* Retrieved on August 27, 2002, from http://www.networkmagazine.com

Lewis, W. F., & Williams, C. (2002). *On estimating the amount of Internet bandwidth needed to support the outgoing and incoming information streams.* Retrieved on November 15, 2002, from http://www.prospect-tech.com/ec/bandwidt.doc

Liu, B., Abdulla, G., Johnson, T., & Fox, A. A. (1998). *Web response time and proxy caching.* Retrieved on November 22, 2002, from http://memex.dlib.ut. edu/~liub/webnet/index.html

Luotonen, A., & Altis, K. (1994). World Wide Web proxies. *Proceedings of the First International Conference on WWW,* Geneva. Retrieved on January 9, 2003, from http://citeseer.nj.nec.com/luotonen94worldwide.html

Nielson, F., Spreitzer, M., Janssen, B., & Gettys, J. (1998). *HTTP-NG overview.* Retrieved on November 30, 2002, from http://www.w3.org/

Passmore, D. (1999, February). The fine print of SLAs. *Business Communications Review,* 16-18.

Phillips, B. (2000, March). Bandwidth management: Whose job is it? *Business Communications Review,* 20-26.

Raunak, M. S. (1999). *A survey of cooperative caching.* Retrieved on December 31, 2002, from http://citeseer.nj.nec.com/432816.html

Rubenstein, R., Hersch, H. M., & Ledgard, H. F. (1984). *The human factor: Designing computer systems for people.* Burlington, MA: Digital Press.

Sevcik, P., & Bartlett, J. (2001, October). Understanding Web performance. *Business Communications Review,* 28-36.

Sridhar, V., & Bhasker, B. (2003, January). Managing information security on a shoestring budget. In *Fifth volume of the annals of information technology.*

Stallings, W. (1998). *High speed networks: TCP/IP and ATM design principles.* Englewood Cliffs, NJ: Prentice-Hall.

Stallings, W. (2001). *Business data communications* (4th ed.). Englewood Cliffs, NJ: Prentice-Hall.

TRAI (Telecom Regulatory Authority of India). (2001). Consultation paper on QoS of dialup access network for Internet. Retrieved on October 17, 2002, from http://www.trai.gov.in

TRAI (Telecom Regulatory Authority of India). (2002). Guidelines for issue of permission to offer Internet telephony services. Retrieved on June 18, 2002, from http://www.trai.gov.in

Tubell, W. J. (2002). *The case for a policy based Web-use management approach.* Retrieved on October 15, 2002, from http://www.wavecrest.net/editorial/include/caseforpolicybased.pdf

Utsumi, T., & Varis, P. T. (2003, January 26-27). Creating global e-learning and e-healthcare through the global university system. *Proceedings of the 3rd eHealth Regional Symposium,* Dubai International Exhibition Centre, UAE. Retrieved on December 12, 2002, from http://www.friends-partners.org/GLOSAS/

Voice & Data. (2002). *Vital statistics on Indian telecom.* Retrieved on October 25, 2002, from:http://www.voicendata.com

Wang, J. (1999, October). A survey of Web caching schemes for the Internet. *ACM SIGCOMM Computer Communication Review,* 36-46.

Wexler, J. (1998, December). ISP guarantees: Warm, fuzzy and paper-thin. *Business Communications Review,* 23-27.

WiscNet. (2003). *Internet access pricing.* Retrieved on January 15, 2003, from http://www.wiscnet.net/internet_access.html

Young, M. (2002, August). Policy-based network management: Finally? *Business Communications Review,* 48-51.

Appendix A. Annual budget for the year 2002-03 for the computer services at IIML

No.	Activity	Budgeted Amount	
		USD	INR
Planned Expenditure			
1	Servers	80,000	4,000,000
2	Uninterrupted power supply units for the computers	12,000	600,000
3	Information security infrastructure and implementation	16,000	8,00,000
4	Campus-wide networking	50,000	2,500,000
5	Videoconferencing facility	14,000	700,000
6	Software & development	14,000	700,000
7	Multimedia machines, furniture, printers, scanners, & projectors.	24,000	1,200,000
8	Training	8,000	400,000
Non-Planned Recurring Expenditure			
1	Annual maintenance contract of computers & facilities management services	34,000	1,700,000
2	Subscriptions for software packages	31,000	1,550,000
3	Internet charges	40,000	2,000,000
4	Annual maintenance contract for information security management	6,000	300,000
5	Consumables and other miscellaneous expenditure	5,000	250,000
Total		**334,000**	**16,700,000**

Appendix B. Comparison of last mile technologies for campus Internet connectivity

Technology	Maximum Bandwidth	Distance Limitation	Characteristics	Reliability	Lead Time for Deployment
1 Wireline Technologies					
Integrated Services Digital Network (ISDN)/ Leased Line	128 Kbps-2 Mbps	Up to 5.5 Km	Use of low-cost telephone lines; most of the BTOs in India, especially in cities, offer the service	Medium	Low (if the nearest exchange of BTO is ISDN capable)
Asymmetric Digital Subscriber Loop (ADSL)	1.54 Mbps-6 Mbps	Up to 5.5 Km	Use of low-cost telephone lines; in India, poor quality of the telephone cables do not permit this service in many areas	Medium	Medium
Fiber in local loop	2 Mbps-100Gpbs	20-100 Kms	Specialized Customer Premise Equipment (CPE) is required; very high capital expenditure (in the range of US$6-10 per meter) for laying down fiber (to be borne either by the BTO or the customer)	Very high	High
2 Wireless Technologies					
Terrestrial Radio	64Kbps-32 Mbps	20-30 Kms	Transmission radio towers at line-of-sight; specialized CPE required (can be rented or bought); low capital expenditure compared to fiber optic	High	Very less
Very Small Aperture Terminal (VSAT) based satellite gateway	400Kbps-32 Mbps	No distance limitations	Specialized CPE required; high capital expenditure; license required from the government for operation; long-term service commitment; very high delay (about 500 ms round-trip time)	Very high	Very less

Appendix C. Sample ISP service level guarantees

Service Level Attribute	Guarantee Levels	Penalty Clauses
Network Availability	99.5%-100%; < 10 consecutive minutes	1/30 of Monthly Service Charge (MSC) refund per 15 minutes downtime per day, up to 10% MSC refund per quarter; one day service credit; one day credit for up to each hour of downtime, up to entire MSC
Maximum Latency	200 ms per month; 85 ms (within U.S.) 120 ms (international)	Remittance of daily charge if each day average doesn't meet guarantee; one day service credit
Maximum Packet Loss	10% in 10-minute period	One day service credit

Appendix D. Incoming and outgoing Internet traffic on IIML network

External - Network traffic - Overview of the day

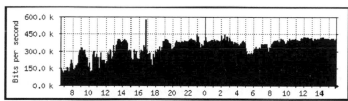

Max In: 72143 Bytes/s Avg In: 41870 Bytes/s Cur In: 51491 Bytes/s
Max Out: 36803 Bytes/s Avg Out: 9140 Bytes/s Cur Out: 22114 Bytes/s

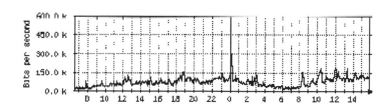

Appendix E. Schematic diagram of the gigabit network at IIML

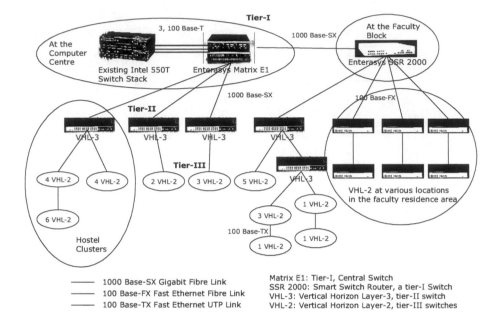

Appendix F. Possible options for improving Internet connectivity for IIML

Connectivity Option	Advantages	Disadvantages
Enhancing Existing 1 Mbp HCL Infonet link to 2 Mbps or more; keeping the same radio link connectivity	No capital expenditure as the existing equipment can be used; no service interruption	No redundancy in case of failure; the service provider does not have enough domestic bandwidth to provide quality of service above 1 Mbp
Explore another vendor for another 1 Mbp (upgradeable later) using radio-link last mile access	Redundancy in case of failure of any one link; traffic balancing can be done on the two links to reduce latency; quick deployment	Need to buy/rent additional radio equipment; limitations on bandwidth of radio connectivity
Very Small Aperture Terminal (VSAT) based international satellite gateway (refer to Stallings, 2001, for details about VSAT technology)	High and scalable bandwidth; direct connectivity from the IIML campus to international POP; better response time and availability; single service provider; quick deployment	Need to get license from the government for gateway operation; pay annual license fee; significant capital expenses for the satellite antenna and indoor equipment; VSAT-based Internet connectivity is very new and evolving service in India; 3-5 years commitment of bandwidth to the service provider
Fiber optic-based last-mile connectivity from the BTO and a terrestrial leased line through fiber optic to the current or any other ISP's POP	The best last-mile solution for increasing bandwidth requirements of IIML; can cater to both voice and data connectivity, and even provide videoconferencing and other enhanced services	Monopoly basic operator demands cost of laying fiber (in the order of US$20 million) to be borne by IIML; even if decided, will take about 6 months to lay the cables

Appendix G. Comparison of squid proxy cache server and appliance-based cache engine

Squid Proxy Cache Server	Appliance-Based Cache Engine
Freeware; can be downloaded from http://www.squid-cache.org; GNU public license.	Proprietary and expensive (about $12,000-$15,000 for the user/host base at IIML); future upgrades are normally charged.
Software based.	Physical appliance, with embedded specialized operating system and high-speed hard disk for storing cache objects; reduces delay in retrieving cache objects.
Limited support through user community found at http://www.squid-cache.org.	Comprehensive warranty and support from the vendor.
Requires expertise for configuring and administration; periodic updates/patches have to be downloaded and installed.	Can be set up for automatic updates of patches/new releases; provides user-friendly configuration and monitoring tools.
Retrieval of objects is serial, which presents delay for the end user.	Certain engines use a technique known as object pipelining to retrieve objects in parallel, thus improving response time for first-time Web page requests.
Audio and video streaming are not well supported.	Provides caching of multimedia streaming objects and automatically replicates a single live stream from originating media server for multiple viewing.
Provides limited support for dynamic access control management.	Typically provides superior access management through dynamic blocking based on statistical analysis of the traffic pattern.
Cooperative caching is supported.	Support scalable cooperative cache configurations.

V. Sridhar is an associate professor of the Information Technology & Systems Group at the Indian Institute of Management, Lucknow, India. He received his PhD in MIS from the University of Iowa. He taught previously at Ohio University, Athens, and the American University, Washington, DC. He has also served as a visiting faculty member at the University of Auckland Business School, New Zealand. His primary research interests are in the areas of telecommunications technology and management, information security management, collaborative technologies, and global electronic commerce. He has published his research articles in the European Journal of Operational Research, Journal of Heuristics, IEEE Transaction on Knowledge and Data Engineering, *and* Annals of Cases on Information Technology.

Piyush Jain is a final-year student of the Fellow Program in Management at the Information Technology & Systems Group at the Indian Institute of Management, Lucknow, India. He completed his undergraduate degree in electrical engineering at the University of Roorkee, India, and was awarded the Science and Technology Entrepreneurship Park Award for best project in the program. He did his post-graduate work at the Indian Institute of Technology, Delhi, India. He had worked at Indian Telephone Industries Ltd. and Military Engineer Services under the Ministry of Defense in various middle-management positions. He is presently working on his doctoral thesis on techno-economic modeling of basic telecommunication Services. His other research interests include network topology design, capacity optimization and management, access network design issues, simulation modeling, and e-governance.

This case was previously published in the *Annals of Cases on Information Technology*, Volume 6/ 2004, pp. 540-560, © 2004.

Chapter VIII

Remote Management of a Province-Wide Youth Employment Program Using Internet Technologies

Bruce Dienes, University College of Cape Breton, Canada

Michael Gurstein, University College of Cape Breton, Canada

EXECUTIVE SUMMARY

A province-wide network of Community Access Internet sites was supported during the summers of 1996 and 1997 by Wire Nova Scotia (WiNS), a government funded program to provide staffing, training and technical support for these centers. The program was managed remotely from an office in Sydney, Nova Scotia (Canada) using a variety of Internet-based technologies, including e-mail, a Web site, conference boards, real-time chat, and mailing lists. Remote management enabled the efficient and low-cost operation of a program involving 67 sites with field placements, plus six regional coordinators and the technical and administrative staff at the hub in Sydney. Effectiveness of remote management was enhanced when employees participated in an initial face-to-face regional training workshop. This training not only familiarized the employees with the communications technologies, but, perhaps more importantly, put a human face and personality to the messages that later came electronically over the Intranet.[Note: For the benefit of those readers who may not be familiar with Internet technical terms, a brief glossary (Appendix A) is provided at the end of this case].

BACKGROUND

Remote management as a key strategy for enabling new kinds of enterprise and other cooperative endeavors has emerged from the necessity of creating networked organizations to deal with the complexity and fluidity of the Information Age society. In the business world, the companies who are able to create effective virtual teams, particularly if they are able to collaborate with other companies that have complementary resources, are the ones who are thriving. For example, in 1995, three "arch-competitors" in the magazine business, *Men's Health*, *Esquire*, and *Rolling Stone* discovered an opportunity to collaborate on a bid for a significant advertising contract, and in three short weeks managed, via rapid and effective use of electronic communication and virtual team building, to put together a bid that beat out the media giant, Time, Inc. for the contract. In Denmark, government policy encouraged small enterprises to create collaborative networks, and this has resulted in a significant increase in the success of these businesses (Lipnack & Stamps, 1997).

Note that it was not just that electronic media created more efficiency or enabled existing projects to be done more effectively. It created the possibility for an entirely different way of doing business. As Marshall McLuhan (1964) has told us, "The medium is the message." Even if our initial motivation in using electronic media is to simplify or speed up existing routines, the nature of the medium inevitably generates secondary effects (Grundy & Metes, 1997), changing the way we work, the way we think, and the way we relate to other people and organizations. One impact of working virtually is that the old hierarchical models of the Industrial Age become transformed into a more interdependent networking model, where access to information and participation in decision making is not limited to the upper echelons.

Instead of asking, "What is the information that matters and how do we most effectively manage it?" companies must start asking, "What are the relationships that matter and how can the technology most effectively support them?" - Michael Schrage, *The Wall Street Journal*, March 19, 1990 (cited in Johnson-Lenz & Johnson-Lenz, 1995)

This creates a need to develop new management and leadership styles, new ways of training workers to collaborate effectively, and a readiness to work across organizational boundaries, whether in commerce (e.g., Sieber & Griese, 1998) or in not-for-profit organizations (e.g., James & Rykert, 1997). Moreover, the network that is built must be constructed in anticipation of rapid change, able to respond immediately to new realities and new opportunities (Metes, Grundy, & Bradish, 1998).

In the case described below, the initial motivation for using remote management via use of an intranet was to create a cost-effective way of coordinating a province-wide program, delivering technical support, and receiving reports. However, once the network was created, it became clear that it also had potential to enable types of collaboration that would not have been practical with conventional organizational structures. Despite the fact that there was a nominal hierarchical structure in place, with both central and regional levels of coordination, the universal access to information and to communications tools enabled both fieldworkers and regional coordinators to operate with a high degree of independence, and were able to use the intranet communications to brainstorm, plan and implement collaborative activities involving multiple sites. The existence of the communications network encouraged and enabled formation of temporary regional action networks, which have the potential to evolve into more permanent Community Enterprise Networks (CENs), described below (see also Gurstein & Dienes, 1998). Such secondary effects of remote management have great potential and should be considered in both the planning and evaluation stages of a project.

Governments, educational institutions and non-profit community organizations are increasingly becoming involved in partnerships for sharing information systems and intranets for collaboration and for delivering services. One excellent example is the Missouri Research and Education Network (MORENet). "The primary mission of MOREnet is to provide collaborative networked information services to its members and customers in support of education, research, public service, economic development and government" (http://www.more.net). Resources on their Web site include technical support, discussion lists, and access to tools such as online databases. Its link to the Missouri Express project for supporting Community Information Networks (CINs) provides an online newsletter as well as contact information, instructions and templates for creating a CIN.

The Wire Nova Scotia (WiNS) case described below incorporates these ideas into its intranet, and adds the element of remote management of staff, including monitoring, reporting, documentation of the project, collaboration on inter-site projects, and technical support.

SETTING THE STAGE

In the summer of 1996, the associate chair in the Management of Technological Change (MOTC) at the University College of Cape Breton (UCCB) in Sydney, Nova Scotia, Canada opened the Centre for Community and Enterprise Networking (C\CEN). C\CEN's function was to research, develop and incubate ways of using the emerging Information and Communications Technologies (ICT) to promote local economic development in non-metropolitan regions, linking university, government and private resources with local needs. C\CEN followed the ongoing traditions of community-based economic development of the Antigonish Movement, which began in the Cape Breton area, and the mission of UCCB, which is to provide support to the economic and social development of the Cape Breton region. (See also the C\CEN Web site, located at <http://ccen.uccb.ns.ca>.)

C\CEN had no core funding, so staffing and facilities would grow and shrink based on the particular projects it was funded to run at any given time. Because of this, its organizational structure was continually in flux. While this was at times confusing, it did create a climate of openness to experimentation with different management and organizational strategies. Base staff included the director, administrator, administrative assistant, systems administrator, and a technical support officer. A project leader would be hired for each new project, along with additional secretarial and technical support as needed, and staffing specific to the project. Project coordinators were given latitude to innovate and implement creative solutions to issues as they arose, generally reporting directly to the Centre's director. As the number of projects grew, a technology coordinator was appointed to oversee the incubation of new projects. Later this function was taken over by two assistant directors, one responsible for project development, the other for project implementation. Often staff roles and job descriptions would change to adapt flexibly to the needs of a new configuration of projects. The ability to share common technical and administrative expertise and resources between projects enabled a more cost-effective solution to staffing needs.

In its first year of operation (June 1996 to June 1997), leveraging an initial $165,000 grant into almost one million dollars via various partnerships and sponsorships, C\CEN was responsible for the creation of 140 person-years of skilled technically intensive work in non-metropolitan Nova Scotia, primarily in Cape Breton.

Cape Breton is suffering from the effects of a depleted resource-based economy, with the former mainstays of fishing, coal and steel no longer able to provide sufficient employment for the local population, resulting in outmigration and increased poverty. C\CEN piloted several information-based employment-generating projects with a view to encouraging the formation of a cluster of activity in this sector in the region, which, once it reached a "critical mass," would be self-sustaining. In particular, C\CEN encouraged the development of the network of Community Access Internet sites initiated by Industry Canada's Community Access Program (CAP), which provides a one-time-only grant on a competitive basis to organizations in up to 10,000 communities across Canada to support public Internet access. (For more information on CAP, see http://cap.unb.ca.)

Each CAP site is operated autonomously by a local community board. In isolation, such sites are unlikely to be economically viable, much less stimulate economic development in their area. However, as a coordinated network of centers (a Community Enterprise Network), they can serve as a distribution system for value-added facilitated access to various government and commercial information services (e.g., Job Banks, health information, online banking) and also as a coordination network for distributed production and joint marketing of local products and services to the global marketplace, using the flexible networking model (Piore & Sable, 1984; Gurstein, 1998a).

In order to stimulate the development of the CAP network in Nova Scotia, and to encourage communication and cooperation between the CAP sites, C\CEN developed a project called Wire Nova Scotia (WiNS) to coordinate and support summer staffing of the sites. (For a description of the WiNS program see the Case Description below.)

One key aspect of WiNS was to demonstrate that a province-wide program such as this could be administered from Cape Breton using appropriate communications technologies, as easily as it could from the central provincial administrative hub in the capital, Halifax. Instrumental to this was the development of an effective, inexpensive system of remote management. It is this system that is the subject of this case study.

Prior to developing the WiNS program, the C\CEN office was working with a small Windows95 Local Area Network (LAN), connected to the Internet via dial-up modem using WinGate software, plus a Unix server, located at UCCB, which hosted a Web server as well as SMTP, POP-mail and Majordomo mailing list services. Internet connectivity for the Unix server was through UCCB's pipeline. Dial-up connectivity for the LAN was donated by MT&T Sympatico. All CPUs were Pentium 100 or 133mhz.

With these simple tools, the Centre managed an impressive array of projects, including creating an online bilingual real-time interface to the Access'96 conference held in Sydney, NS in November, 1996. Participants on the Internet could view real-time summarized transcriptions of ongoing presentations and seminars, simultaneously translated into English and French. Questions and comments from Internet participants would be directed to the speakers for immediate reaction.

Some examples of other projects include:

- **Environmental Web Project:** Web-based index of all organizations in Cape Breton with the capacity to provide services in the Environmental sector.
- **The Cape Breton Music Centre Online:** Web site for the promotion and preservation of Cape Breton music, with an events calendar, resource directory, links to musicians' sites, etc.
- **Technical Support Group:** Offering online technical support to Community Access sites, First Nations sites, etc.
- **Electronic Democracy Project:** Providing a Web site and e-mail lists for open discussion of election issues in Nova Scotia.
- **Student Connection Program:** This was a province-wide program in which students were hired to train small businesses in use of the Internet.
- **ANGONET:** A Government of Canada project, "NetCorps" is being designed to link Canadian youth with technology-based projects in developing countries. As a pilot, C\CEN sent two young Cape Bretoners to provide technical support for ANGONET, Angola's Internet Access Network.

Many of these projects, particularly Student Connection, ANGONET, and the Technical Support Group included a remote management aspect, and experiences with various communications modalities led to a better understanding of what worked and what did not. This experience enabled C\CEN to design an effective remote management system for WiNS.

CASE DESCRIPTION

WiNS'97 Project Description

Industry Canada's Community Access Program (CAP) has supported Community Access sites by providing startup funds as well as funding for summer youth workers. These workers help staff the sites and also encourage involvement of local businesses, government and community groups. The Wire Nova Scotia Project (WiNS), developed and piloted in 1996 by C\CEN, in collaboration with the Nova Scotia Community Access Committee, was an enhancement to this program. WiNS assisted Community Access sites and summer youth workers in Nova Scotia to reach their potential through coordinated training, management and administration, including online technical support. The program was open to any site providing community access to the Internet, not only the CAP sites.

The WiNS pilot was subsequently adopted by Industry Canada as a national model. The following summer, C\CEN was asked to implement WiNS '97, which served 67 access sites across Nova Scotia directly, and many others indirectly. This effort included helping sites to strategize with school boards, libraries, regional development authorities, federal agencies and corporate sponsors to work towards the creation of local collaborative networks, with a view to creating long term sustainability.

In addition to the site workers, there were six regional coordinators with management skills. These staff helped to coordinate the efforts of sites in their area, facilitated provision of training and technical support and provided regular supervision or intervention where necessary.

WiNS is a medium for community capacity building. Community Access and a trained workforce are both critical to the infrastructure for other initiatives that will follow. They provide the groundwork for a new commercial sector — Rural Informatics, and the structures to enable rural enterprises to be both competitive with, and appropriate collaborators for, their urban counterparts.

The development of WiNS involved the investment of many partners including:

- Industry Canada
- Enterprise Cape Breton Corporation
- Human Resources Development Canada
- Nova Scotia Department of Education & Culture
- Nova Scotia Department of Economic Development and Tourism
- University College of Cape Breton
- MT&T/Sympatico
- Nova Scotia Regional Libraries
- Regional Development Agencies
- Acadia Centre for Small Business and Entrepreneurship
- Strait East Nova Community Enterprise Network
- Chebucto Community Network

In order to satisfy the various reporting requirements to these bodies, detailed records of fieldworker activity and site usage had to be kept.

One fieldworker was assigned to each participating Access site. Precise job descriptions would vary with the specific needs of each location, but in general, the fieldworkers' tasks would include:

- Staffing the Access Site in collaboration with local volunteers to maximize hours open.
- Promoting the Community Access site to the community at large and marketing its services.
- Providing training on Internet and related services to representatives of community organizations, business people and individual citizens.
- Providing technical support to the Community Access site.
- Outreach to neighboring communities to promote the Community Access concept and encourage others to develop proposals for the CAP program.

Funding for the fieldworkers' salaries was designated for the 16-30 age group, so most workers were young students with limited experience. This created a greater need for supervision and technical support. Complicating the management process was the fact that the workers had a dual-reporting relationship. The workers were interviewed and hired by the local CAP site committee, and a local site supervisor was assigned to them. Depending on the site, these supervisors were more or less accessible during the summer months. Since payroll was managed directly through C\CEN-UCCB, and the funders had specific expectations for outputs during the work term, the fieldworkers also reported to their WiNS regional coordinator, each of whom supervised between 10 and 25 sites.

When the agendas of the site committees and the funders clashed, it was up to the WiNS coordinators to mediate the situation.

Overview of the Remote Management System

The newest meaning of "virtual" attests to forces that are fast moving teams into an altogether different realm of existence — virtual reality — or more precisely, digital reality. Electronic media together with computers enable the creation of spaces that are real to the groups that inhabit them yet are not the same as physical places. The eruption of the World Wide Web in the last decade of the millennium has allowed virtual teams to create private electronic homes. These interactive intranets-protected members-only islands within the Internet-signal a sharp up-tick in the human capability to function in teams. (Lipnack & Stamps,1997, Chapter 1, Section 2)

Human resource management is a critical component to a successful project. Limited financial resources and a large geographical area require that traditional methods be adapted to adequately monitor, administer and manage a large and widely dispersed group. Using a storehouse of available knowledge and skills, developed from the projects described above, C\CEN was able to institute a form of virtual management. Using a Web site, O'Reilly's WebBoard™ software, e-mail, electronic mailing lists and IRC chat, the project was managed "online," with a formal reporting structure implemented and a variety of support functions available to assist the fieldworkers in their responsibilities. Five regional coordinators and one community network developer were placed across the province. These staff used the virtual management process to coordinate supervision of field workers in their region, as well as communications and cooperation between the regions and with technical and administrative support staff from C\CEN.

Needs

To be an effective system, the following management functions had to be implemented online:

- **Reporting**
  Weekly work summaries of fieldworkers.
  Weekly site usage statistics.
  Weekly regional reports from regional coordinators.
- **Documentation/archiving**
  Permanent record of reports accessible online.
- **Conferencing**
  Several distinct online asynchronous discussions, threaded by subject.
- **Announcements**
  Read-only announcement board.
- **Problem solving**
  Ability to form virtual task forces and workgroups to solve emerging problems.
- **Tech support**
  Online technical support for CAP sites, fieldworkers and regional coordinators.

- **Shared insights/experience/idea building**
  Free flow of information between all workers.
- **Regional collaboration/planning**
  Coordinating local multi-site projects and efforts.
- **Online real-time meetings**
  Real-time chat (IRC) in a secure environment allowing workers to meet "virtually" to brainstorm projects and plan joint activities.

In addition, the communication media had to have the following characteristics:

- **Flexible participation**
  Multi-modal access to communication services (via e-mail, Web and IRC).
  Both synchronous and asynchronous communication modalities.
- **Access control**
  Access to resources contingent on status, function and regional location within the organization. Signal to noise ratio (i.e., information versus socializing) appropriate to each channel of communication.
- **Cost-effectiveness**
  Resources had to be available at a cost easily affordable in terms of both central server and connectivity requirements and local workstation specifications.
- **Ease of use/user-friendliness**
  Even non-technical users should be able to use the system with a minimum of training.

Solutions

When the Web first got started, threaded discussions and live chat were interesting diversions on Internet sites and corporate intranets, but little more. Not any longer. Today online conferencing has moved to the forefront of many Web sites. Some of the Web's most highly trafficked sites are devoted to live discussion areas. And workgroups in small and large companies alike have discovered the productivity advantages offered through Web-based collaboration. (http://webboard.oreilly.com/wb3/ product_info/index.html; July 12, 1998)

The interactive capacities of the World Wide Web give rise to applications that enable a radical shift in communications options for organizations. Initially, the Web was used primarily as a giant bulletin board, but now is fast becoming a set of interconnected virtual communities. Secure interfaces allow for controlled access to information and communication resources, at the same time as allowing access from any terminal connected to the Web, provided the user has the appropriate access codes.

What this means is that organizations with small budgets can take advantage of the existing infrastructure of the Information Highway with only a modest investment in hardware and connectivity costs. Fieldworkers can access the organization's intranet from home, from a work site, or from a public access terminal. Regional coordinators with a notebook computer and a cell phone can travel anywhere in their region and do not necessarily require the expense of a branch office. The head office can be located anywhere, freed of the high rents of urban locations and able to bring jobs to non-metropolitan areas. These are the underlying concepts of the WiNS intranet: flexibility, accessibility, low cost, and effectiveness.

To manage the WiNS'97 project, two new P-150 NT servers were added to C\CEN's network: one to run an expanded LAN at the central office, and one to host the WebBoard and NT-Mail (to provide e-mail service for staff). The main components of the WiNS intranet were:

1.   **WiNS Web Site** (http://ccen.uccb.ns.ca/wins97) housed on the C\CEN Unix server.

This was the "front end" of the Intranet, providing a gateway to the other features and a public face to the world. There were six main sections to the site:

   a.   **Project Information Area:** This was the public interface, which gave an overview of the project and links to the other areas on the site, including contact information. The WiNS general information flyer was available. Prospective employees could find out requirements for qualified applicants and sites could learn how to apply for a summer worker. Success stories from the previous year were highlighted. Press releases and updates about program innovations were posted as they became available. Links to other relevant sites provided information about the broader context of the project. These included links to the Nova Scotia Community Access Committee, the Community Access Program, the Strait East Nova Community Enterprise Network, the main C\CEN page and all other funders and partners of the project.

   b.   **Technical Support Area:** This area gave a description of services that were offered (as well as clear statements about services that were NOT offered!), a Frequently Asked Questions (FAQ) file, based on the highest volume queries dealt with by the technical staff, links to other technical resources on the Web, and, most importantly, a link to the Technical Support WebBoard conference, described below in the section on the WebBoard.

   c.   **Regional Resources:** A clickable map of the province, color-coded by regions, gave quick access to resources of relevance to each particular region. These included a map of the region, showing all the access sites, and contact information for each site, each fieldworker, and the regional coordinator. These coordinators were encouraged to develop their own information resources for their region to add to this area. Graphic components were made available via FTP so that the regional sites thus developed could have a similar "look and feel" to the central site. Some regional sites were hosted on regional servers (such as local Community Network servers) rather than on the C\CEN server. Decentralizing development and hosting of regional Web sites enabled local creative control over content and design, thus giving local workers more "ownership" of the site and making it more likely for them to use it. It also decreased the workload of the central technical team, enabling them to put more attention on technical support rather than on Web design and maintenance.

   d.   **Fieldworker Resources:** This area gave an overview of fieldworker responsibilities, examples of specific tasks that workers would undertake, information on how to get help or technical support, and instructions on using their personal Web conferences for regular reporting purposes.

e.   **Conference Area (WebBoard):** (See description of WebBoard below.)

f.   **Contact Information:** Phone, fax, e-mail and mail addresses for the central office, regional coordinators and technical support were posted. Links also were provided by region to a list of all participating sites, their WiNS fieldworkers and their contact information. This gave a simple, accessible method of reaching anyone in the organization.

2.   **E-Mail:** For those staff who did not already have e-mail accounts, C\CEN provided this service. [Note: With the new availability of free Web-based e-mail services, such as HotMail or EudoraMail, this would no longer be required.] Whereas the WebBoard was used for reporting, announcing and group discussions, e-mail provided a medium for one-to-one correspondence, particularly for arranging meetings, dealing with personnel problems, or getting quick feedback. It also provided a mode for social interaction between the workers, a key aspect of maintaining high morale, particularly in situations where workers were staffing sites alone much of the time.

3.   **WebBoard:** While the Web site was the matrix that integrated the various communications modalities, the WebBoard was the "heart" of the system. (See Peck & York, 1998 or <http://www.oreilly.com/webboard> for a detailed description of the WebBoard software.) WebBoard provides Web-based online discussion conferences that are accessible to anyone with a Web browser and Internet access. A key feature of the WebBoard is the ability to restrict access to each individual conference depending on the status of the user within the organization.

WiNS used four types of conferences: general, regional, staff and personal, incorporating the functions of reporting, planning and coordination, problem-solving, documenting, socializing, idea generation and technical support.

a.   **General conferences** were open to all employees and served a variety of purposes.

   i.   The *Announcements* conference was a read-only conference to which only senior staff could post but all could read. This created a zero-noise board that everyone would check regularly for important updates without having to plow through any chaff.

   ii.   The *Business Development* conference was an idea-generating and sharing medium for people to brainstorm methods of using the access sites to generate income, to support local business, or to create new businesses, using the ICT tools.

   iii.   The *Technical Support* conference provided answers to a variety of technical questions on hardware, software, networking, training, etc. Technical support was made available to any access site worker or supervisor, not just the WiNS worker. Anyone could sign themselves up to the WebBoard and automatically be given access only to the Technical Support and Business Development conferences. A particularly useful aspect of using a Web conference rather than one-to-one e-mail was that, in addition to getting information from the technical support officers, participants could assist each other and share expertise they developed as they solved problems locally. This created a broader pool of expertise and reduced the workload for the technical support staff. It also reduced the tendency for field workers to develop too much dependence on the central technical support, encouraging them to develop their own skills and share them with others. The technical staff would of course monitor the answers provided by other employees to ensure that they were not giving inappropriate advice.

iv.   The *General Discussion* conference provided a medium for anyone to raise issues, talk to groups from other regions, chat, etc. This was an informal environment that facilitated socializing, networking, and resource sharing across the whole project. Providing a forum specifically catering to informal interaction helped to keep the other fora more focussed and productive. It is important to provide a variety of communication venues so that people of a wide variety of personalities and communication styles will feel comfortable using the online medium.

b.   **Regional Conferences** served each region, enabling planning of joint ventures between nearby access sites, region-wide projects, coordination of promotion and advertising campaigns, announcements of regional online real-time chat meetings, and sharing of resources and expertise between people who had met at regional training sessions. Regional conferences were readable by anyone involved in the WiNS project across the province so that others could learn from innovations in other areas, but generally only participants from that region would post to the regional conference.

c.   The **Staff Conference** was restricted to the senior staff: the WiNS coordinator, the WiNS administrator and the regional coordinators. It was used for weekly regional reports and for discussion of any problems involving sites, fieldworkers, or policies. Having this as a Web conference rather than a one-way report to the WiNS coordinator enabled regional coordinators to learn from one another's experiences, and to offer support as appropriate. Coordinators had a variety of distinct expertise. For example, one may have training in community development, another in technical issues and another in business development. The staff conference made these skills available to all as problems and solutions were brainstormed.

d.   **Personal Conferences** were the primary reporting method for all fieldworkers. Each fieldworker had his or her own personal conference that was used to file weekly reports and site usage statistics. This conference was viewable only by the fieldworker, their regional coordinator, and the WiNS coordinator. This gave the WiNS coordinator a log of activities and problems encountered by all staff. He could get a "snapshot" of any given site for any period by simply reviewing the weekly logs of the site worker. The absence of a log would point immediately to a potential trouble spot, and this could be addressed before the problem was exacerbated. These logs (in addition to the regional weekly reports) provided a rich source of data for compiling success stories and for writing the final report on the project.

4. **Online Mailing Lists:** C\CEN provided three e-mail lists of relevance to this project. One was a general discussion list ("center") for those working at the central hub in Sydney, allowing exchange of ideas between people working on different projects; another was a general discussion list for all the Community Access sites in Nova Scotia ("CA-NS"), providing a medium for sharing problems and solutions, new grant or income generating possibilities, upcoming workshops, conferences and training opportunities, etc. In addition, there was a list for general discussion of community access issues across Canada, with a particular focus on universal access to the Information Highway ("UA-C"). Originally run on majordomo software, the latter two are now functioning using LISTSERV. Archives of these lists can be found at <http://ccen.uccb.ns.ca/archives>.

5. **IRC Chat:** Internet Relay Chat (IRC) enables multi-user real-time text-based interactive communication. Individual users' typing is identified by an alias name at the beginning of each line they send, and each participant sees comments from all others as they scroll up the monitor screen. IRC-based meetings work best when there is a designated moderator and a predetermined agenda or set of questions about which participants are prepared to dialogue.

    Chat functions were available for WiNS through the WebBoard or via IRC client software. Most sites used the MIRC shareware software for this purpose, as the chat capabilities in version 2 of WebBoard were limited. [Note: This limitation has been addressed in version 3 and private chat lines can operate within the WebBoard environment.] Use of chat for regular regional meetings was at the discretion of the regional coordinators. Fieldworkers could also set up chat meetings at any time, for example to coordinate multi-site activities (such as creating content for a booth representing area Community Access sites at the local county fair).

    There were mixed opinions as to the usefulness of the medium. Some felt that it was important for fieldworkers to interact in real-time, as this built a sense of community and provided an environment for brainstorming or for sharing immediate issues. Others found the medium somewhat ponderous, especially for those without fast typing skills, preferring the asynchronous modalities of e-mail or Web conferencing, which allow participants to reflect on their messages before posting and to respond at whatever time they happen to be free.

6. **Face-to-face regional orientations and training sessions:**

    a. Prior to the fieldworkers' start date, the WiNS coordinator held an orientation and training session attended by all regional coordinators. In addition to familiarizing these staff with the policies, procedures and communications technologies of the WiNS project, the session afforded the opportunity to get to know one another, identify one another's skills and resources and, perhaps most importantly, put a human face to the e-mail addresses that would be the primary source of communication throughout the project.

    b. Each regional coordinator later held a one or two day orientation and training workshop for their fieldworkers, covering issues parallel to those described above. Those who for whatever reason were unable to attend the training session were harder to integrate into the WiNS team, regardless of the amount of technical training they were given. The lesson seems to be that, for this population at least, it is difficult to build a sense of teamwork and community, or to create a desire for online socialization, without some prior face-to-face contact. Whether this would be true for people who have much more experience working virtually is an open question.

CURRENT CHALLENGES/PROBLEMS FACING THE ORGANIZATION

Funding Cutbacks

Funding for C\CEN expired in March 1998, and at this point it seems unlikely that it will be resurrected. The 1998 summer staffing program for Community Access sites was operated directly from the MOTC office with reduced staff and for a shorter time. While there were insufficient funds to engage in development work to create new opportunities and modalities for rural Community Access, the remote management system (now using WebBoard™ v.3) enabled implementation of a similar program despite the much-depleted central resources.

Community Enterprise Networks (CENs)

One challenge facing emerging CENs is to adapt the remote management model for business purposes. Issues include higher security (encrypted e-mail, secure servers, etc.) and secure Web interfaces to dynamically updated databases such as inventory, contact lists, etc. Depending upon the nature of the particular enterprises fostered by the CEN, unique solutions would have to be designed. The needs of a distributed production network are quite different from that of an information service or technical support delivery network, for example.

SUMMARY

The coordination of the program using remote management via the Internet worked well. The Web-based, password-protected Intranet enabled workers and staff to keep in touch with one another, track activities in other regions, collaborate on group projects, and create an automatic archive of all reports and discussion for future analysis. The online technical support service was particularly useful, as it allowed not only for rapid response from the technical staff at C\CEN, but created an open forum where WiNS participants could answer one another's technical questions and learn from watching the solutions unfold. The WebBoard allowed for management of the information flow such that, for example, fieldworkers' reports were accessible only to themselves, their regional coordinator and the WiNS coordinator, but regional discussion conferences were accessible to all staff. The WiNS Staff conference was accessible only by coordinators, creating a secure forum where confidential issues (such as personnel problems, etc.) could be discussed. One challenge with the system is that it required more training than anticipated, and a more user-friendly interface may be sought for future projects. Overall, the system saved considerable amounts of money on travel and phone costs, and it leaves an archive of all correspondence and worker logs, enabling future programs to learn from previous experience.

REFERENCES

Dienes, B. (1997). *WiNS '97 final report* (Report). Sydney, NS: University College of Cape Breton, Centre for Community and Enterprise Networking. Executive Summary retrieved from http://ccen.uccb.ns.ca/articles/winsum.html

Grundy, J., & Metes, G. (1997, June). *Intranet challenges: Online work and communication*. Retrieved October 8, 1998, from http://www.knowab.co.uk/wbwintra.html

Gurstein, M. (1998a). *Flexible networking, information and communications technology and local economic development*. Retrieved May 31, 1998, from http://ccen.uccb.ns.ca/flexnet

Gurstein, M. (1998b). Information and communications technology and local economic development: Towards a new local economy. In G. A. MacIntyre (Ed.), *Perspectives on communities: A community economic development roundtable.* Sydney, NS: UCCB Press. Retrieved from http://ccen.uccb.ns.ca/articles/ict4led-chapter.html

Gurstein, M., & Andrews, K. (1996, October 31). *Wire Nova Scotia (WiNS '96): Final report* (Report). Sydney, NS: University College of Cape Breton, Centre for Community and Enterprise Networking.

Gurstein, M., & Dienes, B. (1998, June). *Community enterprise networks: Partnerships for local economic development.* Paper presented at Libraries as Leaders in Community Economic Development Conference, Victoria, BC. Retrieved June 28, 1998, from http://ccen.uccb.ns.ca/flexnet/CENs.html

Gurstein, M., Lerner, S., & MacKay, M. (1996, November 15). *The initial WiNS round: Added value and lessons learned* (Report). Sydney, NS: University College of Cape Breton, Centre for Community and Enterprise Networking.

James, M., & Rykert, L. (1997). *Working together online.* Toronto: Web Networks.

Johnson-Lenz, P., & Johnson-Lenz, T. (1995, March). *Humanizing distributed electronic meetings*. Paper presented at Groupware '95, Boston. Retrieved October 6, 1998, from http://www.awaken.com/Awaketech/AWAKEN1.NSF/d4cbbb795713bdee882564640074729d/8441cffd01b521d2882564a9008390dc?OpenDocument

Lipnack, J., & Stamps, J. (1993, August 1). Networking: Not just for big business. *New York Times*. Retrieved from http://www.netage.com/TNI/Publications/Articles/article_nyt.html

Lipnack, J., & Stamps, J. (1994). *The age of the network.* New York: Wiley.

Lipnack, J., & Stamps, J. (1997). *Virtual teams: Reaching across space, time and organizations with technology.* New York: Wiley.

McLuhan, M. (1964). *Understanding media: The extensions of man.* New York: McGraw-Hill.

Metes, G., Grundy, J., & Bradish, P. (1998). *Agile networking: Competing through the Internet and intranets.* Upper Saddle River, NJ: Prentice-Hall.

Peck, S. B., & York, J. (1998). *Conferencing with WebBoard* (2nd ed.). Cambridge, MA: O'Reilly Software.

Piore, M., & Sable, C, (1984). *The second industrial divide: Possibilities for prosperity*. New York: Basic Books.

Sieber, P., & Griese, J. (1998, April). *Organizational Virtualness: Proceedings of the VoNet-Workshop,* 27-28. Berne: Simowa-Verlag.

APPENDIX A: GLOSSARY OF TERMS

These definitions were extracted from the *ILC Glossary of Internet Terms*, Copyright © 1994-97 by Internet Literacy Consultants™. Used by permission.The URL of the full document is: http://www.matisse.net/files/glossary.html, which is where you can look for the latest, most complete version. Permission is granted to use this glossary, with credit to Internet Literacy Consultants, for non-commercial educational purposes, provided that the content is not altered including the retention of the copyright notice and this statement. ILC regards any use by a commercial entity as "commercial use". For permission to use it in other ways or to suggest changes and additions, please contact us by e-mail.: matisse@matisse.net +1.415.575.1156. Last update: 01/28/98

Browser

A Client program (software) that is used to look at various kinds of Internet resources. .

Client

A software program that is used to contact and obtain data from a Server software program on another computer, often across a great distance. Each Client program is designed to work with one or more specific kinds of Server programs, and each Server requires a specific kind of Client. A Web Browser is a specific kind of Client.

E-Mail

(Electronic Mail) — Messages, usually text, sent from one person to another via computer. E-mail can also be sent automatically to a large number of addresses (Mailing List).

FTP

(File Transfer Protocol) — A very common method of moving files between two Internet sites. FTP is a special way to login to another Internet site for the purposes of retrieving and/or sending files. There are many Internet sites that have established publicly accessible repositories of material that can be obtained using FTP, by logging in using the account name anonymous, thus these sites are called anonymous ftp servers.

Home Page (or Homepage)

Several meanings. Originally, the web page that your browser is set to use when it starts up. The more common meaning refers to the main Web page for a business, organization, person or simply the main page out of a collection of Web pages, e.g. "Check out so-and-so's new Home Page." Another sloppier use of the term refers to practically any Web page as a "homepage," e.g. "That web site has 65 homepages and none of them are interesting."

Host

Any computer on a network that is a repository for services available to other computers on the network. It is quite common to have one host machine provide several services, such as WWW and USENET.

HTML

(HyperText Markup Language) — The coding language used to create Hypertext documents for use on the World Wide Web. HTML looks a lot like old-fashioned typesetting code, where you surround a block of text with codes that indicate how it should appear, additionally, in HTML you can specify that a block of text, or a word, is linked to another file on the Internet. HTML files are meant to be viewed using a World Wide Web Client Program, such as Netscape or Mosaic.

Internet

(Upper case I) The vast collection of inter-connected networks that all use the TCP/IP protocols and that evolved from the ARPANET of the late 60's and early 70's. The Internet now (July 1995) connects roughly 60,000 independent networks into a vast global Internet.

Intranet

A private network inside a company or organization that uses the same kinds of software that you would find on the public Internet, but that is only for internal use. As the Internet has become more popular many of the tools used on the Internet are being used in private networks, for example, many companies have Web servers that are available only to employees. Note that an Intranet may not actually be an internet — it may simply be a network.

IRC

(Internet Relay Chat) — Basically a huge multi-user live chat facility. There are a number of major IRC servers around the world which are linked to each other. Anyone can create a channel and anything that anyone types in a given channel is seen by all others in the channel. Private channels can be (and are) created for multi-person conference calls.

LAN

(Local Area Network) — A computer network limited to the immediate area, usually the same building or floor of a building.

LISTSERV®

The most common kind of maillist, "LISTSERV" is a registered trademark of L-Soft international, Inc. Listservs originated on BITNET but they are now common on the Internet.

Login

Noun or a verb. Noun: The account name used to gain access to a computer system. Not a secret (contrast with Password).

Verb: The act of entering into a computer system, e.g. Login to the WELL and then go to the GBN conference.

Maillist

(or Mailing List) — A (usually automated) system that allows people to send e-mail to one address, whereupon their message is copied and sent to all of the other subscribers to the maillist. In this way, people who have many different kinds of e-mail access can participate in discussions together.

Network

Any time you connect 2o or more computers together so that they can share resources, you have a computer network. Connect 2 or more networks together and you have an internet

Node

Any single computer connected to a network.

Password

A code used to gain access to a locked system. Good passwords contain letters and non-letters and are not simple combinations such as virtue7. A good password might be: Hot$1-6

POP

(Point of Presence, also Post Office Protocol) — Two commonly used meanings: Point of Presence and Post Office Protocol. A Point of Presence usually means a city or location where a network can be connected to, often with dial up phone lines. So if an Internet company says they will soon have a POP in Belgrade, it means that they will soon have a local phone number in Belgrade and/or a place where leased lines can connect to their network. A second meaning, Post Office Protocol refers to the way e-mail software such as Eudora gets mail from a mail server. When you obtain a SLIP, PPP, or shell account you almost always get a POP account with it, and it is this POP account that you tell your e-mail software to use to get your mail.

Posting

A single message entered into a network communications system. E.g. A single message posted to a newsgroup or message board.

PPP

(Point to Point Protocol) — Most well known as a protocol that allows a computer to use a regular telephone line and a modem to make TCP/IP connections and thus be really and truly on the Internet.

Server

A computer, or a software package, that provides a specific kind of service to client software running on other computers. The term can refer to a particular piece of software, such as a WWW server, or to the machine on which the software is running, e.g. Our mail server is down today, that's why e-mail isn't getting out. A single server machine could have several different server software packages running on it, thus providing many different servers to clients on the network.

SMTP

(Simple Mail Transport Protocol) — The main protocol used to send electronic mail on the Internet. SMTP consists of a set of rules for how a program sending mail and a program receiving mail should interact. Almost all Internet e-mail is sent and received by clients and servers using SMTP, thus if one wanted to set up an e-mail server on the Internet one would look for e-mail server software that supports SMTP.

Terminal

A device that allows you to send commands to a computer somewhere else. At a minimum, this usually means a keyboard and a display screen and some simple circuitry. Usually you will use terminal software in a personal computer — the software pretends to be (emulates) a physical terminal and allows you to type commands to a computer somewhere else.

UNIX

A computer operating system (the basic software running on a computer, underneath things like word processors and spreadsheets). UNIX is designed to be used by many people at the same time (it is multi-user) and has TCP/IP built-in. It is the most common operating system for servers on the Internet.

URL

(Uniform Resource Locator) — The standard way to give the address of any resource on the Internet that is part of the World Wide Web (WWW). A URL looks like this: http://www.matisse.net/seminars.html or telnet://well.sf.ca.us or news:new.newusers.questions etc.
The most common way to use a URL is to enter into a WWW browser program, such as Netscape, or Lynx.

WWW

(World Wide Web) — Two meanings — First, loosely used: the whole constellation of resources that can be accessed using Gopher, FTP, HTTP, telnet, USENET, WAIS and some other tools. Second, the universe of hypertext servers (HTTP servers) which are the servers that allow text, graphics, sound files, etc., to be mixed together.

Bruce Dienes received his doctorate in psychology from the University of Illinois at Urbana - Champaign. He coordinated the Campus Support Groups Program for five years, developing training models and initiating a local network of online support groups. This program was designated a national model by the American Self-Help Clearinghouse. He has been involved in community development as an organizer, as a trainer and as a consultant. He worked for the Centre for Community and Enterprise Networking (C\CEN) at the University College of Cape Breton as program coordinator for Wire Nova Scotia (WiNS) and later as an assistant director of the Centre.

Michael Gurstein has a bachelor's degree from the University of Saskatchewan and a doctorate in sociology from the University of Cambridge. Dr. Gurstein worked as a senior public servant in British Columbia and Saskatchewan, managed a consulting firm concerned with the human aspects of advanced technologies and served as a management adviser with the United Nations Secretariat in New York. He is currently the ECBC/NSERC/SSHRC associate chair in the Management of Technological Change at the University College of Cape Breton and the founder/director of the Centre for Community and Enterprise Networking (C\CEN) of that institution.

This case was previously published in the *Annals of Cases on Information Technology Applicaitons and Management in Organizations*, Volume 1/1999, pp. 159-173, © 1999.

Chapter IX

Moving Up the Mobile Commerce Value Chain:
3G Licenses,
Customer Value and
New Technology

Martin Fahy, National University of Ireland, Galway, Ireland

Joseph Feller, University College Cork, Ireland

Pat Finnegan, University College Cork, Ireland

Ciaran Murphy, University College Cork, Ireland

EXECUTIVE SUMMARY

In the spring of 2002, a team within Digifone, in the Republic of Ireland, deliberated on how they would generate revenues from a Third Generation (3G) mobile license. Their plan was to facilitate third-party mobile commerce services, and receive a percentage of the resulting revenue. A key challenge would be differentiating the value of content based on transaction context. Mediation technology, based on open standards, was seen as the solution. The case describes Digifone's evaluation of an XML vocabulary as the basis for a mediation and billing solution. The Digifone experience suggests that the key to exploiting 3G technology is inter-industry agreement on measurement standards to effectively manage the value of the associated services — a "bar code" for measuring value and distributing revenue amongst value Web participants.

INTRODUCTION

The sun was shining as Brian Noble looked out over Dublin from his office in Digifone Headquarters in Lower Baggot Street. He was pleased to see the large number of people using their mobile phones as they walked along the street. The "executive dominated" mobile market so evident during the early 1990s had since been replaced by usage patterns that better reflected the general population. The "Celtic Tiger"[1] and market competition ensured that mobile phones were now an affordable item essential to social and business life. The majority of phone users on that day held their handsets to their ears as they held a conversation, but a significant number were using their phones to send or receive SMS text messages: probably arranging where to meet for lunch, he thought! Brian wondered how long it would be before Digifone's customers would regard having a conversation as an "old fashioned" way of using their mobiles.

Turning away from the window, Brian considered the issues facing him in his role as marketing manager, and how these related to the major decisions facing the company during 2002. The Irish government had announced the bidding process for the new 3G mobile licenses in late 2001, and Digifone was widely regarded as a major contender for two of the four licenses on offer. "3G" refers to the third generation of wireless communication technologies, and to the pending improvements in wireless data and voice communications through any of a variety of proposed standards.[2] The immediate goal was to increase quality and speed from current GPRS (General Packet Radio Service) standards. 3G would fully realize multi-media and real-time services, but might still suffer from operational difficulties (3G News Room, 2001). GPRS, 3G and other wireless technologies are explained in the Appendix.

Brian knew that the manner in which Digifone would be able to exploit 3G, and the amount that they could be expected to pay for the license, would depend on their ability effectively position the company within the emerging 3G value Web, and to bill appropriately for the services provided. He needed to consider this complex issue. In particular, he needed to give some thought to the development of intra-industry value definition standards.

ORGANIZATION BACKGROUND

Digifone was established as Esat Digifone in Ireland in 1997, and positioned itself as the main competitor to the state controlled Eircell (now owned by Vodafone). After a series of share purchases, British Telecom (BT) acquired 100% of Digifone in April 2001 (CNN Money, 2000). In November 2001, Digifone, as part of the mmO2 set of companies, completed a de-merger from British Telecommunications PLC and devised plans to re-brand as O2. The group is made up of five national mobile networks — Digifone in Ireland, BT Cellnet in the U.K., Manx Telecom on the Isle of Man, Telfort in the Netherlands and Viag Interkom in Germany, plus Genie, the international mobile portal — and shares the common brand name O2.

By early 2002, the mmO2 businesses served 16.5 million mobile customers in the UK, Germany, Ireland, The Netherlands and the Isle of Man. The group's mobile businesses in these countries were all wholly owned and together covered territories with a total population of over 160 million people. The combined turnover for the year ending March 31, 2001 reached £3,200 million, up 22% from £2,618 million for the preceding year (mmO2, 2001).

Digifone's turnover for the year ending March 31, 2001, was £309 million, with adjusted EBITDA for the same period of £68 million. The company serves 76 of the top 100 businesses in Ireland as defined by Business and Finance Business Information Ltd. in Ireland. In September 2001, the Digifone network had 1.1 million contract and pre-paid customers, representing an overall market share of 41% and approximately 60% of the business market (Digifone, 2001).

In spring 2002, Digifone was widely regarded as a major contender for both category A and category B licenses. The A license would require the operator to roll out a national network giving 53% coverage by 2005 and reaching 80% of the population by 2007. The B license, on the other hand, was targeted to reach 53% of the population by 2008 (ODTR, 2000).

Digifone did not introduce a HSCSD (High Speed Circuit Switched Data) service, but instead invested EUR 45m in establishing Ireland's first consumer GPRS (General Packet Radio Service) or 2.5G network (EuropeMedia.net, 2002). This was released to business consumers in 2001. The take-up was successful with a number of high profile adoptions including Dell, EMC, GE Capital and Logica. Digifone was planning on launching GPRS services including WAP, gaming, Internet and e-mail to pre-paid consumers during the summer of 2002. Digifone's GPRS pricing strategy was based on the volume of information downloaded rather than the length of time that the user is connected to the Internet.

The challenge facing Brian and others in the European telecommunications sector was to find new business models to allow them to exploit the investment they had made. While the 3G license investment is significant, network infrastructure and, more importantly, European-wide brand building would also require a significant commitment. For Brian and others the challenge was twofold:

1.   To position Digifone/O2 as a key player in the complex value Web required for the delivery of next generation mobile services.
2.   To develop new ways of managing the intangible values associated with the consumption of mobile commerce (m-commerce) services.

SETTING THE STAGE

How Much do You Pay for a Piece of Paper?

The Irish telecommunications sector had clearly experienced impressive growth levels in the years preceding 2002. Not only had the mobile penetration rate risen from only 29% in June 1999 to 75% by November 2001, but the Irish Internet market had also experienced a significant expansion, with more than one-third of the population having home Internet access by the end of October 2001 (ODTR, 2001). While the 2002 Irish market was small (see Table 1) by European standards, the potential revenue from its 2.8m subscribers was attracting considerable attention (ODTR, 2001). As part of the process of the liberalization of European telecommunications markets, the Irish government announced the bidding process for the new 3G mobile licenses in late 2001. The Director of Telecommunications Regulation, Etain Doyle, had statutory responsibility for awarding the licenses in question. She launched two competitions for the award of

Table 1. Key indicators in the Irish mobile sector

Mobile subscriptions (Nov 2001)	2.8m
Share of post-paid subscribers (Nov 2001)	Approximately 33%
Mobile penetration rate (Nov 2001)	75%
Total SMS traffic (Q3 2001)	385m per quarter
Average SMS traffic per subscriber	46/month
ARPU (Average Revenue Per User)	€38/month
Number of employees within Sector in 2000	Approximately 2477
Market share of existing operators as of Nov 2001	Vodafone (58%), Digifone (40%), Meteor (2%)
Date of service launch of 2G operators	Vodafone (1985), Digifone (1997), Meteor (2001)

a total of four licenses to provide third generation mobile services in Ireland in December 2001.

During the spring of 2002, Brian Noble and a large team of Digifone executives were working on the bid for the licenses on offer. It was a time consuming and difficult process but one they knew would have a long-term strategic impact on Digifone. The outcome of the bidding process would be the awarding of a 3G license, giving the winning firms the right to exploit a piece of the radio spectrum. The price of the licenses (see Table 2) was to be in the region of EUR 114m and was due to be awarded using a so-called "beauty parade" approach (ODTR, 2001). The beauty parade or comparative selection method sets out measurable indicators, such as speed of roll-out and extent of coverage, which applicants can be judged by. This method has been used extensively throughout Europe and is in contrast to the Auction method where the license is awarded to the highest monetary bidder (ODTR, 2000).

For Brian, the prospect of spending in excess of EUR 100m on a "piece of paper" was daunting, but paled in insignificance compared with the amounts paid by some of the telecom operators in Europe. Table 3 shows the enormous sums that had been paid by some operators throughout Europe (Cellular News, 2002). The European 3G licensing history began in Finland in 1999. Coinciding with strong speculation in telecommunica-

Table 2. 3G Licenses in Ireland

Type of License	Administrative Fee	(3G) Spectrum Access Fee	Annual Spectrum Fee
A	Up to €2.6 m	€50.7 m	€2.22 m
B₁	Up to €2.6 m	€114.3 m	€2.22 m
B₂	Up to €2.6 m	€114.3 m	€2.22 m
B₃	Up to €2.6 m	€114.3 m	€2.22 m

Table 3. 3G Licensing activity in the EU in late 2001 and early 2002

Country	Methodology used and no. of licenses	Licensees or candidates	Prices USD
Ireland	•Beauty Contest •Licenses: 4, one reserved for a new entrant	•Vodafone •O2/Digifone •Meteor	'A' license €50 million 'B' license €114 million
United Kingdom	•Auction •Licenses: 5	• TIW (new entrant) (35MHz) • Vodafone (30MHz) • BT (25MHz) • One2One (25MHz) • Orange (25MHz)	€7.90 billion €10.75 billion €7.26 billion €7.21 billion €7.37 billion Total: €40.48 billion
Spain	•Beauty Contest focusing on quality of coverage. •Licenses: 4 •Winners to supply services to all cities with pop. > 250,000	• Telefonica • Airtel • Retevision • new consortium Xfera	€135 million one-off fee + investment in 'information societies' of the future + 0.15% of annual revenues.
France	•Beauty contest and Fixed fee •Licenses: 4 Commercial launch Jan 2002	•France Telecom •SFR	€631 million + 1% annual revenue share (Was €5.15 billion)
Austria	•Auction, but only for pre-qualifiers. •Licenses: 6	•Connect Austria •Maxmobil •Hutchison 3G •Mannesman •Mobilkom Austria •3G Mobile	€119 million €113.4 million €118 million €112 million €120.3 million €115.8 million
Denmark	•Auction •Licenses: 4	•TeleDanmark •Sonofon •Telia •Mobilix	€135 million per license

tions values, very high fee levels were paid in the British and German auctions during the spring and summer of 2000. In line with the decline in the telecom markets, and in light of reviews of the capability of 3G, interest in the licensing processes declined during the latter part of 2000 and early 2001. This was partly caused by the general slow-down in the international economy, general fall in share prices and lower telecommunications company credit ratings. In addition, the technological uncertainty surrounding 3G grew, as initial testing of the technology did not yield the expected potential of 2 Mbps, but rather an initial speed between 64 and 384 Kbps.[3] However, despite the shakeout in the industry, there were indications that optimism was returning by late 2001. In the 3G auction held in Denmark during September 2001, there were five bidders for four licenses, with a price per capita almost equivalent to the Irish tender (ODTR, 2001).

The investment in what was an intangible asset presented significant challenges for Brian Noble and his colleagues. He believed that the real issue was not what figure to show on the balance sheet for the license, but how they could manage this intangible asset in a way that would generate a return to shareholders, and meet their obligations to customers and other stakeholders.

Table 3. 3G Licensing activity in the EU in late 2001 and early 2002 (continued)

Country	Methodology used and no. of licenses	Licensees or candidates	Prices USD
Germany	•Auction in two stages. •Licenses: 6	• T-Mobil (Deutsche Telekom) • Mobilfunk (Mannesmann) • E-Plus (KPN/Hutchison Whampoa) • Viag Interkom (British Telecom) • MobilCom • Group 3G (Sonera/Telefonica)	€8.80 billion €8.75 billion €8.72 billion €8.77 billion €8.69 billion €8.73 billion. Total: €52.55 billion (Most expensive auction to date)
Sweden	•Beauty contest •Licenses: 4 Ten candidates Commercial launch for Q1, 2002.	•Europolitan •Hi3G Access •Orange Sverige •Tele2	Cost of SKr100,000 (€12.38 million) plus annual fee of 0.15% of turnover.
Portugal	•Beauty contest, •Licenses: 4	•TMN •Telecel •Optimus •ONI	€103 million per license Fixed Fee
Norway	•Beauty contest focusing on quality of coverage. •Licenses: 4 Seven applicants	•Telenor •NetCom •Broadband Mobile (Has handed its license back to the government) •Tele2Norge	NKr18m (€2.3 million) annual fee + one-off NKr100m (€12.62 million) fixed charge per operator.
Finland	•Beauty Contest •Licenses: 4 •First European country to allocate 3G licenses. Commercial launch Jan 2002	•Radiolinja •Sonera •Telia Mobile •Finish 3G Consortium	
Netherlands	•Auction •Licenses: 5 The relatively low totals were raised due to one reason; there were five incumbent mobile operators and five licenses allocated.	• Libertel • KPN • Dutchtone • Telfort • Ben	€0.76 billion €0.75 billion €0.47 billion €0.45 billion €0.42 billion Total €2.85 billion
Belgium	•Hybrid Auction •Undisclosed Entry Fee •Licenses: 4	•Mobistar •Proximus •KPN Orange	€160 million per license
Italy	•Mix of auction and beauty contest •Licenses: 5 (Seven applications received in time)	•Omnitel Pronto Italia •Telecom Italia Mobile •Wind Telecomunicazioni (Enel and France Telecom) •Andala •IPSE 2000 (incl. Telefonica)	Minimum bid of four trillion Lire (€2.12 billion) per license + deposit of five billion Lire (€2.65 million).

The 3G license bidding process demonstrates the growing importance of intangible assets in the emerging knowledge economy. While licenses were issued for free in the Japanese market, European telecom operators invested over EUR100 billion in what one telecom manager termed "little pieces of paper which have no re-sale value and which require an equally large investment in physical infrastructure in order to be of any use."

The fallout from this investment was receiving considerable attention in the capital markets during late 2001 and early 2002. As European telecom operators struggled under growing mountains of debt, many were wondering how they could redesign their business model to extract value from their investment. They took little comfort from capitalized value of the licenses; instead they were struggling to put the business strategies in place to manage their increasingly intangible asset base.

Disruption and Change in the Telecom Sector

The situation facing Digifone was the culmination of a history of disruptive developments, which had begun 30 years previously. The telecom sector has been characterized by periods of intense standard building, punctuated by dramatic techno-logical and operational change.

The traditional telecom sector has a history of standards. In the early days an engineering mindset dominated and telecom standards were primarily based on hardware involved in connecting physical equipment on the network. These standards facilitated international traffic and the reconciliation of those revenues associated with this traffic. This allowed a small number of players, primarily state or semi-state bodies, to operate non-competitive oligopolistic or monopolistic type structures. Many of the international revenue-sharing models were based on "gentleman's agreement" situations where revenue on the international network was shared based on call times, using well-established protocols and conventions. The standards provided a stable low risk, if somewhat low return, environment. Standards were based on simple measurement of traffic and provided the stability and certainty to allow the telecom operators to manage their relationships with their customers. This situation continued right through to the 1980s, when a number of disruptive developments occurred (Figure 1). In particular, the move towards deregulation or privatization of the telecom market meant that many companies were thrust into a new competitive environment. Previously sheltered telecoms had to actively manage their revenue streams, address the more dynamic needs of the competitive market place, and meet the demands of their shareholders.

Other developments included the move from wire to wireless. This paradigm shift meant that the unit of analysis changed from being represented by homes and residential units to a more individually driven market place. It reflected, in many respects, the changing mood from Baby Boomers to Generation X, a more individualistic society. At a less conceptual level, there was a move in traffic from purely voice traffic to data traffic. The crossover point between the volumes of voice and data occurred in the late 1990s. This period also saw a change from a very centralized model of telecommunications in the value chain — from a situation where customers had single relationships with single telecom providers — to a more complex, web-like, relationship with vertical and horizontal connections between participants in the value chain.

Underpinning all of this was the transition from analog to digital and the resulting battle of standards. The importance of standards is reflected in the different experiences in Europe and North America. In Europe, the emergence of the GSM standard allowed the European-wide network of GSM operators to share customers and infrastructures, and facilitate service roaming. This also facilitated the emergence of large-scale mobile telecom players such as Vodafone, DeutscheTelecom and BT. In the U.S., by contrast, the absence of standards led to a very fragmented market, which has persisted.

As such, the late 20th and early 21st centuries witnessed a move from a collection of oligopolistic-type telecoms, operating based on shared hardware standards, to a more chaotic market, based around emerging wireless technologies GPRS and 3G. GPRS

Figure 1. Disruptive developments in Irish telecom sector since 1980

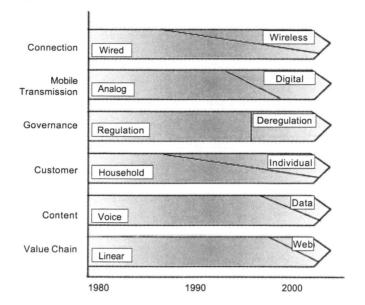

(General Packet Radio Service) is a standard for wireless communication running at speeds of up to 150 kilobits per second, compared with current GSM systems' 9.6 kilobits. GPRS supports a wide range of bandwidths, is an efficient use of limited bandwidth and is well suited for sending and receiving small bursts of data, such as e-mail and Web browsing, as well as large volumes of data. However, it often suffers from sub-optimal performance and operational difficulties (SearchNetworking.com, 2001a). GSM (Global System for Mobile communication) is a digital mobile telephone system that is the de facto wireless telephone standard in Europe. While GSM operates at a higher speed and quality than 1G analog services and it enables roaming, it offers limited data services and is still relatively narrowband (SearchNetworking.com, 2001b). The GSM experience in the U.S. indicated that this new market was likely to fall into chaos unless a consistent standard emerged to allow those within the value chain to share revenues in a meaningful way. In the absence of this, there would be significant barriers to exploiting the licenses that governments had given to Telecom operators in Europe. It was considered essential to develop measuring and reporting mechanisms in order to realize the intrinsic intangible value that the networks had.

The Mobile Internet Device Market

The worldwide Internet device market showed incredible growth in the 1995-2000 period with growth rates of 13% to 20% per annum. According to Cahners In-Stat MDR (2002), the same market was projected to grow at an overall annual rate of 41.6% between 2000 and 2005. Predictions pointed to a growth in wireless Internet devices, particularly in the form of Web-enabled phones/smart phones and other types of hand-held devices. This market was driven by a more mobile population, better Internet access and technology, such as wireless access protocol (WAP), increasing speed and bandwidth in the form of 3G technologies, and Bluetooth wireless connectivity tools. Bluetooth is an evolving short-range networking protocol for connecting different types of devices; for example, connecting a mobile phone with a desktop or notebook computer. Bluetooth devices can communicate by wireless signals within a 35-foot range, and do not need a line-of-sight connection (Computer User, 2002). While market penetration of mobile devices varied across Europe, the relatively recent deregulation of that market segment allowed a number of firms to identify vast opportunities for redeploying traffic in the competitive telecoms environment.

Wireless demand was surging, with the number of wireless Internet users set to grow 18-fold from 39 million in 2000 to 729 million in 2005 worldwide, according to the Intermarket Group. Europe was to have the highest concentration of these wireless Internet users by 2005, with 200 million people going online with a mobile device, up from 7 million in 2000 (Nua, 2002). In comparison, the mobile Internet market in North America had 5.7 million subscribers in 2001, but was expected by to reach over 110.1 million subscribers by the end of 2008 (Frost & Sullivan, 2002). A comparison of population penetration figures showed that mobile penetration in the European Union (EU) stood at 75% of the population, compared with 55% in Japan and less than 50% in the U.S. (Liikanen, 2002).

By 2002, research indicated that the uptake of mobile technology in Europe was likely to see exponential growth in the time period 2003-2005. The associated services were projected to move through a series of steps from the initial period of e-mail access to an information channel, moving on to more transactional and interactive services (see Table 4). New 3G applications include streaming video services, videoconferencing, interactive online shopping, location-sensitive directories, and online banking, stock trading and sports reporting. Applications specifically designed for business focus on extending access to computer facilities to staff in the field, and the use of mobile security to monitor buildings and moving vehicles remotely (Schlumberger, 2003).

It was envisaged that the deployment of 3G services was dependent on the implementation of increasingly sophisticated technologies, both in terms of network infrastructure and handheld devices. Designs for 3G access devices provide evidence of increased integration between Personal Digital Assistants (PDAs) and mobile phones, resulting in a highly interactive multimedia environment.

Table 4. Mobile technology and applications

Year	1998-2000	2001 - 2002	2003-Future
Technology	GSM	GPRS	3G
Applications	SMS	Advertising	Gaming
	Email	Shopping	Video
	Banking	Music	Travel
	Ticketing	E-Billing	Healthcare
	Reservations	Auctions	MMM

Figure 2. Traditional value chain

CASE DESCRIPTION

Understanding the Mobile Commerce Challenge

Industrial sectors can often be described by the interdependency between players in the delivery of value to customers. Traditionally, the telecommunications industry has been characterized as a short sequential value chain as shown in Figure 2. Network operators invested in the physical infrastructure required to operate a telecommunications service. Such operators either operated as the service provider or leased the infrastructure to independent telecommunications companies to provide telephone services to customer.

As Figure 3 shows, the move to 3G services was accompanied by a transition in the telecom sector from a simplistic value chain involving the network operator and the customer, to one which involved content creators, aggregators, packaging solutions, network operators, service providers and ultimately customers (Durlacher, 1999). A configuration such as the one presented in Figure 3 has been termed a value Web as the value-added activities of players is not sequential and the interdependency between players is complex. In addition, new players (e.g., content providers and aggregators) exist in the mobile commerce value Web.

Figure 3. The mobile commerce value Web

There's an old adage that you can only make money *if you know how to bill*. Increasingly, the key players in the mobile commerce arena were focusing not on the business models and infrastructure of mobile commerce services, but on flexible, accurate and timely revenue allocation methods for these services. As carriers in Europe, Asia and the USA raced forward to GPRS and 3G services, the billing and revenue allocation process was becoming inextricably intertwined with the enabling service delivery mechanisms. The next generation billing challenges have been described as follows:

- Facilitating a wide range of pricing structures that differ by type of service as pricing structures evolve;
- Changing requirements as the range of content and application-based services offered over mobile networks expands and operators need to introduce a wider range of prices;
- Communicating charging information to customers in real time in a way they understand;
- Establishing a variety of third-party relationships with the ability to bill for content and transaction-oriented applications;
- Charging for quality of service (QoS) by QoS class, which varies by the type of service (e.g., gaming and video services require different levels of quality) (Schlumberger, 2003).

In particular, two primary factors created a complex challenge. The first was the intangible value associated with the *consumption* of mobile commerce services. The second was the complexity of the value Web needed for the *delivery* of mobile commerce services. Even if these issues could be solved, many saw consumer reaction as the major stumbling block.

Consumption

The telecoms industry's initial reaction to the move towards IP-based,[3] "always-on"[4] data services was a shift to billing according to the amount of data moved, rather than the length of usage time. Unfortunately, this idea failed to take into account the context-driven nature of mobile transactions. Even voice-centered, time-based telecom billing strategies recognized that "all minutes are not created equal," and often used different billing schedules based on when network usage took place. It was quickly realized that data-centered billing systems must take much more into account. For example:

- The kind of data being transferred,
- The transfer direction of data,
- The time-specific value of the data,
- The location-specific value of the data, and
- The signal-to-noise ratio of the data (Cerillion, 2002).

These were just a few of the factors that were recognized as relevant to the billing of a single usage event. Andrew Burroughs (Apogee Networks) saw the key challenge as being able to "collect, identify and rate content on a unique basis." Others agreed. Robert Machin (Logica) argued that billing needed to focus on "event, nature of event, volume of data, length of session" and that carriers must use "different measures for different service types." Campbell Ritchie (EMEA) added that, "the adoption of digital content will put increasing emphasis on the importance of the event type, as the type of service delivered could have an enormous impact on the revenue associated with a particular event" (Helme, 2001).

For example, what is the comparative value of two related pieces of data? The first is a real-time trading quote for a particular company: volume of data moved — 400 bytes. The second, a history of trading values for a six-month period, one year old: volume of data moved — 8,000 bytes. Which is worth more? If the historical data is sent in a file format ready for import into a spreadsheet application, without a change in size, is it worth more than if it is sent for screen consumption only? To illustrate further, are today's headlines worth more than the entirety of yesterday's newspaper? How do we compare the value of a user requested cinema schedule with an automatic flight time update sent to an airline passenger?

Delivery

Assuming that a solution could be found to support the accurate *measurement* of the (mostly intangible) value associated with a mobile transaction event, companies still faced the problem of reporting and allocating this value among the many stake-holders involved in the extremely complex value Web. As mentioned before, the 3G/GPRS value chain was a far cry from the simple telecom-to-customer chain of the past. Instead, it was seen as a value Web of technology enablers, content providers and aggregators, network and service providers, wholesalers, retailers, fulfillment enterprises, etc. (see Figure 8). Many m-commerce services involved selling third-party content, products and services, often *through* several additional mediators, and the compelling question was "how does everyone get paid in a timely and accurate fashion?" Regulators, financial institutions and taxation authorities also needed to be taken into account. This complex value Web required a community of interested stakeholders to come together to develop a common solution for sharing revenue across the value chain.

Consider Susan, an Irish executive based in Paris, who wishes to order flowers for her partner. She roams on the SFR network, using her 3G device. She visits the Yahoo! portal and, subsequently, an Interflora site. She orders flowers for delivery to a location in Dublin, and these flowers are duly delivered by the local Interflora franchise. The total billable value of this transaction needs to be divided between the various telecom operators (SFR who are carrying the roaming traffic as well as the user's primary service provider), the franchisee, Interflora, Yahoo, and the potentially many others involved in financing, infrastructure, etc.

Consumer Reaction to 3G

One of the most worrying factors facing 3G operators was the low level of consumer awareness of 3G technology and capabilities. Many consumers had been very disillusioned with their experience of WAP technology. The technology once described as "the Internet in your pocket" turned out to be a poor imitation of the real thing. Some 3G enthusiasts now feared that hype around 3G would ultimately result in consumer disappointment, and poor return on investment (ROI) figures for those that invested in the infrastructure. Effective communication regarding 3G capabilities was therefore considered important to avoid consumer apathy.

Many believed that the availability of applications for both businesses and consumers would be important to encourage multiple communities of interest. The availability of business applications would be especially important if the handset vendors priced the access devices at the high end of the market. Some profiteering at the beginning was inevitable, but many believed that handsets would have to be priced for the mass market if the fees paid for 3G licenses were to be recovered.

However, the most important issue appeared to be customers' perception of the value added from 3G applications. Most 3G prototypes tended to be faster versions of existing applications, and developers were unlikely to invest much in creating new applications until the infrastructure details were known. Games and travel services seemed the most likely first generation applications, but these provided little value added for business users.

Customers' perception of the value added by 3G applications was invariably linked to the billing issue. Customers were used to "bill by time" approaches; some even were used to "bill by volume." However, this method resulted in complaints when the connection dropped and the customer had to pay for the incomplete download as well as the new download. Also, customers complained that they were not aware of how much a download would cost until it was complete. How customers would react to a "bill by value" approach was unknown. A key element of such pricing is content aging. The concept of aging is used in the film industry. Consumers understand that the cost to see a film falls as it moves from cinemas, to pay-per-view, to video/DVD, to terrestrial TV. However, consumers do not expect to pay more for an Oscar winning film than another film. The manner in which consumers will react to frequent price changes and differential pricing for similar products is unknown.

The Mobile Commerce Challenges at Digifone

The mobile phone market in Europe had witnessed significant consolidation in the years preceding 2002. The emergence of European-wide providers such as Orange, Vodafone and O2 reflected what Brian Noble saw as the increasing reach and range of a small number of key providers. While the development of a European wide network had been an important part of these firms' investment strategies, a more significant development had been their enormous investment in intangible assets. The 18 months to spring 2002 had seen an unprecedented investment in brand building by these firms and a continued investment in 3G licenses. This represented a significant break with the past when the physical infrastructure represented the core of the telecoms' investment strategy. Never before had an industry experienced such a dramatic move to the intangible.

As part of the O2 group, Digifone had significant interest in the European mobile market space. As a holder of several 3G and other licenses across Europe, the O2 brand was faced with the challenge of generating sufficient revenues over the following ten years to justify the large amounts of resources it had invested in 3G to date. As part of the O2 brand, Digifone had implemented a strategy of targeting high value customers and had achieved a strong position in the business market. Digifone's positioning in the market was based on three principles:

1. A state of the art quality network,
2. Customer care,
3. A guarantee of a superior service at a price that represents value for money.

By mid-2001, the Digifone network covered 98.5% of the Irish population and 95% of the country geographically. By this time, Digifone's customers were sending 2 million text messages per day, and the company was keen to move to the next generation of mobile services (Digifone, 2001). The consumer-focused GPRS services planned for summer 2002 included chat, textual and visual information, moving images, Web browsing, document sharing/collaborative work, audio, vehicle positioning applications, corporate and Internet e-mail, Remote LAN access, file transfer and home automation.

Digifone had identified key growth drivers as being carrier growth fueled by deregulation of the market, convergence with the bundling of telecom services, the pervasiveness of Internet packet-based IP/data services, and consolidation of strategic alliances among content aggregators and providers. Digifone was also keenly aware of the emergence of renewed vertical markets where value was moving closer to the customer and further away from network operators. This was recognizable in the number of telecom operators partnering in an attempt to leverage value and increase shareholder value in the face of increasing demands for returns from a more deregulated stock market.

Brian Noble believed that the continued success of Digifone depended on flexible business and product modeling. Specifically, it needed to develop its ability to manage partner relationships more effectively as well as more flexible and dynamic processes for revenue sharing among the multi-participant value chains. In this context, billing systems and billing architectures were a key priority. These systems needed to support second wave CRM and relationship management amongst strategic partners in the value chain, such as content aggregators, content providers and others. Brian believed that this issue centered on the concept of mediation, despite many commentators in the industry describing it as a billing and customer care issue. In particular, he saw mediation of traffic from different devices as the key bottleneck in the move towards more sophisticated pricing approaches.

Mediation and Billing

Brian's colleague Edward Therville, the billing manager for Digifone, advocated the importance of moving away from flat rate billing approaches to one that more accurately reflects the inherent value of the online experience. In particular, Edward stressed the need for firms in the 3G space to be able to mediate and bill for premium services such as news, sports, stock prices, etc. In developing these services, firms such as Digifone needed to develop strategic alliances with content providers, aggregators, etc., to become a significant player in the 3G landscape.

The challenge for Brian Noble, as marketing manager, was to move away from what he referred to as a "black box pricing model" where customers were charged on a flat rate basis for the amount of data they downloaded, to a pricing model that would reflect the value of the service provided. Brian knew that the company needed to move away from charging people the same amount of money for sports results as real time stock prices, which may have a much higher value. Digifone was trying to segment/differentiate the market based on the quality of services provided and other rating characteristics. It was likely that these characteristics would change in the future. Brian required a dynamic solution to the billing and revenue reconciliation problem that would allow the company to maximize the amount of revenue from the value chain, but which also would allow it to be objective and fair in delivering benefits to other value chain participants.

For relatively young firms such as Digifone, the billing and mediation issues facing it were not as severe as those facing the older established telecom companies. In the early days, as government-controlled utilities, older firms developed their own online billing systems. These were proprietary systems developed in-house which took pulse and tone data up off the telephone network and processed this in very simplistic ways. As such, data was gathered in a very raw way to produce bi-monthly bills for residential and corporate customers. These systems initially ran on telephone systems and large mainframe environments. Over the years, they were updated and imported into a client server environment, but the basic underlying architecture remained the same.

Digifone has a much shorter history of billing and mediation, and has benefited from the consolidation in the market for software solutions. By 2002, the market was well served with companies like Keenan, Geneva, Amdocs, ADC and other companies. These provided specialist services to the telecom industry all around the world. In the early days, these specialist solutions were based on simple databases and flat file record systems, with various menus for changing prices and billing of customers. They also included a procedure for reconciling with international operators for international traffic. The primary purpose of these systems was to operate as an accounting system. The systems worked well when the telecom operators enjoyed monopolistic power. As product sophistication increased and market segmentation became more important, telecom operators increasingly recognized the need for more dynamic approaches to billing and revenue management. In particular, the GSM market created new billing problems, and many telecom operators required off-the-shelf solutions for their mobile billing. The existing players in the marketplace typically provided these, and many of them were very impressive products. However, most of these had a unique feature in common in that they still based the billing on time units.

In Brian Noble's view the 3G challenge for Digifone was to "move up the value chain." He was aware that mobile operators did not want to repeat the experience of those in the fixed Internet world, where flat rate pricing and a focus on access speed resulted in ISPs becoming commodity service providers. Mobile operators wanted to offer mobile commerce infrastructures, and therefore collect transaction fees and a percentage of revenue on purchases that customers would charge to their phone bill.

Many telecom operators saw this as a battle for customer ownership. They were determined to have their own portal, create and aggregate (partner for) content and commerce services, and take a piece of the revenue. As a result, these mobile operators needed to manage value chains that stretched from application developers and e-merchants through branded private-label distributors or "mobile virtual network operators" (MVNO).

In the 3G market space, customer care and billing enabled new business models and allowed the operator to broker relationships with everyone in the value chain while retaining absolute control of the customer, and, consequently, control of the revenue stream. For firms such as Digifone the key challenges as the company moved into a new area was to bill for content not time. Digifone therefore required a dynamic billing solution. Brian Noble believed that the "spaghetti collection of underlying systems" that he worked with made it difficult to fully exploit the GPRS licenses which the company had, the 3G licenses that it may buy, and the other intangible assets held by the company.

For Digifone the growing importance of intangible assets, such as licenses, the company brand, the installed customer bases, and the alliances the company had with other operators and content providers required a change in strategic mindset. In particular, it required a move away from thinking mainly about exploiting the physical tangible network infrastructure to focusing on the intangible assets, which constituted the bulk of the firm's investment base.

Like many telecom operators in Europe, Digifone was disappointed with the uptake of WAP services and the overall quality of the WAP customer experience. The low level of availability of WAP services across Europe had been attributed to the lack of effective revenue sharing mechanisms between operators and content providers. If GPRS/3G was to be successful this needed to be addressed immediately.

For Brian Noble, and for Digifone's Billing Manager Edward Therville, this meant that Digifone needed to be able to move towards more consolidated billing solutions and a single view of the customer. In this context, Digifone was faced with a major technological integration issue. In addition, there was an integration issue based on recognizing the importance of intangible assets in operating the business model. In trying to arrive at a solution, key technology and billing decision makers within Digifone spoke to Amdocs, whom they had worked with in developing a GPRS billing solution.

The implementation of Amdocs was the result of a billing problem at Digifone. The company found itself three months behind in billing and payment processing. This resulted in an unprecedented volume of calls to the Customer Care section. Most of the problems stemmed from the legacy systems employed by the company. The IT Department found themselves developing "patch on" systems to try to support customer care and billing. However, the system was so piecemeal that a single call from a customer became four calls.

The IT director at Digifone saw the benefit of Amdocs as having all the customer data in one place. According to Mary O'Donovan, the IT director:

Whether a customer contact comes through our dealers, self service through our portal, IVR or through the agent, it's not only in one database, it's in one system. Everybody is looking at one system — the same Amdocs ecare system. Whenever a customer calls Digifone through the Amdocs environment, we get a full picture of how valuable the customer is to Digifone and we deliver a graded level of service based on that value. We manage customer complaints better because of the follow-up functionality of the Amdocs customer care system.

The solution worked. Within a year after implementing Amdocs, Digifone's customer satisfaction ratings had significantly improved (O'Donovan, 2001).

A Standards-Based Solution

Brian knew that Digifone, along with the rest of the players in the mobile arena, must address the two-fold problem associated with billing for next generation mobile services.

Figure 4. The role of mediation software

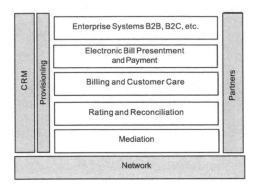

First, a mechanism would have to be put in place that allowed for the accurate measurement and recording of an event's context, thus allowing Digifone to accurately bill based on an event's value, not simply the bandwidth used. Secondly, it was recognized that Digifone would have to forge key alliances with content providers, content aggregators, and others — *and exploit technologies and processes that would support the timely, fair distribution of value among the various partners in the value Web.*

As a solution to the first of these challenges, the IT team at Digifone focused on mediation software and processes. Mediation refers to the processing and delivering of information between a telecommunications network, for example a GSM or 3G network, and an operator's business support systems, for example a billing or customer care system such as Amdocs (Comptel, 2002). The relationship between mediation software and other organizational systems is shown in Figure 4. Thanks to mediation, the network elements and business support systems need to communicate only with the mediation software. This single point of interface reduces the number of connections needed and facilitates change management. It also reduces time-to-market periods for the introduction of new services to customers, which will be of extreme importance to Digifone and other Telecoms in the near future, when numerous new services will be introduced after the adoption of 3G.

Mediation systems track real-time information from the hundreds of packets and myriad of servers making up an IP transaction, creating the equivalent of the CDR (Call Detail Record) in traditional systems. The call detail record is the fundamental unit of data exchange used by circuit-switched vendors (Lucas, 1999). The mediation software collects the subscriber usage data (all transferred bytes; for instance, the dialed calls and the sent short messages for the past month) from the telecommunications network. Its key function is to collect and transform this network utilization and service usage data into a common format for the billing system to enable customers to be billed (Openet, 2002). Due to mediation software, Digifone can be sure that its billing system is based on reliable usage data.

Mediation software also enables other operations, such as managing and charging various content-based services, controlling the subscribers' usage charges and ensuring reliable revenue sharing between operators and content providers. This will be of particular importance to Digifone in the future with the launch of its 3G services and it will help Digifone with the difficult task of deciding how to fairly and accurately distribute revenue between the various parties involved in the complex value chain.

Initially it will be essential for Digifone to focus strictly on the billing-related applications of mediation and ensure that they get the billing mechanism right from the start. However, over time, the full value of mediated data will be realized as innovative uses of the data are developed. For example, the software easily interfaces with applications that can provide a unified, real-time view of each customer and their usage habits. This will enable Digifone and its strategic partners, such as content providers, to target strategic sales and marketing campaigns that lead to increased revenues and improved speed-to-market when introducing high-margin, value-added services.

IPDR

Digifone's decision to continue to use Amdocs as their solution provider in addressing these billing challenges was based not only on their positive track record with the company, but also on the fact that Amdocs' technology for mobile services billing is built on top of an established, non-proprietary industry standard for describing network usage events. The standard, known as IPDR,[5] is the product of an independent consortium which includes Amdocs, some of their competitors like Openet, and many other companies including Accenture, Hewlett-Packard, Intel, Lucent Technology, Apogee and Nortel (IPDR.org, 2002a). IPDR is essentially an agreed upon list of the various contextual characteristics that can be used to describe data exchange events within networks, and an agreed upon format in which to represent these descriptions. IPDR thus makes it possible to access, measure and record the information needed for billing context-driven events in complex value Webs — the key problems facing the mobile arena. At present, IPDR is in its third version release, and is already quite mature. The current version supports the description of many different types of next generation mobile services — Video on Demand (VoD), Application Services (ASP), Voice over IP (VoIP), E-Mail Services, Streaming Media, Authentication and Authorization Services (AA), Internet Access (including wireless), Content/Service (including wireless) and Push Delivery (including wireless), to name a few. More importantly, the specification is fully open — a service provided to the business sector by the members of that sector in a win-win collaborative effort.

The goal of IPDR is to create a framework of open standards to facilitate the exchange of usage data between network/hosting elements and operations/business support systems. This framework provides a mechanism for various network and service providers to seamlessly exchange information, and for that information to eventually be passed on to various business support systems (IPDR.org, 2002b).

For example, many mobile services will depend upon authentication and authorization events (the act of verifying the identity of a user and then determining whether the user will be allowed access to a particular resource). These events are information-rich, and must be carefully recorded for the purpose of trend analysis, auditing, billing, or cost allocation. The NDMU[6] extension for authentication and authorization services includes mechanisms for recording:

- Type of authentication/authorization service that is invoked.
- The date and time of both requests and responses.
- The location of users, network access providers, and other service providers involved in the event.
- The user's identity.
- The authentication and authorization service provider's identity.
- The amount of data transferred, etc.

Although immature, IPDR has made considerable gains in terms of industry adoption. ACE*COMM, Amdocs, Convergys, Hewlett-Packard, Narus, NEC America, and TSI have all demonstrated their ability to exchange IP usage records utilizing the NDMU protocol. Proof of concept demonstrations have been performed in which data associated with next generation services such as streaming media, Voice over IP, and wireless applications has been seamlessly exchanged between different software products.

The IT team at Digifone understands the importance of the IPDR standard in enabling the Amdocs solution and, in turn, Digifone's future plans. They also realize that IPDR was only made possible by yet another, broader, industry standard, namely XML (eXtensible Markup Language). XML is an open, standardized data format — only a few years old — that has had an impact on nearly every aspect of the business use of computers. The XML data format plays a key role in delivering digital content in multiple formats to multiple agents (desktop Web browsers, mobile devices, printing devices, etc.); it enables more intelligent information discovery and retrieval and allows computers within and between organizations to seamlessly exchange data. Most importantly, XML is not a language in itself, but a tool for building languages — for modeling the information in a specific business sector. Using XML, companies participating in a particular industry vertical or functional horizontal can come together and agree upon a common language for exchanging information. IPDR is just one example of such a community agreement. Further details on XML are available in the Appendix.

Thus, Digifone's solution to the mediation challenge is doubly reliant on technological standards. Firstly it relies on IPDR, an intra-industry standard for describing network usage and this in turn relies on XML, an inter-industry standard for building sector-specific languages. This situation was unsurprising to Brian. The telecom industry had always relied on standards, and as the technology increased, the standards were needed more than ever before. And, just as the roll out of next generation business models were seen to depend on collaboration between many different companies, implementing these business models depended on these same companies, and others, collaborating to build a common information infrastructure.

CURRENT CHALLENGES

For Brian Noble, the challenge facing Digifone means moving from a very simplistic view of value based primarily around large scale infrastructure and tangible assets towards one where the participants are faced with making investments in much more intangible assets and trying to manage these assets. That means moving away from time as a mechanism for calculating revenues to looking at the inherent characteristics of the transaction from a value point of view, using these to charge value across the value chain, and allocating revenue to the various participants.

The Digifone experience highlights the importance of industry communities in developing standards. In particular, the IPDR experience represents a community of stakeholders coming together to recognize the need for a standard. In the case of IPDR the standard is not just a static vocabulary, which characterized the standards in the early days of the telecommunications industry, but a much more dynamic approach where a negotiated standard continues to evolve and meet the needs of the community. In an increasingly intangible world, firms such as Digifone need to develop capabilities for measuring, managing and reporting intangibles because the substantial investments they are making in organizations are not in physical infrastructures, and indeed many of the participants in this new emerging 3G environment will be resellers who don't own infrastructure. The big challenge is for these to develop the capability to manage their intangibles.

In the new world of intangibles the role of billing and mediation has changed dramatically. In the past billing and mediation were seen as back office accounting activities where revenue was calculated and collected. In the new intangible world of global brands and 3G licenses, billing and mediation are key critical success factors. In particular, the ability to mediate and bill for value across the multi-participant value chain will be a core competency within firms such as Digifone. Firms that fail to put flexible mediation and billing solutions in place will be constrained in their ability to deploy new and more profitable services. As the value of telecom operators becomes increasingly dependent on their intangible assets, the role of information is shifting to center stage. The old adage "if you can't measure it you can't manage it" will become increasingly important. In the future, the key resource will be customers' brands and relationships, and the measurement of these relies upon effective mediation and billing solutions.

While the focus in recent years has been on Customer Relationship Management (CRM) and Billing and Customer Care (BCC), the real challenge lies in effective and efficient mediation. The measurement and management of intangibles presupposes a mechanism for tracking and understanding the online customer experience, and that requires standards-based mediation of the highest quality. In the absence of standards the industry measurement and management of intangibles becomes little more than guess work.

Concluding Thoughts

Brian Noble considered the 3G challenges as he waited in line for the checkout at the corner shop. At the register, staff were busy scanning the barcodes on the products that customers purchased. To his left, two staff members were discounting some products that were approaching their "best before date." Further down the aisle, one assistant was setting up a "two for one" promotion for some new products. Barcodes dominated the activities as value was promoted, discounted and paid for. It didn't matter where the products came from, or what retail channel supplied them; the bar coding system using the universal product code (UPC) worked. Discounting or promoting just required a different bar code. He began to think of the 3G billing issue as bar coding for value. The IPDR standard might work within the telecom sector to address this challenge, but he knew other industry sectors were developing XML standards to facilitate their work. That is what 3G needs, he thought, a way of categorizing value across industrial sectors and customer segments.

As Brian left the shop he considered the issues facing Digifone in the context of the 3G challenges. To his mind, the issues all hinged on one question: how to position Digifone within the emerging 3G value Web? The bar-coding for value: idea was neat, but how could it be achieved? In addition, other issues needed to be addressed. Specifically, these were:

1. How would Digifone partner with other 3G value Web participants (e.g., content providers, application developers, and perhaps even other 3G license holders)?
2. How would Digifone build on its brand image and customer base within the 3G value Web?
3. How would customers perceive the value added from 3G networks and applications? In particular, what would the "killer app" be?
4. How would customers react to value-based billing, and how could Digifone best help them make the transition?

ACKNOWLEDGMENTS

This case was made possible through the generous co-operation of Digifone. The case is intended as a basis for class discussion rather than to illustrate either effective or ineffective handling of management situations. The research for this case study, as part of a portfolio of 15 cases, has been funded by the European Commission via IST project 2000-29665 "RESCUE." For further information see http://www.EUintangibles.net.

REFERENCES

3G News Room. (2001). *What is 3G?* Retrieved January 10, 2003, from http://www.3gnewsroom.com/html/what_is_3g/index.shtml

Cahners In-Stat MDR. (2002). *Internet access device market to grow 41.6 percent in five years*. Retrieved January 10, 2003, from http://www.canvasdreams.com/viewarticle.cfm?articleid=1144

Cellular News. (2002). *3G licenses and contracts*. Retrieved January 10, 2003, from http://www.cellular-news.com/3G

Cerillion. (2002). *What will GPRS and UMTS mean for billing?* Retrieved January 10, 2003, from http://www.cerillion.com/frame.html

CNN Money. (2000). *BT is Esat's white knight*. Retrieved January 10, 2003, from http://money.cnn.com/2000/01/11/europe/bt_esat/

Comptel. (2002). *Comptel*. Retrieved January 10, 2003, from http://www.comptel.com/p10012.html

Computer User. (2002). *Definition: Bluetooth*. Retrieved January 10, 2003, from http://www.stlcu.com/resources/dictionary/popup_definition.php?lookup=7912

Digifone. (2001). *Frequently asked questions*. Retrieved January 10, 2003, from http://www.digifone.ie/faqs.jsp

Durlacher. (1999). *Mobile commerce report*. Retrieved July 10, 2002, from http://www.durlacher.com/downloads/mcomreport.pdf

EuropeMedia.net. (2002). *Digifone launches Ireland's first consumer GPRS service.* Retrieved January 10, 2003, from http://www.europemedia.net/shownews.asp?ArticleID=7837

Frost & Sullivan. (2002) *North American Mobile Internet access markets.* Retrieved from http://www.frost.com/prod/servlet/fcom?ActionName= DisplayReport&id=A313-01-00-00-00&ed=1&fcmseq=1039454224237

Helme, S. (2001, November/December). Top ten trends for 2002. *Billing, 18,* 48-53.

IPDR.org. (2002a). *Membership list.* Retrieved January 10, 2003, from http://www.ipdr.org/membership/list-member.htm

IPDR.org. (2002b). *Network Data Management-Usage (NDM-U) for IP-based services, version 3.0.* Retrieved January 10, 2003, from http://www.ipdr.org/documents/NDM-U_3.0.pdf

Liikanen, E. (2002) *The vision of a mobile Europe.* Retrieved January 10, 2003, from http://www.m-travel.com/20322.shtml

Lucas, M. (1999, July/August). The IP detail record initiative. *Billing World,* 30-32. Retrieved January 10, 2003, from http://www.rate integration.com/articles/ipdr.pdf

mmO2. (2001). *About mmO2.* Retrieved January 10, 2003, from http://www.mmo2.com/docs/about/company_facts.html

Nua. (2002). *Wireless Web population to soar.* Retrieved January 10, 2003, from http://www.nua.ie/surveys/index.cgi?f=VS&art_id=905357560&rel=true

O'Donovan, M. (2001). *Amdocs customer interview.* Retrieved January 10, 2003, from http://www.amdocs.com/customers/CustomersStories.asp? CustomerID=38

Openet. (2002). *The value of mediation* (Openet Telecom White Paper). Retrieved January 10, 2003, from http://www.openet-telecom.com/

OTDR. (2000). *Introduction of 3G mobile services in Ireland* (Office of the Director of Telecommunications Regulation Document No 00/48). Retrieved January 10, 2003, from http://www.odtr.ie

OTDR. (2001). *Four licenses to provide 3G services in Ireland* (Office of the Director of Telecommunications Regulation Document No. ODTR 01/96). Retrieved January 10, 2003, from http://www.odtr.ie/docs/odtr0196.doc

Schlumberger. (2003). *Billing for 3G services.* Retrieved January 10, 2003, from http://www.schlumbergersema.com/telecom/ind_insight/3gbilling3.htm

SearchNetworking.com. (2001a). *GPRS.* Retrieved January 10, 2003, from http://searchnetworking.techtarget.com/sDefinition/0,,sid7_gci213689,00.html

SearchNetworking.com. (2001b). *GSM.* Retrieved January 10, 2003, from http://searchnetworking.techtarget.com/sDefinition/0,,sid7_gci213988,00.html

APPENDIX

From 2G to 3G: The Evolution of Mobile Services

In this section, we will look at three platforms for mobile services as shown in Table A. First, we'll discuss the GSM (2G) systems that emerged in the early 1990s and dominate today's market. Then we'll examine the roll out of GPRS (2.5G), which took place in 2001-2002. Finally, we'll look at possibilities of true Third Generation (3G) mobile services, which are to be implemented in 2002-2003.

Table A. Three generations of mobile networks compared

	Strengths	Weaknesses
2G (GSM)	Security. Roaming. Higher quality and speed than 1G analog services.	Still relatively narrowband. Limited data services.
2.5G (GPRS)	Faster than GSM. Broader band services. Real-time services. True Internet access.	Network impact. Sub-optimal performance. Operational difficulties.
3G (EDGE/WCDMA)	Higher quality and speed than GPRS. Fully realized multi-media and real-time services.	Operational difficulties.

GSM

As of early 2002, the dominant platform for mobile services is a digital system known as GSM (Global System for Mobile Communications). Developed in 1991, by 1997 this open, vendor-independent standard had become the *de facto* platform for mobile services in both Europe and Asia. By 2001, GSM was in use in over 170 countries. Like all mobile services platforms past and future, GSM is an open standard rather than the proprietary architecture of any particular company. Intra-industry cooperation, standardization and vendor independence have all been critical to the widespread adoption of GSM.

GSM is a digital technology, and thus supports faster, more reliable and more sophisticated transmission methods than the first generation (1G) mobile systems that were based on analog technology. Even so, while GSM has been able to offer high quality, highly secure voice services, and impressive "roaming" capabilities, it is still quite limited in terms of data services.

GPRS

At the time of writing, we are witnessing the roll out of GPRS (General Packet Radio Service) systems. GPRS improves upon GSM in many ways, and supports the delivery of much more sophisticated data services than its predecessor. It is seen as the stepping-stone to 3G mobile services, and for this reason, GPRS enabled networks are often referred to as 2.5G platforms.

One significant change represented by GPRS is the industry move towards building mobile data services on top of the Internet Protocol (IP) — the standard that supports the Internet itself. GPRS promises greater data transmission speeds (in ideal circumstances up to three times the faster than landline speeds in 2002 and up to 10 times faster than GSM) and better connectivity (GPRS data services are "always on" and no dial-up connection is needed for a particular transaction). These changes, among others, will enable GPRS users to consume and produce broader band media and interact with more sophisticated real-time services.

It is hard to separate marketing hype from sober prediction, but GPRS data services are likely to include real-time, text-based Internet chat, dispatch services, mobile e-mail services, mobile financial/commercial transactions, mobile Web browsing (limited), text-based messaging services (superior to those offered using GSM), graphics-based messaging services, audio-visual data transmission (limited), document sharing and file transfer (limited), remote collaboration and remote computer network access (limited), home device monitoring and control (limited).

GPRS is, of course, not without limitations. First, GPRS will have an impact on current levels of network capacity. Second, the transmission speeds in practice are likely to be far less than the ideals mentioned above — most likely the same speed as land lines and about three times that of GSM (still an improvement). Third, and perhaps most importantly, many operational details associated with the delivery of GPRS services have not been fully resolved.

3G

3G mobile systems will not constitute a substantial *conceptual* shift beyond GPRS — although they will be based on more sophisticated technologies (the most popular standards are called EDGE and WCDMA) and make more extensive use of the Internet Protocol. Rather, 3G can be thought of as GPRS in its maturity. As with the shift from GSM to GPRS, the move from GPRS to 3G will include an enormous increase in bandwidth and speed — 3G data transmission rates are predicted to top out twenty times faster than GPRS! Even an increase factor of 10, rather than 20, would mean that the various GPRS services listed previously as "limited" will come to maturity, and new services such as full-streaming video, Voice over IP, and "always on" electronic agents will roll out. 3G will, in many ways, be GPRS plus full multimedia, entertainment, and enterprise class business services. Like GPRS, 3G is likely to be plagued by both operational and technical challenges, as companies figure out how to effectively exploit these new capabilities.

One of the interesting changes expected to take place with the emergence of 3G services is the shift to new physical form factors for mobile devices, or terminals. Because 3G devices will need to accommodate video input and output hardware, they will begin to look more like palm-top computers than like today's GSM handsets.

Self-Describing Data with XML

XML is an international standard data format that promises great things in virtually every area of contemporary business computing. For a technology which is only beginning to mature, XML has enjoyed widespread industry adoption. Dozens of consortia made up of international corporations, professional organizations, and public bodies have formed to develop XML-based applications, called "vocabularies," for various horizontal and vertical industry sectors, and the major Independent Software Vendors (ISVs) — Microsoft, Sun, IBM, Oracle, SAP, etc. — have all made XML a core part of their technological strategies.

XML is revolutionizing several key application spaces. These include:

* **Enterprise Application Integration** — One of the greatest challenges in business computing is getting the various hardware/software systems within an organiza-

tion to effectively communicate with each other. XML makes this significantly easier.

- **Inter-Organizational Systems** — The cost of the "manual gulf" created by the need for human intervention and physical activity in data exchanges between organizations cannot be easily measured. XML helps to automate data exchange between industry partners.
- **Content Management** — XML allows content providers to store information in a single format, and then to transform that information to address the specific requirements of the device that is consuming the information — whether that is a Web browser on a PC, a mobile phone, a printer, or any other application.
- **The "Semantic Web"** — XML enables the development of a more richly described, and thus more accurately searchable, network of digital information.
- **The "Web of Trust"** — XML is at the heart of solutions to address the need for privacy, identification and authentication in Internet systems.

The above list only scrapes the surface, but is impressive enough in its own right. How does the XML data format do all these things? There are four inter-related answers.

1. **XML is a highly interoperable, pan-industry standard.** XML enjoys its growing role as the "glue" of the world's information systems because it is an open, widely accepted standard (developed by the members of the World Wide Web Consortium, or W3C) and is highly interoperable (it can be easily exchanged between computers both within and between enterprises). The reason XML documents are so easy to exchange is that, physically, they are nothing more than plain text files, generally using the now ubiquitous UNICODE character set. As UNICODE text, XML documents are in the "universal currency" of the data exchange world.

2. **XML is self-describing.** If XML were nothing more than plain text, we wouldn't be discussing it now. In reality it is much more. The core XML specification represents something intangible and very powerful: a worldwide *agreement* on what a self-describing document should look like. By self-describing, we mean that XML documents contain both data and meta-data (data about the data). Take, for example the number 9415551212. This is a bit of data but without any context — there is no way to "make sense" of it. Within an XML document, it is possible to add meta-data to the string — margin notes, if you like — which describe what the data is about:

```
<phone-number region="US">
<area-code>941</area-code>5551212
</phone-number>
```

The bits within the angle-brackets (< >) are XML "tags," the meta-data describing the whole number as a U.S. phone number, and the first three digits as an area code.

3. **XML is flexible.** The example provided above hints at probably the *main* reason why XML is enjoying so much success. XML is flexible or, in its own words, extensible. As noted above, XML is an agreement on what self-describing documents look like. It does not attempt to dictate what those documents are actually about. XML is essentially half of a language. It has a hand-full of grammatical rules,

but it has no vocabulary — that is up to the developers and business domain experts of the world. Any organization, industry or field of endeavor can use the XML grammar to create a vocabulary specific to their needs. And they have. Our little "phone number" vocabulary above is a trivial example. Numerous industry vertical sectors (steel, pharmaceuticals, architecture, advertising, etc.) and horizontal sectors (human resources, business reporting, etc.) have formed massive vocabularies to represent the data handled within and between the members of their respective value chains. These vocabularies are changing the dynamics of procurement, customer relationship management, and every other aspect of daily business. IPDR, which we'll discuss soon, is one such example. IPDR is an XML vocabulary for creating self-describing records of mobile network usage — a key information need for the mobile sector.

4. **XML is not just XML.** The final reason behind XML's success is that the core technology (the agreement on what self-describing data looks like) is only part of a wide family of technologies, most created by the international membership of the W3C but some by other independent organizations. The rest of the family is briefly discussed.

- **Schema technologies.** There are several sophisticated tools/frameworks available for building an XML vocabulary for a particular information space. The three most popular are DTD (simple to learn but limited), XSDL (more difficult to learn but very powerful), and RELAX NG (somewhere in between the two in both complexity and power). RELAX NG is the product of an important standards agency called OASIS; the other two are from the W3C. Without schema tools, XML's extensibility and interoperability are compromised, since these tools are used to create the common languages in which organizations can exchange data.
- **Processing technologies.** As was probably evident from the phone number example, XML is human-readable in a pinch, but is machine-readable by design. Again, there are several technologies that enable different computer programming languages to read, modify and create XML documents. There are also tools that enable different software packages to transform an XML document into other forms (for example into a Web page, a mobile text message, or a printable PDF file), and various technologies for searching XML documents, linking them together, etc.

The XML infrastructure space is an impressive display of standards building in itself. What's even more impressive is the huge number of sector specific vocabularies that have emerged based on these standards. In addition to IPDR described in the case, here are just a few examples:

- XBRL (eXtensible Business Reporting Language) is a vocabulary framework for describing financial statements in the public and private sector. XBRL creates value in the form of accuracy and efficiency for every member of the financial information supply chain — financial statement creators, analysts, investors, regulators, taxation authorities, financial publishers, data aggregators, and financial software vendors, among others. XBRL is the product of an international consortium with over 140 member companies, and has enjoyed significant adoption in Australia, Denmark, Singapore and the U.S., with other countries planning adoption in the near future.

- P3P (Platform for Privacy Preferences) is a vocabulary that enables Web sites to express their privacy practices in a standard format that can be retrieved automatically and interpreted easily by user agents. P3P user agents will thus allow users to be informed of site practices (in both machine- and human-readable formats) and to automate decision making based on these practices when appropriate.
- ebXML (electronic business XML) is the product of UN/CEFACT (United Nations Center for Trade Facilitation and Electronic Business) and OASIS (Organization for the Advancement of Structured Information Standards) and is a suite of specifications that enables enterprises of any size and in any geographical location to conduct business over the Internet. ebXML provides companies with a standard method for exchanging business messages, conducting trading relationships, and communicating data in common terms. ebXML thus facilitates the creation of a "Single Global Electronic Market" open to all parties, irrespective of size.
- Resource Description Framework (RDF) is an XML vocabulary for creating metadata for information resources. RDF can be used in resource discovery (better search engines), cataloguing, knowledge sharing and exchange, content rating, etc.
- NewsML is an XML vocabulary for supporting the creation, transfer and delivery of news items. NewsML is maintained primarily by The International Press Telecommunications Council (IPTC), which is comprised of major news agencies, newspaper publishers, news distributors and vendors from around the globe.

ENDNOTES

[1]   A name used to describe the phenomenal growth in the Irish economy during the mid-late 1990s.

[2]   Also referred to as Next Generation Mobile Networks in the U.S.

[3]   Internet Protocol (IP) is the underlying standard for exchanging data between Internet hosts.

[4]   3G offers always-on or constant connection to the Internet eliminating the need to "dial-up" to connect to the network.

[5]   IPDR (Internet Protocol Detail Record) facilitates the exchange of IP usage data between network elements, operations support systems and business support systems.

[6]   NDMU (Network Data Management Usage) defines the essential information for any network service usage, but can be extended to support new information only applicable to a particular type of service (e-mail, etc.).

Martin Fahy is a senior lecturer in accounting and IS at the National University of Ireland, Galway. He is a chartered accountant and holds a PhD in informatics from UCC. Prior to joining NUI, Galway he worked as a management consultant with KPMG. He has written extensively on the areas of business information systems, ERP systems and emerging issues in finance. He is currently an adjunct professor at the Universite d'Auvergne in France and is involved in research in the areas of shared service centres and strategic enterprise management.

Joseph Feller is a college lecturer in business information systems at University College Cork, Ireland, and is director of the University's master's program in electronic business. His research on XML has been presented at the European Conference on Information Systems 2003 and he has published more than 50 practitioner articles on the topic. His related research on open source/free software has been presented in a number of international journals and conferences, and he has edited four journal special issues on the topic. He has published three books, and was awarded "Best Paper on Conference Theme" at the International Conference on Information Systems 2000 for his paper "A Framework Analysis of the Open Source Software Development Paradigm" (with Brian Fitzgerald).

Pat Finnegan is a senior lecturer in MIS at University College Cork, Ireland, and is director of the University's MSc in MIS. His research on electronic business models and inter-organizational systems has been published in international journals including Information Technology and People, Information Systems Journal, Journal of Electronic Commerce, Journal of Information Technology, International Journal of Electronic Markets, International Journal of Technology Management, *and the* Australian Journal on Information Systems. *He has also presented his research at international conferences, including the International Conference on Information Systems (ICIS), European Conference on Information Systems (ECIS) and Hawaii International Conference on Systems Sciences (HICSS).*

Ciaran Murphy is Bank of Ireland professor of business information systems at University College Cork, Ireland, the head of the BIS group, and director of the Executive Systems Research Centre. He has more than 20 years of research and commercial experience, and has acted as a consultant to a wide variety of organizations in Ireland and abroad. His research work has appeared in number of international journals and books.

This case was previously published in the *Annals of Cases on Information Technology*, Volume 6/ 2004, pp. 128-156, © 2004.

Chapter X

The Integration of Library, Telecommunications, and Computing Services in a University

Susan A. Sherer, Lehigh University, USA

EXECUTIVE SUMMARY

Today many IS departments and individuals are attempting to transform from technical groups and specialists to user oriented functions and customer support personnel. The major responsibility of the traditional IS department has evolved from the development, operation, and support of technology to the management of information. In the university environment, managers of information have traditionally been librarians. Librarians have increasingly become users of electronic information resources. A merger of the library with computing and telecommunications brings together technical expertise with information management skills. This case study describes the process of integrating the library, computing and telecommunications services in a university. Within the last two years, a new manager in the newly created position of Chief Information Officer merged these diverse organizations. We will describe the techniques used during the first year to foster communication, develop new strategic direction, and create and implement a new organizational structure. We will focus on establishing leadership, the organizational change and operational planning process, and the initial implementation of the new organizational structure. We will describe some of the problems and obstacles that needed to be addressed, including new management's establishment of trust and control, creating an environment for change, managing change amid strong time pressures, human resource issues, and resource constraints. It is expected that many of the issues that arose during this merger will be addressed by organizations in other industries as they attempt to evolve from technical IS groups to more customer oriented organizations. Today it is imperative that IS functions provide client support, which requires different types of skills from those traditionally nurtured among technical experts in traditional IS departments.

BACKGROUND

The University is an independent, coeducational institution located in northeastern United States. Approximately 5000 undergraduate and 1500 graduate students are enrolled in programs in the arts and humanities, business, education, engineering, natural and social sciences. The University employs approximately 1400 faculty and staff.

The University's mission is to advance learning through the integration of teaching, research and service to others. It is committed to integrated learning, promoting the discovery, integration, and communication of knowledge. The University's mission statement embraces a commitment to the intellectual, physical, social, ethical, and spiritual development of all members of the academic community.

To support the discovery and integration of knowledge, the University built extensive library resources, including a collection of over 1 million volumes. The advent of the campus wide network enabled the library to supply all users with electronic as well as traditional services. In 1995 the University library staff numbered approximately 60, having just been significantly reduced from 75. The organization of the libraries in June 1995 is shown in Appendix A.

The University also built extensive computer and communications resources to support its mission. The computing center served the needs of students, instructors, researchers, and administrative users. A total of approximately 90 people worked in this department, whose structure is shown in Appendix A.

SETTING THE STAGE

The library and computing center at this university, as at most universities, evolved separately, with different roots and cultures. We describe here the evolution of these two groups, similar integration efforts at other schools, and the University's decision to integrate these functions.

Library Services vs. Computing and Communication Services: Evolutionary Differences

Historically the library's primary function has been to provide information to faculty and students who are in the pursuit of knowledge. Initially when computers were introduced into universities, their primary purpose was the storage and dissemination of data. Whereas data are simple observations of states of the world that are easily structured, captured, quantified, and transferred, information is data endowed with relevance and purpose which requires a unit of analysis, needs consensus on meaning, and requires human mediation (Davenport, 1997). The role of the librarian has generally been to aid the user in finding meaningful information that can be transferred to knowledge, information that has been reflected upon. As computers have become more pervasive and user oriented, they have evolved as a source of information rather than simply data. As a result, data processing centers have evolved to encompass information management.

While the library and computing and communications services both provided support for information gathering and processing, there were significant cultural differences within these professions. These differences were embedded in different approaches to problem solving. For the computing people, new technology was often the key to solve problems because it enabled more efficient and effective processing of data and information. For the librarians, the information itself provided answers.

Gender differences, the nature of the educational training, and the problems that were solved by librarians as opposed to computing personnel contributed to cultural differences. The library profession is predominately female while computing is male dominated. At the University, the library's director and four out of six associate directors were female. In computing and telecommunications, only one associate director was female and she was two levels down within the hierarchy. The demarcation between exempt and nonexempt staff was much greater in the library as compared to the computing center, most probably because of the greater importance of formal education in the library.

Librarians were generally excellent oral and written communicators. They used their skills in meetings where they joined together to plan and discuss various issues. The computing people had strong analytical skills and analyzed and solved problems often without as much discussion. Whereas the librarians worked best in a very collegial atmosphere, computing people often stressed doing the best individual job, i.e., developing the best program, solving the problem quickest, etc.

The librarians interacted primarily with faculty and students. They were comfortable working with novice users, often providing general solutions to meet the needs of large groups. Computing people entered the profession when communication with external constituents was minimized through the segregation of the computing center. "Many computer people were attracted to the IT profession because they were happier communicating with machines than with people" (Caldwell, 1996). As the computing environment evolved to a more distributed model, computing people generally supported users with some degree of technical literacy, providing one-to-one interactions and customized solutions. Communication between the computing center and telecommunications personnel and faculty and students continued to be fairly limited to help desks.

Not only were the cultures of library and computing different, but both professions had different subcultures as well. In the library, there were differences between public services and technical services librarians. Media specialists had their own background and differences. In computing, subcultures existed between academic and administrative computing, between hardware and software support personnel, and between user services and programmers or technicians (Hirshon, 1998).

These departments also had very different structures. One year prior to her retirement, the director of libraries flattened her organization, resulting in the six associate director positions shown in Appendix A. The computing and communications groups remained more hierarchically organized. There was no comparability in the management titles.

Mergers at Other Universities

During the mid- to late 1980s a number of universities integrated libraries and computing. Few of these mergers survived. They did not effectively merge the departments below the top level because the technology had not yet advanced to enable full integration. However, in the 1990s, with the advent of the World Wide Web, the need to access large amounts of interactive information, replacement of administrative systems with enterprise information systems, and increased incorporation of technology in the curriculum, many schools revisited the need for integration. Universities pursued several different integration models ranging from total integration to limited integration of one or two key service areas such as help desk, or joint management of some projects. Of the approximately 100 schools that have merged their functions, more than 80% integrated within the last five years and only 17% have attempted full integration (Hirshon, 1998).

The University's Decision to Merge

In 1994, the University's president decided to merge the library with computing and communications services. The vision was to integrate information systems, resources, and services in support of the teaching, learning, and research missions of the University. The decision was based upon the following recent trends in the environment:

- Library resources were increasingly electronic.
- Technical computing personnel were increasingly required to support end user computing as new technology enabled the transformation from centralized to distributed information processing and users became more technically capable.
- The objective was to manage information, not just systems.
- Both were support functions, providing information resources to the academic community.
- Content and technology were merging.

Most members of the community believed that this was a reasonable strategy. There were some who thought that economic considerations were driving the decision as well, suggesting that the ability to reduce administrative costs was a driving factor. The opportunity to merge was created when the director of libraries retired and the assistant vice president for computing and telecommunications returned to the teaching faculty.

CASE DESCRIPTION

This case will focus on the *process* of organizational change. It describes the establishment of new leadership, the six-month process of planning for the new organizational structure, and the initial implementation of that structure, including the importance of training personnel to meet the new requirements (Agarwal et al., 1997; Keen, 1991). A time line of major events is shown in Appendix B.

Leadership

In 1994 a search committee was constituted to select a chief information officer (CIO). Since this was a newly created position, the committee not only selected a candidate, but also formulated the job responsibilities and qualifications detailed in Figure 1. Since this was a new type of position, it was recognized that it would be difficult to find an individual with experience in all areas: computing, telecommunications, and libraries. This was recognized as a potential problem, as members of the group whose background differed from that of the new leader were concerned that the new CIO might not be as supportive of their areas of operation. However, given the fact that few universities had previously combined these areas, candidates with experience in any one of these areas were considered.

After reviewing the qualifications of numerous candidates, an individual was chosen who provided a vision for the use of information technology in higher education. This candidate understood the opportunities to improve support of the teaching, learning, and research missions of the University presented by the merger of the information management resources of the University. The candidate understood the challenges of merging the different cultures and offered a management style that was expected to recognize and deal with these challenges. The successful candidate had previous experience primarily in the area of library management with a background in library technology.

The new chief information officer came on board in late 1995. The first weeks were spent meeting the staff, with scheduled interviews with every professional and offers to meet any individual support staff member. In addition, the CIO visited every department and held group department meetings.

Interim Management

Given the significantly different organizational structures of the two component units, an interim management group was established, consisting of 15 individuals pulled from the management of the existing organizations. Since the library had six direct managers, the CIO went down to the next reporting level to provide similar representation from computing and communications. This resulted in a group of nine managers from the computing and communications group, shown in Figure 2. The purpose of the initial management group was to help the CIO manage solely during the transition for the six-month planning process, after which a new management would be established in a new organizational structure.

The interim management group met twice a month throughout the six-month planning process. Several participants noted that in early meetings, library and computing people typically sat separately and talked among themselves before meetings.

Figure 1. Chief information officer responsibilities and qualifications

Responsibilities
- Integrate library and computing services and resources in support of the teaching, learning, and research missions of the university.
- Create an organizational structure and a working environment that encourage creativity, cost effectiveness, and change.

Qualifications
- Possess a vision for incorporating innovative information technologies in higher education.
- Understand the changing paradigm in scholarly communication.
- Understand the use of information technologies to support instruction, research, and administration.
- Understand the administration and operation of libraries, computing, and telecommunications services.
- Experience and expertise in at least one of the areas: libraries, computing, and telecommunications services
- 7-10 years of increasingly responsible management experience.
- Demonstrated commitment to participative management style and open decision making.
- Excellent interpersonal and communication skills.
- An advanced degree in a relevant field.

Figure 2: Interim management team

Associate Director, Library Technical Services
Associate Director, Library Automation
Associate Director, Library Special Collections
Associate Director, Library Reference Services
Associate Director, Library Access Services
Associate Director, Library Media Services
Director, Computing User Services
Director, Administrative Systems and Telecom
Associate Director, Computing and Networking
Associate Director, Computing Facilities
Associate Director, User Services
Manager, Computer Operations
Manager, Network Facilities Support
Manager, Systems Programming
Manager, Academic Programming

Gradually, however, the managers began to mingle and communicate with others in different areas.

Interim Organization: Getting to Know Each Other

The combined organization was composed of approximately 140 people. At its inception, many people did not know the responsibilities of their colleagues in the other component units. Several initiatives were instituted to enable personnel to become better acquainted. First, in January each group was asked to provide an overview presentation of their area. Everyone was obligated to attend these sessions. Some consideration was given to making this a competition, asking groups to compete for the best presentation, to elicit more excitement about the process. However, it was thought that this might be unfair because some groups, by the nature of their jobs, had much higher presentation skills than others, e.g., media productions. Attendees believed that they learned some new things by attending these group orientation meetings. While these sessions provided a forum for presenting information, they did not necessarily engage people in cooperative efforts nor was this its purpose. If the group were to coalesce, they needed to have opportunities to work together. A retreat and cross-functional planning team structure were instituted to accomplish this task.

Planning for Organizational Change: The Retreat

Prior to instituting any organizational change, an intense planning process was begun. This was necessary to create motivation for change or "unfreeze" the set of different assumptions, values, and beliefs in the library and computing cultures (Schein, 1997). A one-day retreat was scheduled in early 1996. The entire organization was invited; approximately one half attended. The focus of the retreat was to develop mission, vision, environment, and values statements, which were discussed in a series of small group sessions.

Planning for Organizational Change: Working Teams

The major goal for the new CIO was to develop a plan for an integrated organization, and thereafter to restructure the functions to implement that plan. To develop the plan, four working teams were set up to focus on technology, information, financial and staff resources, and client services. In addition, a lead team coordinated the activities of all the teams. The lead team was made up of the four team leaders and four at large members. Each of the four working teams included 8-10 people. Each team was managed by a member of the interim management group and included at least one other member of that group. Teams were organized so that there could be a fresh look at the problems and extensive charges were prepared for each team to outline some issues. Figure 3 shows the leadership and composition of these teams.

To identify appropriate people throughout the organization to participate in this planning process, the interim management group was consulted. They were asked to identify non-parochial individuals in the organization who would have the vision to help move the combined organization forward. Teams were to mix individuals who represented the different subcultures within the combined group. The CIO consulted with others concerning team appointments, but ultimately decided who would lead the teams. As unanimity in such cases is impossible, in some cases, the selections were inconsistent with some of the interim management group recommendations.

After the work teams were initially formed, there were some changes to the composition of the teams to incorporate further advice from the interim managers. In some cases, a team lacked experience in an area assigned to that team. Changes were made to add or change team members so that "experts" were assigned to each team, albeit not always the ones the group might have selected. Several people felt that lack of inclusion of a particular person who was considered an expert in a particular area caused a lot of

Figure 3. Planning team composition

Lead Team
Leader: Chief Information Officer
Members: Team leaders (listed below), 4 at-large members from computing and libraries

User Services Team
Leader: Associate Director, Computing User Services
Members: Libraries (5), Computing and Telecom (3)

Information Services Team
Leader: Associate Director, Library Technical Services
Members: Libraries (4), Computing and Telecom (4)

Financial and Staff Resources Team
Leader: Associate Director, Library Reference Services
Members: Libraries (3), Computing and Telecom (4)

Technology Services Team
Leader: Associate Director, Computing and Networking
Members: Libraries (2), Computing and Telecom (7)

unnecessary time to be spent in team meetings. However, this was purposeful, as change was a primary objective, and teams had the opportunity to call upon additional expertise as needed.

The team structure directly involved approximately 25% of the organization. Some who were involved found that time pressures were very uncomfortable because they could not get their regular jobs done, which raised stress levels. On the other hand, those who were not involved were anxious because they did not know how the outcome would affect them. The team structure seemed to have accomplished several objectives:

- It brought together individuals with little prior relationship and got them to work well together.
- It brought together different and oftentimes new perspectives for looking at issues, getting people working intensively together in a cooperative environment while developing intra-organizational alliances not previously in place.
- It got staff involved in the planning process so that they would "own" the resulting strategy and organizational structure, building a base line of support for the outcome.
- It provided an opportunity for the new CIO to observe the actions of personnel in order to identify future leaders in the new organization, providing a testing ground for personnel.

Communication During the Planning Process

Throughout the planning process, there was an attempt to provide open communication both internally and externally. Traditional channels of communication were altered. The sensitivities of the process coupled with time constraints lead to the structuring of much of the communication.

Teams met from January through April to develop the strategic plan draft. During the process, there was an attempt to make this process more open via listservs and discussion groups providing everyone with input and access to the deliberations. As work progressed, however, information became more filtered. For example, as the teams got closer to defining organizational structure, there was concern that sharing too much raw information might cause panic among individuals. This sensitivity prompted the limitation of team postings to minutes that eliminated some of the detail, but resulted in the promulgation of significant rumors.

As the planning process evolved, traditional flows of communication were no longer satisfactory. With multiple teams working on issues simultaneously, managers could not provide answers to their own staff if they were not on certain teams. Channels of communication were altered. During this period, many individuals missed the cama-raderie and felt a loss of community when, for example, librarians no longer met together, but they also bemoaned the significant amount of time they were spending in planning meetings. However, given the ambitious schedule and requirements on individuals, there was little time to mourn this loss. Open meetings and informal brown bag lunches were held periodically throughout this six-month process to share information. In fact, some members of the organization felt that there was too much communication and, in fact, they were overloaded with information.

During this time period, external communication was limited. Focus groups were brought together to gather input from faculty and students. Although these sessions were short (typically 1-1.5 hours) and very structured, many attendees noted that this had been the first time that they had been asked for their opinions. Some participants felt that the structure limited their ability to provide information on real problems but some structure was essential if useful information was to be gathered in a short time period.

Centralization of Internal Business Functions

During this time period, the CIO centralized the internal business functions that were previously in three separate structures into a central location to track and control financial resources. With a large and complex budget, the CIO believed this was necessary to manage effectively. Some managers who were now required to go through a central process to acquire needed items resented this act because they perceived that local support was removed. This caused some hard feelings during a time period when trust was still being established and staff were still very anxious and fearful about any type of change.

The Process of Introducing the New Organizational Structure

In mid-April an open meeting was held to discuss the strategic plan draft. After the first draft of the strategic plan was sent out, each lead team member was given a week to propose the rough outlines for a new organizational structure that would advance the strategic plan. Extensive discussions ensued within the lead team, and there were many iterations of the structure. In May, the first draft of a new organization was unveiled at a brown bag lunch. No individuals were named on the chart, only the broad outlines of the organizational structure. The purpose of this meeting was to allow people to ask questions. As anxieties were very high, the meeting was referred to facetiously by some staff members as a "body bag," rather than a "brown bag" lunch. Most individuals were trying to determine whether and how they might fit into this new structure. A concerted effort was made to use new terminology to communicate whenever possible. The intent was to help break away from the old patterns of communication and ideas and create a common language for the new integrated culture (Schein, 1997). This also caused some confusion and discomfort. In fact, one participant commented that "it is a bit like a puzzle to see where you fit in."

Within two weeks after this lunch, the organization chart was revised and the CIO identified the high level managers who would populate it. All others were given a formal opportunity to request in writing the types of jobs they desired and their placement in the organization by completing a job request form developed by University Human Resources. It was recognized that not everyone could be matched with his or her desires, but ultimately a substantial percent (90%) were placed in one of their top three choices. The newly identified managers selected people for their organizations after reviewing these forms. There were some cases where two managers wanted the same person in their group, which resulted in some discussions of what was best for the entire organization.

The New Organizational Structure

Major recommendations from the planning process were the need for improved management of technology in classrooms, and the creation of client service teams. It was recognized that client needs varied broadly across campus, from sophisticated researchers requiring heavy library resources and computing power, to those with very basic needs. The organizational structure that evolved to address these needs combined elements of the old structure with new requirements (Appendix C).

The old "glue" behind the organization that was retained in the new structure was the familiar functional organization or infrastructure group. What was radically different was a new matrix user (client) services organization. Each of the colleges was assigned a team of consultants in four areas: computing and networking, library assistance, instructional technology support, and administrative information systems support. While each consultant's primary assignment was to a College team, s(he) also was to report to a "functional team" leader who headed each of these four functional areas. For some individuals, these secondary assignments created conflicting loyalties.

Personnel and Job Changes

A key issue was finding personnel to fill new positions, particularly in the new user services area. In many cases, these jobs required very different skills from previous job responsibilities. Particularly among computing people, there was a concern that technical expertise was no longer a valued skill, but that management now primarily valued the ability to support and work with users. For some, this change was not welcome. Not unexpectedly, some individuals, particularly computing personnel who were most mobile in the job market, left to find employment elsewhere where they could continue to apply primarily their technical skills. Shortage of personnel also lead to the sharing of consultants between more than one college, which was an additional cause of stress.

All job descriptions were re-written within two months to reduce job ambiguity and insure that no tasks would fall between the cracks. This was particularly important in the creation of the new user services organization. The University Human Resources Department audited all of the jobs. As a result, some salary levels were changed. Approximately 25% of all staff saw an increase in salary while a small percentage, about 5% of staff, received decreases in salary because managerial responsibilities were reduced.

Physical Changes

By early the next year, employees began to move their offices so the new teams would be physically housed together. User services teams were consolidated by their college rather than by their functional departments. For some individuals, this physical move was as upsetting as the organizational change. It meant giving up "homes" to which they had become accustomed. Moreover, it meant physically leaving behind traditional sources of information that had supported functional capabilities. Some individuals experienced a loss of identity as they were moved away from their functional colleagues, and were concerned that they would not continue to learn and grow professionally.

Technological Change

Another item that had a big impact on the ability to implement these changes was the need to undertake simultaneously major technological changes. In particular, beginning in the second half of 1996, installation of Windows 95 began across campus. This occurred at the same time that the organizational changes were put into place. Some users blamed the new organization for difficulty in meeting this challenge, but it should also be recognized that in a technology-based organization there is always a major technological change that needs to be implemented.

Implications for Users

When the organizational change went into effect, it took some time for users to learn whom to contact in the new IR organization. Many users had to change old habits, as they were accustomed to contacting specific individuals in the old departments to obtain help. Users now were provided with specific contacts for their colleges. Initial user response was a function of the quality of their assigned user services team. Users who were assigned consultants who were primarily dedicated to their college and who possessed a good customer service ethos saw service improvements. Some users complained because service was not as promised. In some cases the clients may have been getting used to better service. In other cases, they may have been promised but did not receive improved service because individuals in the new structure were either not yet trained to provide it or were attempting to perform several jobs at the same time. The need for providing additional training for the new jobs, particularly in user services, became very clear. It is not possible to just change organizational structure; people need to be trained to work within a new organization (Agarwal et. al., 1997). As staff moved into their new positions and had to keep old operations going at the same time, there was not time to get all people trained at once. Some training occurred, but more was needed.

CHALLENGES/PROBLEMS

We will now review some of the specific problems and challenges that arose through the planning and implementation of the new organizational structure. Throughout the planning process, major challenges were associated with new leadership and the aggressive time schedule. Choices often had to be made that would satisfy some people and address some concerns, but would leave others unsatisfied. The challenges of implementing the new organizational structure included human and financial resource issues.

Leadership: Establishment of Trust

This traumatic change was instituted under the direction of a new leader. New leaders are often brought in from the outside to create change. They often have a "honeymoon" period to do so, but for the change to be accepted new leadership must build trust during this period. Initial respect is based upon the experience and credentials of the new leader. In this case, the computing people, in particular, were skeptical about the new CIO because of the perceived lack of computing experience. Since these people expected that he would not become involved in technical decisions, any efforts by the CIO to do so were considered "micro-management."

To help build trust not only of the CIO but throughout the organization, the collaborative and collegial planning process was put into effect. However, some staff believed that the CIO did not have the opportunity to learn about the capabilities of all managers and staff prior to the constitution of the interim management group and the working teams. Since the management structures in the library and computing organizations were so different, the CIO initially created a larger and more inclusive management group rather than attempting to select a few from the existing structures. Several staff felt that, while this larger group enabled more personnel to have input, the size made it more difficult to build trust. As previously noted, the CIO's decision to centralize the internal business functions prior to the completion of the planning process also contributed to further difficulty in establishing trust.

Aggressive Time Schedule

The aggressive time schedule led to numerous challenges. A very ambitious and strict time line was established, announced, and adhered to. At the January 1996 retreat, the CIO announced that the organization would definitely change, and that all organizational changes would be announced within six months. It was felt that the schedule should be fixed not to exceed six months to minimize the stress of the traumatic changes, resolving doubts and ambiguities for staff as quickly as possible. However, the schedule needed to allow time for staff to participate while continuing with the work of the traditional functions. This meant that stress levels for these six months were exceptionally high. Staff were continuing with their own jobs and working very diligently on the new team assignments. There was often not a lot of time for exploration of actions. For example, when the management team was chosen, there was little explanation of its composition. This led to initial feelings that further input from staff might not be considered.

While the communication was intended to be very open, and very often was, the aggressive time schedule and the high stress level made open communication more difficult. Some staff found too much to review and too little time to digest all the material. Moreover, it was felt that there was too much at stake to publish everything in such a short time period. Opportunity for small group discussion was limited because of the time pressures. Some staff felt uncomfortable about their instructions and the level of detail required in the process, while others felt that they wasted time on issues that went away after the new organizational structure was announced. Communication with external constituents primarily occurred through job interactions, structured focus groups, and an open request to the campus for feedback on the published draft plan.

Resources

Other major challenges of the new organization were related to resources, both human and financial. The new organization required major changes in human resources. A change management plan and a human resources program were imperatives. The human resources plan included new job descriptions, performance measurements, and some training, but more training was required. "Changes in roles and performance appraisal criteria have to be supported by individuals who are appropriately trained to fulfil these roles. It is not simply a matter of calling an erstwhile systems analyst a project consultant if she is not equipped to handle the new job" (Agarwal et. al., 1997).

Existing staff had to learn to redefine long standing culture assumptions and develop new behaviors that would be reinforced by the external environment (Schein, 1997). A few technical personnel were immediately moved directly into user services positions. Some personnel did not have the training while a few others did not have the interest to redefine their own values and focus on user services. These staff revered their technical skills over their service skills. With a limited number of staff from which to choose, it was not always possible to place each person into their ideal position, but it was necessary to select those willing to acquire new skills for these new roles, provide training in new skills, and mentor those individuals.

The initial implementation was exacerbated by the lack of personnel. There were some vacancies to be filled when the new organizational structure went into effect, caused in large part by the significant University reduction in staff. These vacancies were not filled as quickly as hoped. In the meantime, some individuals performed more than one job, limiting their abilities to deliver user services and mentor others. There were also some shortages in critical areas with technical people assigned to cover areas for which they did not have all the requisite skills, which created some user frustrations. However, as these positions were filled, the new employees were generally better oriented toward their jobs than the people who were already on staff and had to adjust from their previous to their new positions.

In this case, increasing hardware and software demands had to be evaluated along with demands for new staff in a period of constrained financial resources. This was further exacerbated by a complete turnover of the senior management of the University, including the person to whom the CIO reported. These changes occurred in 1997 just after the new information services organization was implemented, and as individuals were beginning to develop new operating processes to test the effectiveness of an organizational structure that was driven by the strategic vision of the former top management of the University. Transitions at the top levels of management often alter institutional priorities and allocation of financial resources. As this top management turnover occurred sequentially and not at one time, it made it particularly difficult to secure commitments for additional resources.

SUMMARY

While this case illustrates the planning process used by a university to reorganize its library, computing, and telecommunications functions, it has applicability to information systems organizations in many other industries who are developing a new vision to provide end user support while supporting infrastructure investment. Most information systems organizations need to re-deploy human capital to allow for careers in business services and support as well as development support and technical services (Keen, 1991). Organizations must develop human resource strategies to develop new skills while providing opportunities to advance with traditional skills. IS functions who want to evolve from a technical oriented to a customer oriented approach must employ change management processes and provide training for client service functions.

REFERENCES

Agarwal, R., Krudys, G., Tanniru. (1997). Infusing learning into the information systems organization. *European Journal of Information Systems, 6,* 25-40.

Caldwell, B. (1996, October 28). We are the business. *Information Week,* p. 36.

Davenport, T. (1993). *Process innovation: Reengineering work through information technology.* MA: Harvard Business School Press.

Hirshon, A. (1998). Integrating computing and library services: An administrative planning and implementation guide for information resources. *CAUSE Professional Paper Series, No. 18.* Boulder, CO: CAUSE.

Keen, P. (1991). *Shaping the future: Business design through information technology.* MA: Harvard Business Review Press.

Schein, E. (1992). *Organizational culture and leadership.* San Francisco: Jossey-Bass.

FURTHER READING

Huber, G., & Glick, W. (1993). *Organizational change and redesign.* Oxford University Press.

Knights, D., & Murray, F. (1994). *Managers divided: Organisation politics and information technology management.* Wiley.

Stoddard, D., & Jarvenpaa, S. (1995, Summer). Business process redesign: Tactics for managing radical change. *Journal of Management Information Systems, 12*(1), 81-107.

Zuboff, S. (1988). *In the age of the smart machine.* Basic Books.

Appendix A. June 1995 organization chart

```
                        Vice President for Academic Affairs
                                        |
        ┌───────────────────────────────────────────────────┐
Assistant Vice President
Computing and Telecommunications                    Director of Libraries

    -Director Administrative Systems & Telecom        - Associate Director, Library Reference Services
        -Manager, Systems Programming
        -Assoc. Director, Computing and Networking    - Associate Director, Library Special Collections
        -Manager, Network Support
        -Manager, Computer Operations                 - Associate Director, Library Technical Service
                                                      - Dept. Head, Acquisitions & Cataloging

    -Director, Computing Services                     -Associate Director, Library Automation
        -Associate Director, User Services
        -Manager, Academic Programming                -Associate Director, Library Access Services
        -Associate Director, Computing Facilities
```

Appendix B. Schedule of events

1994	Director of Libraries announces retirement and Assistant Vice President of Computing and Telecommunications announces intention to return to faculty Search committee established
Fall 1995	New Chief Information Officer arrives Interim management group established
First quarter 1996	Group orientation meetings The planning retreat Working teams established Working team meetings Client focus groups Centralization of internal business functions
Second quarter 1996	Open meeting to discuss strategic plan draft Discussion of proposed organizations Brown bag lunch unveiling first draft of broad organizational structure Revised organization chart released High level managers identified Staff completes job placement request forms
Third quarter 1996	Implementation of new organizational structure begins Installation of Windows 95 begins
1997	Physical relocation of employees

Appendix C. July 1996 organization chart

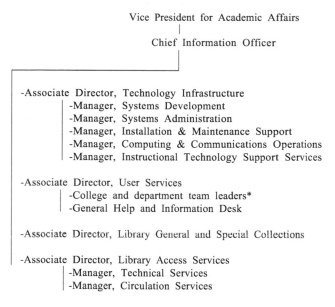

Vice President for Academic Affairs

Chief Information Officer

-Associate Director, Technology Infrastructure
    -Manager, Systems Development
    -Manager, Systems Administration
    -Manager, Installation & Maintenance Support
    -Manager, Computing & Communications Operations
    -Manager, Instructional Technology Support Services

-Associate Director, User Services
    -College and department team leaders*
    -General Help and Information Desk

-Associate Director, Library General and Special Collections

-Associate Director, Library Access Services
    -Manager, Technical Services
    -Manager, Circulation Services

Susan A. Sherer is an associate professor of business at Lehigh University where she serves as director of the Information Systems Curriculum and vice chair of the Business Department. Research interests include software failure risk management, information systems risk management, and risk in manufacturing networks. She is the author of Software Failure Risk: Measurement and Management, *as well as articles in* Journal of Systems and Software, Software Engineering Journal, IEEE Software, Information and Management, *and* Journal of Information Systems. *Sherer received a PhD in decision sciences from the Wharton School of the University of Pennsylvania.*

This case was previously published in the *Annals of Cases on Information Technology Applications and Management in Organizations*, Volume 1/1999, pp. 200-212, © 1999.

Chapter XI

Nation-Wide ICT Infrastructure Introduction and its Leverage for Overall Development

Predrag Pale, University of Zagreb, Croatia

Jasenka Gojšic, Croatian Academic and Research Network, CARNet, Croatia

EXECUTIVE SUMMARY

This paper describes 10 years of efforts in introducing the state-of-the-art information and communication technologies (ICT) and development of ICT infrastructure on the national level. The aim of the project was to build Internet in Croatia and to foster its leverage in the broad range of activities of public interest in the society as a whole. The prime target group was academic and research community, as a vehicle for the overall development in the society. Croatian Academic and Research Network (CARNet) had been started as a project in 1991, and, after five years, it was transformed into a government agency. A broad range of activities had been started, from building and maintaining private nation-wide communication and computer network to information services, user support, education, pilot projects and promotion. The academic community has been treated not only as the main customer, but also as an active partner in developing and providing services. CARNet has been fully funded by the state budget for 10 years, without any participation of the commercial sector, domestic donations or international financial support. Although CARNet is treated as Croatian success story, recognized inside and outside of the country, the question is whether the initial goals have been realistic and achievements sufficient, considering the low penetration of ICT into the Croatian society. Likewise, budget cuts, continuous struggle for political recognition and authority, as well as fights with national telecommunication monopoly, have created an array of questions to be answered at the beginning of the second decade of this highly ambitious endeavor.

BACKGROUND

The late eighties of the 20th century had found Croatia as a part of the former Yugoslavia, with relatively poorly developed national telecommunication infrastructure and absolutely no academic network infrastructure. Due to the extremely difficult economic situation, the academic and scientific community had almost no access to the international scientific publications as well as scarce resources for traveling.

CARNet initiators perceived the Internet, and computer networks in general, as the possible way around this crucial obstacle to scientific and professional activity and development.

In 1990, Croatia had declared its independence from the former Yugoslavia, which triggered military intervention of the former Yugoslavian army and eventually led to a full-blown war.

CARNet initiators had three guiding principles regarding the future of the country. Firstly, the future Croatian independence was to depend significantly upon the strength of its economy. Secondly, the modern economy was to be information-based and future industry was to heavily depend on the scope, level, intensity and quality of application of information technology. Thirdly, much as in developed and progressive countries, implementation and deployment of new technologies were to be trusted to scientific community.

Those three principles led to a natural conclusion that Croatia needed a change agent. As a step forward, the national computer network was to be built in the academic community. The community was supposed to use it for its own education and work as well as to gain experience in pilot projects in various areas of human activities, and then to use the gained knowledge, skills and experience in helping industry and society as a whole to embrace and leverage the information technology for the development and strengthening.

This conclusion had been made by a small group of very young scientists already involved in computer networking development and deployment on the small scale. They prepared a simple proposal and approached Ministry of Science and Technology (MST), basically advocating establishment of national educational and scientific computer network. The Ministry accepted the proposal, the initial group and project director had been appointed and the seed money of $1 million was allocated. The project was dubbed "Croatian Academic and Research Network - CARNet."

In the first year of the operation, basic computer infrastructure and connectivity for about 40% of the community were established and were included in the Internet. From that point on, the project grew significantly, not only in the number of institutions to be connected, but also in introducing new activities and services like education, information services, pilot projects, etc.

This required technological and organizational changes in the project, as well as repositioning the whole project within a more operational institution than the Ministry was.

SETTING THE STAGE

In order to understand the environment in which the project CARNet has been launched and developed over the course of a decade, basic information on the political situation in Croatia, its market, telecommunication market and academic community seems to be required. In addition, the development in academic and research networking in Europe needs to be kept in mind as well.

Croatia

Croatia is a Mediterranean state located in the central Europe. It covers 57,000 square km of land with 2,000 km of land borders and 6,000 km of coastline along the Adriatic Sea, its 1,185 islands being its special geographical beauty.

Croats had their own state already in 9th century. After that, they had been the constituent nation in various states from Austro-Hungarian Empire, Kingdom of Yugoslavia and, finally, Socialist Federative Republic of Yugoslavia (SFRY). SFRY had been constituted of six federal republics. By constitution, those republics had a right to decide on separating from SFRY. In 1990, in Slovenia and Croatia referendums were held and a vast majority voted for their respective independence. However, Croatia did not decide to separate from Yugoslavia immediately; it would have rather sought for more autonomy within SFRY, especially in independent self-deciding how to spend large sums of money it was to donate to less developed areas of Yugoslavia. It was only after Yugoslavia had launched into a military intervention in Croatia (September 15, 1991) that the Croatian parliament declared complete autonomy and separation from the rest of SFRY (October 8, 1991). European Union and United Nations soon recognized Croatia as a sovereign state during 1992. However, Croatia, who had to continue fighting off Yugoslavia, was pushed to defend its independence, and the war lasted until 1995.

During the war, one third of the territory was occupied and the population of barely 4.5 million had to accommodate over 700,000 displaced persons and refugees. As much as 30% of companies were destroyed either directly in the war operations or indirectly having had transportation routes or electricity supply cut off for several years.

The present-day population of Croatia is 4.3 million. About one-fourth lives in the capital, Zagreb. GNP fell from 24.4 billion USD in 1990 to 11.86 billion USD in 1993, rising to 24.9 USD in 2000 (Croatian Bureau of Statistics).

Market

Croatia is a parliamentary democracy with guaranteed private property and a market-oriented economy. However, 50 years of planned economy in Yugoslavia cannot be erased overnight, certainly not from the heads of people nor from the ways of doing business, which is offering very little readiness to integrate into global trends.

Previously, state-owned companies had been privatized, although it did not automatically bring changes in their product and services portfolios, internal organization or working practices. The Yugoslav market had been lost, and the buying power of domestic market had been tremendously weakened in the years of war.

Croatia inherited a monopoly in telecommunication market. By law and in effect, all telecommunication was in the hands of a single, state-owned company: Croatian Post and Telecommunications (HPT). Yugoslav Telecommunications were not in bad shape, though still far from developed, especially with regards to the services. During the war, Yugoslav army had been advised to destroy telecommunication infrastructure. Although it was having a monopoly, HPT had heavily invested in development of infrastructure. In early 90s, they decided to rebuild a major infrastructure with new technology: fibre optical cables. Despite the war, Croatia soon had the whole national telecommunication infrastructure rebuilt and upgraded, and it was all optical.

In 1999, HPT had been divided into two companies: Croatian Post and Croatian Telecom ("Hrvatske telekomunikacije-HT" = «Croatian Telecommunications») with government declaring intention to have them privatized. Soon after, Deutsche Telecom became the strategic partner in HT, but the state kept controlling the significant package of stocks. In 2001, the government sold more stocks to Deustche Telecom, which made them the majority stockholder. The first deal was kept secret by both sides, and the public never got to know under what conditions their most profitable company and asset was sold. The situation was further mystified by the fact that government prolonged its monopoly period for an additional year (until December 31, 2003) as a part of the second deal. Neither was it ever clear whether Deutsche Telecom had bought the "holes in the ground," i.e., the infrastructure which could easily accept additional cables potentially from other providers, once the telecomm market in Croatia got fully de-monopolized. This is of particular importance as Croatia does not seem to have any other telecommunication infrastructure at present, not even for military, police or other public sectors of special interest. As a potential market entrant, Croatian electricity and power grid company did lay some fibres along their power lines but they have been far from ever representing a network, as it is far from any possibility of commercialization.

In contrast to the public expectations and political rhetoric, Deutsche Telecom did not enlarge investments in HT. Actually, they have been minimized. All investments seem to have even been stopped from the very first day. The prices for the services have remained very high despite some initial understanding of the possible reduction.

Academic and Research Community

The Croatian academic community consists of four universities: Rijeka, Osijek, Split and Zagreb. The universities are weak and formal unions, while real power lies in individual constituent Faculties (schools). Faculties are legal bodies with their own property, status and autonomy. There are 20+ public research institutes, but the majority of research is performed on the Faculties. The community counts some 12,000 staff and 100,000 students. The cooperation within the academic and research community is very weak.

The cooperation with commercial companies does exist, but it is far from the required level. This is partly the result of generally very low level of investments in research and development within the industry and partly due to enterprises' intentions to have their own research facilities not trusting the competence of academic and research community, in general.

As a consequence, the largest financing of the academic and research community comes from the state budget. Only 3% of the budget, or approximately US$200 million, is spent on all activities in the academic and research community each year.

In higher education, the law (1996 Act) allows privately owned educational institutions at all levels. However, the majority is still owned by the government and the higher education is free for all citizens of Croatia. Students do pay for textbooks, food and lodging. Still, those expenses are partly subsidized by the state budget. The overall quality of higher education is considered to be traditionally high, and proof can be found in the fact that Croatian diplomas are readily recognized in most developed countries. The Croats who have graduated in Croatia easily get employed and prosper in the developed world. This is especially true for technical and natural sciences fields. However, there is a growing opinion that Croatian higher education needs redefining and restructuring in order to prepare students better for the information age and global economy. Unfortunately, the solution has to rely on slow formal processes, and it means mostly waiting for changes in legislation. There is no exploitation of possibilities for fast changes in spite of targeted developmental programs and projects, campaigns and experimental facilities.

Research is being financed by the budget through about 1,500 research projects with very low participation of industry. In addition, the tendering process run by Ministry of Science and Technology, typically re-launched every three to five years, does not specify practical problems to be solved but rather invites researchers to propose topics they find attractive to deal with.

There is almost a decade old program in place, aimed at rejuvenating the academic population by financing 1,000+ young scientists, in the period of up to 10 years, to achieve their master and doctoral degrees and attend the postdoctoral studies. Their salaries have been financed directly by the state budget through the Ministry of Science and Technology. They have been chosen among 10% of best graduates in their respective field and assigned to existing research projects, on the request of the project leaders.

Government Initiative

Croatian government consists of about 20 ministries that are to administer and divide the budget of approximately $6 billion. The budget is allocated predominantly to activities not projects. Although ministries are supposed to propose their budgets based on projects, once funds have been allocated there is virtually no project follow-up, and financing is based on activities rather than on results. In fact, in projects that last for several years, financing is provided for the next year, even if no results had been achieved in the previous years, largely due to the lack of monitoring.

Although officially CARNet project is a government initiative, it was actually an initiative of five engineers who were supported by a university professor, recently appointed deputy minister of science, at the beginning of the war. The project quickly (in the course of one year) generated large and visible results, and from that point on, it had been progressing using the so-called "avalanche" effect. Thus, despite overall lack of vision and guidance on the part of the government, and ministries of education, culture, and public health total lack of interest, the project was growing and remained present maintaining influence over a decade.

Academic and Research Networks

The idea of academic and research network was not invented in Croatia. It was spontaneously launched in late eighties in the developed countries of Europe and coordinated in the collective effort under the overall Framework of Research Programs of European Union.

EU recognized the importance of information technology and the need not to lag behind the development in the USA. Therefore, significant funds had been provided with the aim to develop new technologies and to build national but also pan-European academic and research computer network. In the later stage, funds had been provided to eligible countries from the Central and Eastern Europe (CEE) for connectivity of their national networks to pan-European infrastructure.

Although specific and different, most of national academic and research networks (ARNet) were similar in their basic goals and operations. A typical ARNet was set to establish international connectivity and national backbone. Connectivity of individual institutions to the backbones was to be left to ambition and finances of individual institution. A minimum of services would then be provided just to support the basic activity, help desk, basic communication and information services, targeted research as well as information packages. Extensive information services, databases, large scale educational activities, pilot projects and promotion were not considered to be part of their tasks and duties. This view remained throughout the major part of 90s. As a contrast, the CARNet initiative, from the beginning, had a broad range of services in vision. Main differences between the typical academic and research network (ARNet) development scheme from a country in transition and Croatian are summarized in Bartolini (2000).

In addition, a typical ARNet would be strictly focused on academic community leaving the rest of population to commercial developments. Since, academic community was only the first phase of CARNet vision and, in many instances, a tool for overall national development, CARNet believed there is much more to just providing communication infrastructure.

CASE DESCRIPTION

As initially envisioned CARNet role was twofold: to provide infrastructure, services and support to the academic and research community (ARC) as well as to act as a change agent for the society as a whole. In order to fulfill the first role, CARNet initiated full range of activities of an ARNET (Bekic, 2000). The milestones are listed in Appendix 1. The second role could be fulfilled in two ways: by increasing the ICT and managerial competence of individuals (not only in academic community) and by CARNet's active involvement in implementation of ICT in national projects and systems.

Network Infrastructure

The technical concept of CARNet had been created at the end of 1991 (Pale, 1992). It was the time of the European Community's COSINE and IXI projects. TCP/IP protocol was considered too American and too old to be used in future European network infrastructure. X.25 & X.400 were foundations of European efforts (CISCO, 2002).

Table 1. Comparison of CARNet and an academic network in a country transition

Typical ARNet of the country in transition	CARNet
Majority of efforts put in international connectivity	Majority of efforts put in high speed domestic backbone
International connectivity financed by EC and Open Society Institute	International connectivity financed by the state budget
A&RN maintains only backbone	Maintenance of the whole network
Support limited to connectivity problems	Support covers host administration, LAN designing, courses, promotion
From the beginning limited to academic community	Free access to everybody, pilot projects with the whole community
Most of the work is done by the A&RN itself	Outsourcing and project cooperation

However, CARNet designers recognized that X.25 and X.400 products were still scarce and thus expensive. X.25 required specific interface in computers. Users needed more services from their computers and networks. It was recognized that TCP/IP was old, meaning reliable, that it was available for virtually every computer platform, that it would operate via RS-232 interface available in just any computer. It provided all services users needed e-mail, file transfer, network file systems, remote terminals and many others. It fully erased the difference between local and global networks from the user's and computer's point of view. The fact that it was completely license-free only sealed the first decision. Despite European trends and recommendations, CARNet was going to be an Internet network.

National public network infrastructures in Europe were built on PSDNs (Public Switched Digital Networks) mostly using X.25 infrastructure. Although such a network was available in Croatia (CROAPAK), it was scarce and expensive, and it only offered user speeds of up to 4,800 bps. CARNet was aiming at higher speeds: 9,600 bps. at least and 19.200 bps. preferably.

Thus, the second decision of the first phase was to build CARNet as a private network based on leased (copper) lines.

Phase One (1991-1994)

CARNet communication nodes were established in all university cities acting as local centers of a hierarchical, star-shaped network topology. Due to the lack of funds, speed of deployment and public telecommunication network being the frequent target of the wartime operations, planned redundant lines were never established. Thus, the established network topology was hierarchical with one major node (in Zagreb) and three regional nodes (in Rijeka, Osijek and Split). Despite that, the network operated with surprising reliability and availability. Within the first year of the project, the national backbone had been established and connected with the rest of Internet. In two years 60% and in three years 100% of academic and research institutions had been connected in CARNet.

Phase Two (1994-1996)

At that time (1994), the connecting speeds of 19,200 bps became insufficient, and, in some instances, bottlenecks were showing up. In addition, a single major node of the network (at Zagreb University Computing Centre - SRCE) was a continuing source of concern due to its vulnerability. As a single point of failure, it could bring the whole network down. It was clear that new backbone was required.

Most of European Academic and Research (but also other) networks had already switched to Internet technology and used 64 kbps and 2 Mbps leased digital lines, mostly via TDM (Time Division Multiplex) technology (Behringer, 1995)

It was difficult to obtain such resources in Croatia, and they were largely expensive. There was also one other concern: CARNet designers were fully aware that upgrade of the backbone would take between 18 months and two years. They did not want their new backbone to become obsolete again, before or at the time it becomes operational. It was already clear that multimedia was going to be in demand, that voices and moving pictures were going to take up the major part of network traffic in the middle of 1996 when new backbone was going to be operational. Thus, the network backbone for the future, not for the current needs, needed to be established. Such a network had to be capable of transferring both, packet data and isochronous signals, like video and audio.

Fortunately, there was some good news. Firstly, new technology, called ATM (Asynchronous Transfer Mode) was emerging, aimed at unifying transfer of packet data (e-mail, file transfer, etc.) and of isochronous signals (audio, video, etc.) in one communication infrastructure. Secondly, Croatian telecommunication monopoly HT had been rebuilding public communication infrastructure using fibre-optical cables. They had plenty of raw bandwidth («dark fibre») and virtually no customers.

As a consequence, two strategic deals were made. The first one with HT, allowed CARNet, in the future, to use "dark fibre" (fibre optical cable between two CARNet nodes without any HT equipment in between) for a small and fixed charge. The second agreement, with CISCO, delivered CARNet the first available ATM equipment at a very favorable price which included education, replacements with next generation of equipment and other important benefits. In this way. CARNet, in the second phase, built a new broad band backbone at the speed of 155 Mbps with ATM technology that enabled audio and video conferencing throughout the country. The cost was 60% of the price that needed to be paid for the technology used on the backbones by other academic networks, offering speed of only 2 Mbps and no ability to do video conferencing.

In addition, the core of the backbone has been fully redesigned (Appendix 2). Instead of a single node, the core of CARNet backbone is now an "unfinished" cube (Appendix 3). Major academic and research institutions act as nodes, interconnected at 622 Mbps. Other, regional parts of national backbone are connected each to another node of the cube, at 155 Mbps. The core of the backbone is now fully redundant and reliable at the utmost.

Phase Three (1998-Present)

In the third phase, the connectivity of individual institutions (Appendix 4) to the backbone needed to be significantly upgraded. However, the dark side of telecommunication monopoly started to get prevalence. HT did not want to sell cheap copper-leased lines any more and allow CARNet to install xDSL modems thus effectively boosting up connectivity to 2Mbps or more. They forced CARNet to buy expensive 2 Mbps digital connections, but even with the contract signed, they did not deliver service. CARNet ended in an unacceptable situation: state of the art high-speed multimedia backbone, and obsolete connectivity of many members at speeds of 19,200 bps, sometimes even slower ones. Besides that, despite the signed contract, HT did not want to connect new institutions nor new locations of already connected institutions. The main reason was that HT perceived CARNet as a competition. Unable to attract other customers, HT wanted academic and research institutions as their customers at prices they could freely set.

Services

Initially, at the time when appropriate communication infrastructure on national scale was not available, the focus was on establishment of the backbone and on international lines. However, as the backbone deployment was well under way it became clear that it was not the only task, more needed to be done.

Connectivity Services

Deep in the foundations of the project was the goal to have as many institutions and individuals using the network as soon as possible.

Other national ARNETs, especially in developed countries, concentrated on establishment of backbones. They relied on the institution's motivation and financial resources to buy a connection from a telecommunications operator to the backbone. However, in Croatia, the situation was significantly different. Because of the weak and war-torn economy, academic and research institutions were almost exclusively financed from the budget. This financing was insufficient even for the basic operations. Other major unresolved problems in the academic community were outdated equipment, brain drain (towards the commercial sector and other countries), and the physical infrastructure (buildings) damaged. Besides, the Internet was a fairly unknown term in 1991, even in the academic community.

Therefore, CARNet couldn't count on their motivation, much less on their money. CARNet had to reach out much further than most other ARNETs, so its connectivity service included permanent communication line from an institution to the backbone. If an institution had multiple locations, multiple lines were needed. Services included purchasing and installing equipment for the institution's central node were communication (modem, router) and computing (UNIX server) equipment offering mail and Web services to all students and staff of the institution. If the institution had no system administration capabilities, CARNet could offer such services (limited to the basic functionality of the central node), as well.

Individuals, who were members of the academic community, had rights and possibilities to access the Internet through modem pools distributed throughout the country paying only the minimal local communication cost.

In this way, CARNet provided "connectivity to the door" both to institutions and individuals, and it was free of charge for them, as end users. This initial infrastructure deployment was performed much in a centralized, planned fashion, almost without end-user involvement. The Internet and its services were simply given to them, regardless of whether they asked for it or not. However, this model proved to be very successful in bypassing the traditionally slow reaction and conservative approach of university managements. Pioneers and early adopters in community were embracing the "gift" and quickly spreading the "gospel." It was a kind of bottom-up approach with (extensive) external help.

Communication, Information and Data Services

As expected, a number of pioneers were found in the academic community who were to discover new communication services early on and to implement them for the sake of research, curiosity or prestige. However, CARNet quickly learned that services born in such spontaneous way would have the form, quality and lifespan according to the interests of individuals who started them, not according to the needs of those who would use them. Industry was fully unaware of the Internet and its (commercial) potential. The first commercial ISP started operation in 1996. Thus, the fundamental CARNet goal, to make the latest communication technologies and services available to every member of the community in the sustained way with guaranteed level of quality of service, cannot be fulfilled if it means letting "someone else" establish and run the service, i.e., hoping that someone would do it and do it in the proper way. Much more deliberation, planning and larger resources are required.

Therefore, a strategic decision had been made early on — that CARNet has a duty to take care that such services do exist and that they are available to everyone. CARNet should also do its best to make the services free of charge to end-users (discussed in the "Finance" section) and to guarantee a level of quality of services.

It was felt that this approach will be chosen in other aspects of academic networking as well. As much as creators, financers and executors of this policy were convinced of the nobility of the goal itself and sure of the method to pursue it, they also believed that it would be wrong to try to provide those services from within the CARNet organization. Instead of increasing the size of the organization (potentially endlessly) and competing with (imperfect but innovative) services provided spontaneously by innovators and pioneers, it was decided to build on cooperation.

As a consequence, CARNet had encouraged individuals and institutions in the academic community to explore new services and to propose them for support by CARNet. CARNet was, then, jointly with them and potential users, to define the service, provide necessary equipment and money for sustained provision of the service. CARNet was also to promote the service and monitor its quality. In this way, the provider of the service was to get substantial supply of money for additional education of staff, salaries and other needs. They were also to get promotion of their work and thus visibility in domestic and international community. The equipment could be used for other academic purposes as well. Perhaps the most important benefit academic entities would get was the experience in providing a service and cooperation under "commercial" contract — something they did not have and very much needed in order to gain survivability in market economy.

If a partner for desired service could not be found in academic community, it would be sought in the commercial environment. It was believed that in such a way CARNet would serve as an active agent in modernizing Croatian economy, supporting development of the new services. CARNet's activities were to be focused on precise definition of service, tendering, financing, promotion and quality control.

In this way, CARNet had always established new communication services like news, list server, IRC, video conferencing, etc. The aim was to provide the academic community with new services as soon as they were introduced somewhere in the world and to provide at least one service in the country that would be impartial, non-commercial and public, at least to the ARC.

Communication services enabled users to communicate among themselves, with international community and to create virtual communities thus enhancing their work and increasing its efficiency. Similarly to communication services, a range of information services had been established like directory services, public ftp, Web hosting, search services, PGP, media on demand, etc. They improved not only group but also individual work.

Communication and information services, amazing though they may be, left many potential users with the question "and what now?" unresolved in their minds. Majority of users, especially from non-technical areas were actually seeking data. Relevant (scientific) data had to be provided to start the process of acceptance of new technology and recognition of its benefits which would hopefully lead to later overall and universal leverage of the Internet and related technologies in all aspects.

Communication services in their essence offered communication means not content, or at least it was not provided by CARNet. Information services did contain data, but they were provided mostly by users. In addition, CARNet launched a range of data services, and they were all about third-party data. They contained scientific databases (Current Content, MEDLINE, Inspec; ...), referral services or portals (www.hr).

Providing a data service requires a technological base (computer servers and high speed connections), data itself and data maintenance. These components are usually provided by different parties. The content was either purchased by the Ministry of Science and Technology (MST) or produced by the service contractor. CARNet role was to organize all the parties in a homogeneous service, to promote the service and to offer user support.

User Support

People usually expect other people to be like them: to share values, attitudes, believes and to behave in the same way. CARNet was launched by pioneers, and they expected everybody else to behave like one: to grab the opportunity, to use new technology and to figure out how it works and how it can be used in most part by himself (Moore, 1999).

Soon, it was discovered that pioneers constitute a very small fraction of academic community and that a whole new track of activities needs to be established in order to attract and involve at least a major part of, if not the entire, academic community (Bates, 1999). Information in the form of brochures, manuals, interactive CDs, Web materials explaining benefits and usage of ICT were provided for different users groups. Merely providing infrastructure and services did not guarantee its successful implementation and users due to continuous upgrading and changing of tools and technologies needed training and support. Training of end users was found to be crucial for adoption of technologies and development of skills, and it was organized by CARNet. It was accompanied by a general purpose helpdesk.

Institutions needed on-site technical support and maintenance. The technology was new and academic salaries were low, so it was very difficult to get eligible technical staff: system engineers. Therefore, CARNet developed dedicated educational track for system engineers and organized suitable separate helpdesk support. To assist system engineers in their work and relieve them of some common activities shared with others, CARNet contracted development and maintenance of standardized set of operating systems, server software and other tools intended to be used in every CARNet node. System engineers were supplied with regular updates of these packages.

Cisco Networking Academy, as the first program for broad professional audience, was introduced, in order to raise the number of skilled technical staff and lower the entry barriers for broader implementation of ICT.

Again, the actual delivery of individual courses was left to educational professionals, but CARNet staff was identifying users needs, defining course outlines, recruiting trainers or contracting institutions and providing funds.

Information technology was new in society and was not part of the curriculum in the formal education. Even when it was, it was in rudimental form and very theoretical. In the first four years, more than 35,000 people were educated in more than 50 different courses and the demand for the participation in courses grew. It became apparent that CARNet couldn't provide sufficient training for the community of 100,000 students and 12,000 staff with educational activities in their present form, especially considering new students enrolling in the universities each year (30,000 students in 2001).

In 1998, it was decided to continue with courses as before but introduce the limit on the number of participants from each institution. A parallel activity, training the trainers system was established, enabling institutions to train their employees and students as future trainers to run courses for all other employees and students in their institution. In such a way, if an institution wanted to have larger number of staff and students trained, they had to make some effort: to find future trainers, to motivate them, to devote some space for a special classroom. CARNet and Ministry of science and technology helped them to get appropriate equipment.

In addition to infrastructure, services and education, users needed a variety of software tools and in order to efficiently use them, they also needed continuous assistance. Typical and widely used tools (statistics, modeling, math libraries, etc.) were to be identified. Institutions were sought which had substantial expertise in leverage of the tools. Contracts were, then, signed with them, making them referral centers to provide support to users community. In such a way, user community would serve itself in an organized manner. Referral center would organize education for the specific software tool, provide helpdesk, organize workshops, seminars and conferences and also negotiate with vendors for community-wide licenses. CARNet would be monitoring the quality of service, promoting the service and providing funds.

In implementation of new information technologies in a variety of applications and in a complex infrastructure like the CARNet, there arose a huge number of problems that needed to be resolved or agreed upon. Usually, this is a task for special interest groups. CARNet invited users to conceive such groups and offered assistance in the form of meeting space and support, travel expenses for representatives to respective international meetings, publishing and promotion of results. However, up to the present moment the response was close to none. Except for a low, activity loose and informal association of system engineers, there had been no user interest group formed.

Visibility of user results, exchange of experiences, checking on new ideas and an easy way to get introduced in "whys and hows" were the goals to be achieved by the CARNet User Conference, the annual event initiated in 1999. The attendance had been continuously growing and the satisfaction of attendees had been very high. Despite the low support from taxation laws, conference managed to attract significant sponsorships. They enabled the organizers to award best papers, presenters and presentations with prizes like PCs, palm computers and travels to international conferences. The conference started to act as a hub for other related events that were to take part immediately before or after the conference or would run in parallel.

In the world of the Internet, it seems to be very easy to convey any information to a large number of people. So, one would not expect much trouble for CARNet to announce new services, products and opportunities to the members of academic community. However, there is a catch: how to tell someone about e-mail, by e-mail if they do not use e-mail yet? There is a problem in the reverse communication as well: if a number of people from an institution are suggesting or demanding one type of service and the other group is advocating something exactly opposite, what should CARNet do?

In order to assure appropriate dissemination of information to end users in institutions a network of CARNet coordinators has been established. Every institution appoints an employee as a CARNet coordinator whose primary role is to act as a liaison officer. News, plans and other information sent from CARNet are relayed to employees and students. Likewise, problems, suggestions, needs and events in an institution are consolidated within the institution and communicated back to CARNet. All coordinators together constitute "Users Council" who has an advisory role to CARNet management influencing annual programs and strategic plans.

Since 1997, CARNet member institutions have been reporting annually about the usage of resources available through CARNet services (Appendix 5). In 2000, the total of 164 institutions submitted reports. Comparison across years indicates speed of ICT penetration and the level of its utilization and is differentiated across specific user groups. It also addresses importance of individual CARNet services and activities to end-users. The annual reports about the usage of CARNet infrastructure and services and institution's needs have been filed by CARNet coordinators and endorsed by top management of individual institutions. Thus, the reports are considered as official feedback from users.

Change Agent

The far-reaching goal of the CARNet activities was to make impact on the national level, outside academic community. Academic community on institutional and on the individual level was to be the partner, CARNet acting as a coordinator and organizer.

CARNet aim was to collect knowledge and experience in the field of information technology and implement them into academic network. The use of global information infrastructure such as the Internet and the Internet-based information services was the prime interest.

It was expected that after graduation students would leave the academic community, trained to use information technologies, and be willing to build or require that type of infrastructure at their working places. It was also expected that academic and research community would gain experience in implementation of ICT in their respective professional areas and thus be capable to act as consultants or contractors in implementation outside of academic community in areas like education, judicial or health care system, public administration or culture. However, in order to gain the competence, academics needed hands-on experience on real-life problems.

Therefore, in order to support pioneers in implementation of ICT in different areas, to solve a particular problem using ICT, or to make a first step in a big project, CARNet ran a range of pilot projects (www.CARNet.hr/projects) in a broad area as a complementary segment of its activities. Their goals were to prove and measure the benefits of ICT implementation, to discover the limits and estimate the costs, while building the knowledge, experience and thus competence of academic community. Projects were performed by groups and institutions from academic and research community on the contractual basis. CARNet was tendering assignments or accepting proposals, financing, monitoring and promoting projects and results.

Every occasion was used to promote usage of information technology. Upon request from organizers, CARNet participated in various events, including non-academic, providing infrastructure like Internet access, videoconferencing or streaming, technologies that were not commercially accessible. Presenting overview of the technology and trends, gaining experiences in different projects or new services, CARNet was promoting ICT to different professional and user groups.

As an independent, non-commercial agency, CARNet had initiated and maintained several national services important for whole Croatian Internet community. With this act, CARNet ensured specific infrastructure service, common for the Internet service providers sector, enhancing their operation while lowering the costs. The most important services are Domain Name Service (DNS), Croatian Internet Exchange (CIX) and Computer Emergency Response Team (CERT).

Defining the policy of Croatian Internet top-level domain ".hr," administration of domain names (DNS) under it, coordinating operation, promotion and legislation, is the role for a non-profit, impartial body, and thus CARNet assumed it. To motivate the industry to create national information space, domains are assigned to users free of charge.

Exchange of the traffic among Croatian ISP's through Croatian Internet Exchange (CIX), lowers the burden for networks outside Croatia and decreases the communication costs.

CARNet CERT (Computer Emergency Response Team) ensures cooperation among ISPs, users, legal bodies and international community, on the topics of education, prevention and response on security problems in the network on the national scale.

More active methods were possible and expected as CARNet tasks in the further development of public information systems: strategy development, project design, project management, executors supervision, etc.

Organizational Development

During the 10 years of CARNet, its organizational form (Mintzberg, 1993) was continuously changing (Appendix 6). However, the metamorphosis can be grouped in four phases.

First Phase (1991-1995)

CARNet started as a project. In the first phase of stable, ongoing operation, CARNet was fully run by University of Zagreb Computing Centre (SRCE), financed and coordinated by the Ministry of Science and Technology.

Young enthusiasts worked in SRCE, dedicated to provide good service to academic community, performing all communication, computer, user-support and information services. State-of-the-art technology and noble mission, in the time of war and overall depreciation, made them eager and curious to show they could make a difference. That was a time for learning and cooperation, without strong organization, planning and sustained financing.

CARNet project was initiated by a young (age 31) engineer who was immediately appointed the project director. Two years letter, in 1993, he was invited and appointed Deputy minister of science in charge of information technology. This gave the project the next boost in importance and financing. He continued to be in charge of the project.

Second Phase (1995-1998)

SRCE was the computing center of one of four Croatian universities. Although SRCE was implementing CARNet in all universities political conditions did not allow to change its constitution and broaden its mission or "jurisdiction."

Only one-third of all SRCE employees were engaged in CARNet operation. The company as a whole was not supportive to further challenges, such as international cooperation, public relationships and customer management. The project was growing and spreading. Management in terms of project management and human resource management was becoming the predominant part of CARNet activities and SRCE management was neither competent nor ready for these non-technical types of activities. Marketing and promotion were on the edge of blasphemy in the low-salaries, engineers dominated culture. Therefore, managing tasks were spontaneously organized and performed by the Ministry of science's newly formed department for information technology. Young engineers also populated this department but they started understanding that the key of future success of the project lies in professional management and completely new working culture. It was clear that a major change needed to take place.

In 1995, CARNet agency was put in operation. It was fully owned and financed by the government but largely independent from other state authorities. The idea was to create an organization in charge of organizing the potentially huge and endless numbers of projects. The actual work of running communication infrastructure, services, education and other activities was to be outsourced to academic community or to the market (Kowack, 1995).

A young engineer was appointed for CEO. He was a good engineer but without any managing experience. It was believed that he will learn flying on his own. He was very systematic and professional. His inclinations laid in strong hierarchy and a rigorous financial control, which were established.

CARNet operations were run by CARNet Executive Committee (CEC) formed from CEO and four deputies, leading four departments: infrastructure, services, R&D and special projects. CARNet was still very much relying on enthusiasm and learning by doing. It was impossible to find such people outside of community, so three deputies were appointed from the "inside": SRCE and Ministry, joining CARNet on the contractual basis. It was clear that this dual role (and sometimes conflict of interest) would pose the problem but it was also hoped that it would generate some benefits. Besides, it was expected to last only a few months, a year at most until "real" deputies were found.

The Deputy Minister, founder and godfather of CARNet, was the chairman of the CARNet Board. He was always present at the CEC meetings, and all strategic, especially technological decisions, were strongly influenced by him. He was deeply involved, and was sheltering young agency from the outside problems.

All those dual roles produced continuous tensions between development and operation (CARNet and SRCE), future and present priorities (Ministry's and CARNet's), "doing right things" and "doing things right" cultures.

Third Phase (1998-2000)

CARNet didn't have it's own building. It was dispersed in four locations, instead. Management was caught by surprise when four different subcultures evolved within one organization. In 1997, it was becoming obvious that CARNet organized in four divisions, with four different cultures, priorities and strategies, and only about 30 people altogether could not deliver what had been required from it. Besides, full time management staff was required, among other things, to employ fully all the employees and partners potentials. The first CEO, after three years of building an organization almost from scratch, got tired from management and decided to leave.

For the expected development of organization and nation-wide role, to fight scarce resources and passive environment, for further pushing of ICT into the public sector, the position was offered to chief of operations in the Ministry. She was already responsible for special projects, corporate culture and human resource management in the period of strong divisions. Engineer by education, manager by her aspirations and leader by the nature, she was the choice.

She moved from Ministry to CARNet and was appointed CEO soon to be followed by the deputy for services who was transferred from SRCE to CARNet to become vice CEO. The new, flat organization was chosen (Appendix 6) believing it could be better accommodating for project type of work and intensive knowledge sharing it required. Almost everybody was given responsibility for his or her segment of job. All of a sudden there were no (big) bosses anymore and everyone was the (small) boss.

Employees belonging to more structural departments, like infrastructure and research and development, had a rough time. Their safe haven was gone, and teamwork and responsibility has put a lot of burden on them.

At the same time, project organization was emerging, providing the platform for the multidisciplinary teamwork and more structural involvement of non-employees (partners, part-timers). Strategy was to outsource and contract all the operation, even ("hard core") research and development. That meant, "real work" and success were given to other organizations, and only "dirty administrative tasks," like running the projects and writing project documentations and contracts, remained in CARNet.

Many good engineers left CARNet in this period (Appendix 8). Only those who liked fast drive and high risk stayed in CARNet. That was time of learning new skills, large investment in non-technical education of employees, transformation from all-engineers organization in multidisciplinary organization. It was also the time of building alliances and fighting for the sufficient budget to preserve the pioneering position among national networks.

Fourth Phase (2001-Present)

Numerous, complex, different activities and lack of strict formal structure imposed project work as the natural way of doing things. Those who resisted "administrative project management work" left. Support functions were developing. More people began to work for CARNet as contractors, employees of partner organizations or part-timers.

In the flat organization, a small management group became a bottleneck. Many parallel projects required distributed responsibilities, but still intensive cooperation.

It was time to introduce new rules of the game — adhocracy (Appendix 6). Adhocracy (Waterman, 1992) with its inefficient mutual adjacent coordinating mechanism, with lots of liaison functions, with no operational core, had become the new stage of CARNet evolution.

Majority of senior employees became managers, mentors and coordinators. Even more, responsibility was distributed among employees, especially those who were in positions of project managers.

Most of the services were formally outsourced, in the first place to SRCE, the oldest and still the most important partner. Skills in negotiation, contracting and coordinating became natural requirements for CARNet employees.

Acquisition of the new skills, like knowledge management, alliance building and fund-raising had progressed. Unfortunately, emphasis on the non-technical competences caused animosity and lack of purpose and focus by engineers. A strong chief technical officer became necessary, to add technical view and priorities to the management team, as well as to link technical teams and their goals with overall goals.

Financing

CARNet was a governmental agency, financed directly by the state budget. The benefit it offered was a planned and secure "income" allowing efficient planning of activities. CARNet budget was to be also stable over the course of years, which would make long term planning possible.

However, experience had shown that execution of the budget rarely reached the planned sums and that it varied throughout the year. In addition, every new government administration needed to be familiarized with CARNet role and needs. This took time and usually caused disruptions in financing (Appendix 9).

Budget expenditure limits also limited salaries while not recognizing particularities of ICT professionals and high market demand for them. Expenditures were also limited in case of education, travelling and other items characteristic for young and new types of organizations.

Despite the problems, CARNet managed to increase, although not continuously, its budget over the years. However, a major part of the budget went for telecommunication services. Legislation was still supporting the monopoly of the national telecom operator, which allowed it to keep much higher prices than those in deregulated markets. CARNet was not successful in making its activities a state priority, which would force the monopoly holder to treat CARNet differently from other consumers.

From the day one, all CARNet services were completely free of charge for end users, both institutions and individuals. Other national networks often charged academic institutions for access to the Internet. CARNet considered there would be no benefit from it since, almost all academic and research institutions were government owned and financed from the budget. If they had to pay for services, they would put pressure for the increase of their budget. Thus, the money would come from the same source but encumbered with more administration, control and accounting expenses while giving up benefits of economy of scale.

Croatian economy had barely survived the war losses but was on the way to recovery. Companies and universities alike were still trying to figure out what the market economy was all about. Many decision makers still lived in the past, in the concepts on the benefits of the planned economy. Tax regulation did not favor donations and sponsorships for educational or scientific purposes.

In addition, advanced services were not always the priority for expenditure decision makers in many academic institutions. Thus, the plan was made to finance communication infrastructure and services centrally from the budget and then to transfer the responsibility of financing on to users and their institutions at the moment when the critical demand had been created in users community (Jennings, 1995). This would mean that CARNet would have to start charging for those services.

There were some suggestions, mostly from the Ministry of finance who is in charge of state budget, that CARNet should increase its budget by selling its services on the market to non-academic users as well. There were examples of similar behavior in some Central European countries. However, CARNet was opposing and resisting these suggestions for several reasons. Firstly, CARNet had much broader range of activities than any other national networking organization. Most of those services were oriented strictly to academic users, which chronically lacked funds to pay for services. Secondly, based on its non-profit and academic status, CARNet was eligible for significant discounts in purchase of hardware, software and services, both on domestic and international markets. If CARNet was to buy them at commercial prices, it would never manage to sell advanced services at an economical price. Thirdly, CARNet role was to be a pioneer and to cater to pioneers. This is almost always not profitable.

While most of Central and Eastern European countries enjoyed benefits of EU funds for development of academic networking, due to the war and political relationships, Croatia did not appear to be eligible for them. Thus, in the whole period, CARNet did not receive any significant international donation or support of any kind. In addition, international connectivity, sponsored by EU for other countries, had to be acquired, from the national monopoly holder, at extremely high prices despite big education discounts.

CURRENT CHALLENGE/PROBLEMS FACING THE ORGANIZATION

Network Infrastructure Consolidation

CARNet's 1995 projection was that, after deregulation of telecommunication market (in 2003), major part of its connectivity services will become commodity products and, due to the economy of scale, would be available at a price on the market. This, however, meant that CARNet would have to cease to operate its communication infrastructure as a private network, and be buying the service. The only exception would be much smaller experimental test bed used for piloting new technologies and services not yet available on the market.

However, this did not and will not happen in the next several years (to 2004 and beyond). There are several reasons for that, all mutually intertwined and interdependent:

- Usage of ICT did not grow as expected and penetration to various activities was still very low, making the market still weak;
- Government has not initiated informatization of public services, which would stimulate consumption and development of public networking infrastructure; and
- Monopoly has been prolonged (to 2004) allowing high prices.

Therefore, CARNet has to continue operating and enhancing its private network. However, this is becoming increasingly difficult due to the fact that:

- After privatization of the national telecomm operator (HT) the monopoly period has been prolonged. Private owner (Deutsche Telekom) has increased prices and shows no interest for special arrangements with CARNet. Instead, CARNet is treated as its competitor;
- HT is technically and organizationally not able to provide advanced services;
- HT does not want to sell low level services like copper lines or dark fibre (unbundling the local loop) which would allow CARNet to install its own advanced equipment; and
- HT is forcing CARNet to use its medium level services like 2 Mpbs digital lines, but even with the contract signed, HT does not deliver them at all, or does so with unacceptable delays of more than six months.

Continuation of this situation is making CARNet network obsolete fast. A solution must be found and implemented quickly. Currently, CARNet has been seeking its own way out in two directions: technological and legislative.

Technologically, CARNet has been piloting wireless LAN/MAN sub-networks (2.4 GHz spread-spectrum de-regulated solutions). On the legislative side, CARNet is exploring cooperation with other would-be providers after complete de-monopolization of telecommunications that is expected to occur in the coming years, as well as possibility to partner with owners of eligible infrastructure even before de-monopolization. For example, the town sewage company owns drainage system connecting every building in the town. Thus, a pilot project is running exploring technical, organizational and legal aspects of using sewage system to deploy CARNet's own fibre infrastructure within the city. Further, the national power company has already laid fibres in the cables of the power lines connecting cities. Partnering with them would enable CARNet to have alternative supplier of connections on the national level.

These potential alternative telecommunication providers appear to be showing some interest, though it does not seem to be strong enough probably because they do not yet know where to start. In addition, all of those companies are still state owned and are waiting for the privatization decisions from the government.

On the international level, in these ten years, CARNet has been a user of European networking infrastructure, being connected to the node in Vienna. It has always been CARNet's vision to become the connecting network between the neighboring countries. However, it has not been possible due to the lack of traffic interest among the western neighbors as well as due to the lack of political will to approach the eastern neighbors.

Fortunately, new EU project GEANT has decided to establish a POP (Point Of Presence) in Croatia, thus connecting CARNet with the Austrian and Hungarian Networks. This gives hopes and represents a foundation for possible establishing of connection between CARNet and other networks in the eastern countries.

The Level of ICT Usage in ARC

Deploying a national networking infrastructure and establishing a wide range of services was a huge enterprise with no previous example in the country. CARNet was concentrating on fulfilling those tasks and believed that all users will eagerly embrace and use them as soon as they were available.

It did not seem to be the case, so CARNet had decided to shift the emphasis to facilitating and stimulating the usage and implementation of ICT in the academic community's life and work. In a number of surveys, CARNet was asking users about their ideas of innovative usage of ICT in what they do. The response was more than weak. It seems they may lack knowledge and experience to answer the questions regarding their primary needs or problems that ICT can fulfil or solve, not yet being able to consider how to plan to use ICT.

However, CARNet cannot fulfill this assignment alone. Students are the key alliance, because their requirements towards universities will create the demand and need for ICT. University administrations and Ministry can and should influence the change by launching projects and imposing various standards and requirements on level and quality of education and research.

Change Agent

The soft, passive role by promoting, influencing market, students and graduates and educating project leaders have been assumed by CARNet. This role should be intensified by increasing the number of employees, organization partnerships and omnipresent promotion.

However, the active form basically has not been used. Government did not launch into "informatization" of public systems like health, education, government administration or judicial system (European Commission, 2000). Those who initiate similar projects on institutional level seem not to understand the importance of project preparation and management and/or to recognize CARNet as eligible partner and resource of knowledge and experience.

As an example, primary and secondary school system is not only very similar to academic community but also naturally connected. CARNet experiences, infrastructure and services could be easily used, multiplied, cloned for the educational community. So far, there have been no requirements towards CARNet from the authorities despite CARNet showing willingness to take part and sending active messages regarding it.

The issues are:

- How to raise awareness of authorities for the needs of huge ICT systems and CARNet's possible role in their establishment;
- When awareness becomes present and demand for CARNet participation significantly outgrows CARNet's current capacities, should a new agency be formed, commercial spin-offs stimulated or CARNet repositioned and reshaped; and

- All activities performed by CARNet so far are only a small fraction of the change agent's activities. In Croatia there is no other example of a change agent and even on the global scene there are few with the nation-wide role. Thus, the question is: where to gain required knowledge in order to become a true and successful change agent on the national scale.

Human Resources

From the very first day of CARNet, people were the primary resource. CARNet always looked forward to "people flow" through the organization since it helps influencing society and transferring knowledge and organizational culture. However, to run efficient services and projects and to create, develop, maintain its own culture and transfer it to the novices, an organization needs a core of "old" professionals. Market demand for such people is tremendous and their price is rocketing. CARNet has not even begun to fill all the job vacancies, especially in top management positions.

CARNet, being financed by the budget, has serious limitations not only in the area of salaries but also in a number of other expenditures like education, travel, equipment, office comfort, etc.

The issue, at present, seems to be how to attract and retain key personnel. So far, motivation was based on challenging projects, learning and education, warm-hearted atmosphere. However, the key personnel that grew with CARNet and is getting older, forming families, thinking about the career seems to be shifting emphasis on the importance of financial compensations.

This need can be met in two ways:

- CARNet should ensure some kind of additional income that could be used for increasing salaries and other expenses, or
- Key personnel should have reduced working hours and be allowed to earn additional income working on projects both in CARNet and outside.

Financing

There is a range of reasons to commercialize some of CARNet services:

- Getting all required finances from the budget is becoming increasingly difficult with growing suggestions to commercialize some of the services;
- There appear to be emerging potential customers outside academic community interested in some of the services like education, connectivity, consultancy, project management, etc.;
- There are views that if academic community were to pay for some of the services, they would value and use them more; and
- Some of human resources requirements could be fulfilled with additional, non-budget income.

The negative sides of commercialization cover:

- The fear that budget sums might be decreased even more because of false expectations that everything could be commercialized, which is not true for the backbone and international connectivity;
- Discounts for educational and non-profit organizations currently used might no longer be applicable;
- Tensions among employees working on profitable projects and those on budget would be developing; and
- The basic role of academic and research community is centered on the area which is rarely profitable which transposes to corresponding ARNET activities.

There is evidence and examples on international scene that there exists interest in charitable and non-profit financing of activities similar to CARNet's. This might prove to be a significant source of income and replacement for expected budget cuts. The issue is whether to establish a fund raising department or to form a separate trust or foundation that would primarily finance CARNet's activities.

REFERENCES

Bartolincic, N., Pezelj, I., Velimirovic, I., & Zigman, A. (2000, October 10-14). The implementation of broadband network technologies in CARNet. *Proceedings of SoftCOM 2000*, Split - Rijeka, Croatia, Trieste - Venice, Italy (pp. 937-946).

Bates, T. (1999). *Managing technological change: Strategies for college and university leaders*. San Francisco: Jossey-Bass.

Behringer, M. (1995, May 15-18). Technical options for a European high-speed backbone. *Proceedings of the JENC 6*, Tel Aviv, Israel (pp. 513-522).

Beki, Z., Gojši, J., & Pale, P. (2000, October 10-14). The role and strategy of an ARNET in a developing country. *Proceedings of SoftCOM 2000,* Split - Rijeka, Croatia, Trieste - Venice, Italy (pp. 955-960).

CISCO. (2002). *Internetworking technology handbook*. Retrieved from http://www.cisco.com/univercd/cc/td/doc/cisintwk/ito_doc/index.htm

Croatian Bureau of Statistics. (n.d.). Retrieved from http://www.dzs.hr

European Commission. (2000). *IST 2000: Realising an information society for all*. Luxembourg: Office for Official Publications of the European Communities.

Jennings, D. (1995, May 15-18). An Internet development model for Europe. *Proceedings of the JENC 6*, Tel Aviv, Israel (pp. 522-523).

Kirstein, P. T. (1995, May 15-18). A user's view of the EU research networking programmes. *Proceedings of the JENC 6*, Tel Aviv, Israel (pp. 531-532).

Kowack, G. (1995, May 15-18). Concepts for provision of Internet services to academic and research community. The EUnet perspective. *Proceedings of the JENC 6*, Tel Aviv, Israel (pp. 523-524).

Mintzberg, H. (1993). *Structure in fives; Designing effective organizations*. New Jersey: Prentice-Hall.

Moore, G. (1999). *Crossing the chasm*. New York: Harper Collins.

Pale, P., Bulat, D., Maric, I., Simicic, L., & Vujnovic, V. (1992, September 15-18). Concept and development of CARNet. *Proceedings of the 14th International Conference on Information Technology Interefaces*, Pula, Croatia (pp. 265-272).

Seršic, D., Pale, P., Vucic, R., Zigman, A., Bartolincic, N., & Maric, I. (1996). A nationwide 155 Mbps ATM academic network reach to the desktop. *Mipro 96*, Opatija, Croatia.

Tesija, I., Vucic, R., & Zigman, A. (1997). Applications of broadband digital networks. *DORS'97*, Zagreb, Croatia. Retrieved from http://cn.carnet.hr/arhiva/1997/970417/pred1/index.html

Teteny, I. (1997, May 13-16). Paving the highway by PHARE. *Proceedings of the JENC7*, Budapest, Hungary (pp. 242-243).

Waterman, R. (1992). *Adhocracy: The power to change.* New York: W. W. Norton & Company.

FURTHER READING

CARNet. (2000). CARNet submission for Croatian strategy for 21st century. Available at http:/www.CARNet.hr/strategy

CARNet Information Service. Available at http://www.CARNet.hr

CARNet Network. Available at http://www.CARNet.hr/network

CEENet. (1997). CEENet Tartu Declaration. Tartu, Estonia. Available at http://www.ceenet.org/ceenet_tartu.htm

GEANT Network. Available at http://www.geant.net

Senge, P. (2001). *The fifth discipline: The art and practice of the learning organization.* Zagreb: Mozaik knjiga Internet Sites.

TERENA Compendium. Available at http://www.terena.org/compendium

UCAID. (2002). Available at http://www.internet2.edu/ucaid/

APPENDIX

Appendix 1: Milestones (www.FER.hr/Predrag.Pale/publications/ACIT/Appendix1)

Appendix 2: CARNet Network Backbone (www.FER.hr/Predrag.Pale/publications/ACIT/Appendix2)

Appendix 3: CARNet Network Backbone in City of Zagreb (www.FER.hr/Predrag.Pale/publications/ACIT/Appendix3)

Appendix 4: CARNet Network Structure (www.FER.hr/Predrag.Pale/publications/ACIT/Appendix4)

Appendix 5: Users Feedback (www.FER.hr/Predrag.Pale/publications/ACIT/Appendix5)

Appendix 6: Organizational Structure (www.FER.hr/Predrag.Pale/publications/ACIT/Appendix6)

Appendix 7: Employees (www.FER.hr/Predrag.Pale/publications/ACIT/Appendix7)

Appendix 8: Finance (www.FER.hr/Predrag.Pale/publications/ACIT/Appendix8)

Predrag Pale obtained his BSc and MSc degrees in electrical engineering from the University of Zagreb, Faculty of Electrical Engineering, where he is lecturing today. He has been designing computer hardware, operating systems, applications and computer networks, but most of his twenty years of professional experience lie in the area of ICT applications: from medicine to civil engineering, commerce, financing, libraries, news media, education, to government administration. He started Internet in Croatia in 1991 with CARNet project, Scientific Information System in 1994 and National Information System for Libraries in 1998. From 1993 to 2000 he had been appointed a deputy minister of science and technology. His primary interests are in the application of ICT in education, medicine and other public systems. He is frequent speaker at international events about the future of the cyberworld and on issues of privacy and security of information systems.

Jasenka Gojšic obtained her BSc in telecommunications and has had 10 years of experience in networking. She worked in Croatian Telecom in information services department for two years. She joined CARNet project in Ministry of Science and technology in the early phase, in 1993. Since then, she has been the project leader, deputy chief executive officer, and for the last four years chief executive officer of CARNet. Her prime interests are organization and human resource management. She has obtained MBA degree in those fields. Ms. Gojsic strongly believes in ICT being a vehicle to the knowledge society.

This case was previously published in the *Annals of Cases on Information Technology*, Volume 5/ 2003, pp. 585-607, © 2003.

Chapter XII

The Telecommuting Life:
Managing Issues of Work, Home and Technology

Gigi G. Kelly, College of William and Mary, USA

Karen Locke, College of William and Mary, USA

EXECUTIVE SUMMARY

As the 21st century approaches, the work environment is transforming, driven in large part by technology. For example, technology is challenging ideas about where and when work needs to take place. Technology allows employees to work from home. However, this new distributed work arrangement brings with it many new challenges for both the employee and employer. In this case, we introduce one company that has decided to experiment with the telecommuting arrangement. Through the eyes of one teleworker, many of the benefits and challenges of telecommuting are explored.

INTRODUCTION

One doesn't have to look vary hard at the business world to figure out that something significant seems to be happening with the current workplace. Look at the language we are using to describe our situation (emphases added): "A tidal wave of change is sweeping across the American workplace (Ehrlich, 1994, p. 491); "The last decade, perhaps more than any other time since the advent of mass production, has witnessed a profound redefinition of the way we work (*Business Week*, 1994, p. 76); and, the very notion of a job itself, is being questioned (Ancona, Kochan, Scully, Van Maanen, & Westney, 1996a). Whether one views the terms in which change is described as dramatic hyperbole or a reasonable representation of what is happening, it does seem that change, dramatic or evolutionary, is indeed taking place. Furthermore, the impetus to change is being felt across a wide variety of organizations, from Fortune 500 companies to government bureaucracies and even to the military.

One innovation in how work is accomplished is telecommuting, and it is increasingly being adopted by organizations. Telecommuting constitutes a fundamental change in where, how, and even when work is accomplished. After this brief introduction to telecommuting, we will present the case of Glenn Smith, an account of one individual's and organization's experience with creating this new work arrangement. We hope that the issues raised will offer insight into the implications of this new work arrangement both to individuals and organizations that might be thinking about the telecommuting option.

Telecommuting or teleworking, as defined by U.S. General Service Administration (1995), "refers to a means of performing work away from the principal office-typically at home or at a nearby telecenter." Telecommuting increases separation from the principal office while simultaneously increasing connection to the home. Telecommuting is not new; it has existed for several decades. In the late 1970s and early 1980s, businesses projected that the physical location of the worker would shift from the central business building to the home. Although there were organizations that pioneered this movement, it was not until the 1990s that telecommuting became a viable, acceptable, and, occasionally, a preferred work option.

A survey conducted by Telecommute America reported 11 million U.S. teleworkers in early 1997, an increase from 4 million in 1990 (Murphy, 1998). Furthermore, conservative estimates for the number of teleworkers in the year 2000 range from 11 to 15 million (Piskurich, 1997) with other estimates indicating growth rates of 18% a year resulting in 20 million U.S. telecommuters by the year 2000 (Nilles, 1997). The GartnerGroup predicts even more involvement of U.S. workers in teleworking situation — by the year 2000, 30 million with more that 137 million involved in some form of teleworking by the year 2003 (Langhoff, 1998). These estimates suggest a growing recognition that telecommuting is a viable work option that organizations must acknowledge. Indeed, as the case will present, the forces driving telecommuting clearly express the interests not only of individuals seeking increased flexibility to reconfigure home and work but also of organizations focusing on economics, environmental regulatory pressures, and technological advancements. With this alignment of organizational and individual interests, the increasing growth in the population of teleworkers is not surprising. Nevertheless, it remains a "new work arrangement," very much in its initial adoption phase.

Let's now turn to Glenn's experience to help us better understand what might happen when an individual completes his work at home, electronically connected to but geographically distant from his principal office. This case portrays the real-life experiences and reflections of one teleworker. It is worth noting that this case tells a telecommuting story from one demographic perspective — a manager-level teleworker who belongs to a dual career family with children. Although some issues raised in this case are unique to this particular situation, many of the issues are faced by the mass of teleworkers.

BACKGROUND:
AMERICAN BANK CORPORATION

Kelly Watson is the vice-president for information systems (IS) at American Bank Corporation (ABC) in Chicago. ABC is a large financial institution that has typically been characterized by its conservative culture. However, over the past decade mergers and acquisitions in the financial industry have required ABC to quickly change from an ultra-conservative, "change very slowly" mentality towards new and more progressive ways of doing business. Fortunately, the president and CEO, who are both in their mid-60s, realized the need for change and have been instrumental in setting the precedence for making change happen. One recent change in the IS division was the hiring of Kelly Watson three years ago as Chief Information Officer (CIO).

In this capacity, Kelly is in charge of the migration of systems from the mainframe legacy environment to a new client server environment, and she faces a number of challenges. First, the IS division is under considerable pressure from its customers to get new systems up and running. Second, even though the IS professional resources in her division have grown tremendously (see Figure 1 for Kelly's staffing structure), she has employee retention problems. The market for IS professionals is so competitive that Kelly faces the constant problem of having to replace staff who have been lured away by competitors, consulting companies, and vendors.

SETTING THE STAGE:
FINDING A SOLUTION TO A PROBLEM

Kelly's most recent expression of her retention problems is the imminent loss of one of her most senior development supervisors, Glenn. Glenn's wife, Vicki, had received a very attractive job offer in Cincinnati, where Vicki had grown up. Though Glenn really enjoyed his work at ABC and has developed satisfying work relationships with his manager as well as with Kelly, he feels that it wouldn't be fair to Vicki to deny her this opportunity. Given the number of people who have recently been hired away from the IS division and the fact that he receives calls from headhunters all the time, Glenn knows he'd have little problem finding a comparable position.

Kelly, anxious not to lose yet another experienced development professional (especially someone with Glenn's track record with clients) was prompted to try something new. Kelly had recently come across a number of interesting articles that

Figure 1. Organization chart

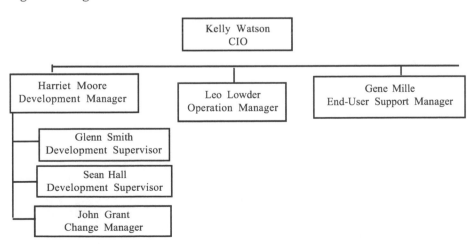

highlighted the benefits of creating a virtual office. Apparently, companies were increasingly allowing knowledge workers to telecommute. In the financial industry, specifically, companies such as Bank of America, Citibank, and HomeFed bank were all offering some form of telework to their IS personnel, and they were reporting benefits in employee productivity and in reduced operating costs. She called Glenn into her office and broached the subject with him, "Would you be interested in working out a telecommuting arrangement with ABC?" Glenn was at first taken aback, but the idea did have some appeal. Organized and hardworking, he felt that he would have little problem working in a less structured format. Additionally, in his role as a development supervisor, most of his work revolved around projects. Glenn believed that during some phases of the project life cycle he would require close contact with the users; however, his current project was a long way from that point. And, right now, at least, his physical presence was not really a necessity. Certainly, he wouldn't miss the two hours he spent commuting through city traffic every day. On the other hand, Glenn was somewhat concerned with the implications of his telecommuting with his staff, and he knew this would have to be addressed. Kelly asked Glenn to think it over, to speak with Vicki about it, and they would talk again soon.

In the meantime, Kelly met with Harriet Moore, her development manager. It was Harriet who had first informed Kelly that they might be losing Glenn and had urged her to "do everything possible" to keep him. So, while Harriet was uncertain as to how telecommuting would work out — she had never worked under an arrangement like this before - she was anxious to give it a try. She did suggest to Kelly that it would be important to discuss this with the other IS managers. Kelly agreed and convened a meeting with her three managers: Harriet from development, Leo Lowder from operations and Gene Mille from end-user support. She also asked a representative from human resources (HR), Bob Loskins, to attend.

CASE DESCRIPTION: THE DECISION TO TRY TELECOMMUTING

Kelly began the meeting by emphasizing the retention challenges the IS division faced. She informed everyone that they were confronting the imminent loss of yet another experienced professional, this time a development supervisor. She told them that she was giving very serious consideration to allowing the supervisor to telecommute in order to keep him at ABC.

Leo immediately asked Harriet if she knew about this and if she supported the idea. Harriet responded, "Yes, Glenn is really respected by his clients, and I can't afford to lose him right now ... and, who knows, if this works out, it may give us an advantage in attracting and keeping the best people."

"That's right," interrupted Kelly. "I wouldn't be proposing this if I didn't see it as a strong initiative for not only the division but also whole company. The opportunities with distributed work are really considerable," insisted Kelly, passing around a sheet that outlined the benefits of teleworking (see Table 1).

"But, you're only offering it to Glenn, right now?" Gene retorted, looking at Kelly. "Yes," Kelly replied, "We've never tried anything like this before so I want to move slowly on it...I'm sure there will be some bugs we'll work out as we go along."

Table 1. Benefits of teleworking

BENEFITS TO ORGANIZATION	BENEFITS TO INDIVIDUAL
* Improve office productivity	* More retained income (reduce personal expense)
* Increase use of new technology	* Increase job satisfaction
* Enhance recruitment and retention of personnel	* Productive and balanced lifestyle
* Increase competitive advantage	* Reduce stress and illness from commuting
* Reduce central office costs	* Integrate work and personal life
* Optimize personnel performance	* Improved family functioning
* Minimize cost and disruption of relocation	* Effective time management
* Improve environmental reputation	* Increase personal safety
* Strengthen Disaster Planning	* Control and design of workplace

Source: Hodson, 1997; U.S. General Service Administration,1995; ITD Telecommuting Task Force, 1997

"Right!" Leo jumped in, "How are you going to maintain control over his work quality? What happens when he has to meet a deadline and he's got a sick child at home with him who's lying around on the sofa? For that matter, what is he going to do at home, set up a separate office or is he going to be working from his kitchen table?"

"And," Gene chimed in, "I know that Glenn's really reliable, but what are you going to do when everyone starts wanting to try telecommuting?"

"Hold on!" Kelly interrupted, "These are all really important points and they illustrate what I was just saying, ... we'll have a lot of issues to work out. But, for right now, can we agree in principle that this is something we want to try and that we'll begin offering it to Glenn?" Gene nodded.

"It's definitely something that is in our interest as a division to explore," agreed Leo looking at the list of benefits Kelly had passed around, "Heck, I might try it myself!"

"What do you think, Bob?" asked Kelly, turning to the HR manager, "You've been pretty quiet during all this?"

"Well, I'm all for trying something new, especially if it's going to put a dent in our hiring and retention problems," replied Bob. "But, our discussion here has made it clear that there are going to be a lot of issues we'll have to deal with. We'll need a policy on telecommuting — I'll start looking into what other companies are doing with it."

"Good, then we have agreement," said Kelly, "We'll move on offering it to Glenn as an experiment. In what kind of time frame do we want to revisit this?"

"How about three months?" Bob offered.

Kelly looked around the room; the others all nodded. "O.K." said Kelly, "Now I only hope that Glenn wants to move ahead with this."

Glenn Tries Life as a Teleworker

Glenn let Kelly know that he would like to give telecommuting a try. The juggling of their respective professional careers was a recurring topic of conversation in the

Copyright © 2006, Idea Group Inc. Copying or distributing in print or electronic forms without written permission of Idea Group Inc. is prohibited.

household. Glenn and Vicki agreed that they should do this if for no other reason than this was a working arrangement that they would definitely have to explore at some point in their careers. So why not now?

Initial Set-Up: The First Couple of Weeks

Setting up their new home in Cincinnati, Vicki was thrilled with the prospect of Glenn working from home. Now he would be home when the kids got off the bus and to help out with all the "taxi" and other family duties that had always been her responsibility to manage. Their lives would change in other ways too. Glenn needed an office. Since the living room was rarely used, this space was designated his office. With his own money, Glenn purchased used office furniture. Its cost was considerable, $1,500.00, but since it was to go in the living room, Glenn and Vicki had decided that it was worth buying quality pieces. Also, Glenn wanted a good chair, not a spare from the kitchen.

The phone lines were a little more of a challenge. After spending $120.00 for the initial set-up and an additional $30.00 a month (for voice mail, call waiting, caller ID, and call forwarding), Glenn was ready to connect to ABC. They already maintained two telephone lines for their house (one for the computer and one for voice), but a month into the project, Glenn's work conversations were so monopolizing their voice line that Vicki insisted that they get a third line designated for the family. They carried the cost of this additional phone line. After complaining about the costs that he was carrying, ABC did come through and provided Glenn a laptop computer (56k modem), and they provided Glenn a calling card for the voice long distance. ABC already had an "800" number to support the long distance connection for data. Glenn had asked for an ISDN line, but ABC was not set-up to receive ISDN communications. Actually, ABC only currently supports data connections at the speed of 28.8k. And, Leo Lowder in Operations was not ready to make any additional purchases to support teleworkering at this time.

The hassles of getting set up continued. Connecting to ABC's system from his home was more difficult than he originally thought it would be. His first week of telecommuting was pretty unproductive as the kinks were worked out of connecting his computer to ABC's network. Glenn spent a lot of time trying to get the technical support he needed to get set up. But, the technicians at ABC were swamped with service requests at the home office; and when they did find the time to call Glenn back, they had difficulty understanding and responding to his particular needs. They just hadn't had any experience with helping someone set up a distributed office. Then, Leo Lowder raised some concerns about security because Glenn was dialing directly into ABC's network. While Leo was uncomfortable with the current arrangement, he allowed it to continue for the duration of the experiment. Further, the connection problems persisted. Sometimes the local telephone company was to blame; at other times it was ABC's corporate telecommunication network.

Managing his time productively would be an issue, and this required some structure. Thinking that it would have been helpful to have had some specific guidelines or training courses for structuring his day and for maintaining effective communications links with the corporate office, Glenn, nevertheless, proceeded on his own to make changes he thought would be helpful to his situation (certainly, ABC offered no such programs). Glenn decided to set up a daily agenda of tasks to complete. Because he was a "night owl," he found himself completing tasks in the late evening hours when the house was quiet and he was not disturbed by the phone. Glenn had always been a hard worker and this continued dedication to work did not change with his new work venue. Glenn also set up daily lists of the people he needed to contact. He would spend a considerable amount of time on the telephone as well as maintaining contact through e-mail.

Back at ABC, the initial reception of Glenn's telecommuting highlighted just how much of a departure from traditional work practices this new arrangement was. On the phone, Glenn received a lot of friendly ribbing about how it must be great being able to get up when he wants and do what he wants. Ads running on the TV of teleworkers at home in their pajamas only added to the barrage of stereotypical comments (e.g., "Your golf game must be improving!" or "What's on ESPN now!"). Some of these comments about the informality of working at home were correct. Glenn did enjoy the relaxed dress code and was often on the golf driving range during lunch; however, these offhanded comments concerned Glenn. The monetary costs associated with establishing himself as a teleworker were clear, but the friendly ribbing made him wonder what other price he might be paying?

The Next Two Months: A New Set of Issues

The project Glenn was working on had up to this point been conducive to letting him work at home. During this time frame, he had returned to ABC twice for four days. ABC picked up the costs for travel and lodging. These interim visits to corporate office were going to be necessary at least for the short term and maybe long term. In addition, the project would soon be starting implementation and Glenn realized that he would have to be on-site more; although, he wasn't sure how much more. He could have done with some help to develop a corporate on-site schedule. Needing to be on-site was not all bad. Sometimes, Glenn felt the isolation and wanted to simply go out to lunch with his colleagues. He was also concerned that he was out of the day to day loop. Though, Glenn was fortunate that Harriett, his boss, acted as an advocate for him in his absence.

The clients Glenn supported also raised some issues. Over the last month or so, there had been times when end-users wanted to have Glenn on-site. To maintain his communications with end-users, Glenn audio-conferenced in on meetings, however, no video was being used. And, while audio conferencing provided a connection, Glenn did not always feel as an equal participant. One observation Glenn made was that if all participants were meeting face-to-face in a room with Glenn present on the speakerphone, he did not feel a real part of the exchanges taking place, regardless of how frequently he "spoke up." By contrast, if everyone "met" on a conference call from their separate offices, the meeting dynamics were more equal and satisfactory.

A more daunting issue for Glenn was learning how to manage from a distance. Glenn's prior management style had been "by walking around." He would stop by his staffs' cubicles to chat and keep abreast of their work. Now, this was impossible to do from his home. Some of Glenn's employees had no problem adjusting to Glenn's physical absence; however, one employee was having difficulty completing his work. Glenn knew that a more formal management routine was going to be necessary; so he set-up weekly audio staff meetings. What else could he do to better monitor his employees' performance? What technology might further facilitate his managerial responsibilities? Glenn was very interested in exploring the available groupware technologies to support virtual teams, but ABC had initiated little use of this type of software (except for e-mail). Furthermore, his ongoing management and project responsibilities kept him so busy (he was hiring additional staff) that he simply didn't have the time to investigate the technology. Glenn also felt he needed a fax machine, but at this time no money had been budgeted for such purchases. Frustrated at not having all the means of keeping in contact that he needed, Glenn reached into his own pocket to buy a fax machine.

On the home front, many of the benefits Glenn believed he would have by working from home did come to fruition. He did have more flexibility and was able to do things with the family that he simply did not have the opportunity to do under his previous working arrangement. Occasionally, he had been able to meet with the children at lunch; he had cared for the children when they were ill; and he was able to be more active in the neighborhood.

The adjustment to working from home, however, was not without problems. A number of issues arose with his family. Glenn started receiving an increasing number of work related calls in the evening when the family was eating dinner or when they were relaxing together. Time and time again, Glenn would have to leave family activities in order to deal with a work-related problem. Glenn and Vicki decided that he would not answer his work-designated phone after 6:00 p.m. That worked for a while until his manager and clients figured it out. If Glenn didn't pick up on his work line, they simply dialed the home number. Sometimes, it felt like he was "on call" 24 hours/7 days a week.

Additionally, late afternoons when the kids came home from school and Vicki returned from work proved difficult. When the kids got off the bus in the afternoon, they were thrilled that Dad was home and were anxious to talk about their day. Glenn loved being available to hear all the school stories, but, occasionally, the children would rush into his office excitedly shouting, "Dad! Dad! Guess what ..." only to interrupt a telephone call or an urgent task. This disruption was annoying and would lead to conflict as the kids were told to, "Leave the room; Daddy's working!" On rare occasions when Glenn was dealing with a particularly frustrating problem, family members would bear the brunt of his frustration.

Tensions also arose between Glenn and Vicki in the allocation of responsibilities for managing the household. At those times when Glenn had to return to Chicago to interface directly with the home office, Vicki had to single parent, and this put considerable pressure on her. On the other hand, Glenn's apparent availability at home had also lead to some conflict. On one occasion, Vicki arranged to have some landscaping done. That morning was one of those difficult times when Glenn was having problems getting connected just when he was under a time crunch. The door bell rang five times in the space of two hours as truckers wanted to know where they should dump the mulch, the gardener wanted to know if Glenn had any oil for the tractor because it was low, and everybody wanted to get paid. Glenn called Vicki at work furious that she had presumed on his time in this way. At moments such as this, Glenn wondered whether the new work arrangement really was working out satisfactorily.

Back at American Bank Corporation: The Experiment Receives Mixed Reviews

Results Inside the IS Division

As the three months experimental period drew to a close, the reviews on the telecommuting experiment were mixed. It certainly had taken Glenn a little longer to get set up then either he or Harriet anticipated, and the downtime associated with that put pressure on everyone to catch up the project. Still, once the technology-related bugs were worked out, Harriet and Glenn had both been happy with the latter's overall productivity and work quality. Also, Glenn's current clients were beginning to adjust to Glenn's physical absence and realized that quality products were still being delivered. Some clients did believe that Glenn was going to have to be at corporate more for the upcoming implementation and they were asking Harriet for her assurances that he was going to be physically available.

Unfortunately, the experiment also created some tensions. Several other development professionals approached Harriet. And, when they were told that it was only being offered to Glenn as an experiment for the time, they wanted to know why Glenn was being singled out for this special privilege. Harriet also received a few complaints from some of the division's other internal clients who wondered why they couldn't find Glenn despite the fact that they had dropped by on several different occasions to ask him questions. Additionally, the concern around "special privilege" also cropped up in the other IS functions and highlighted some stresses in the division. For example, operations staffers who as a group were prone to feel that they always had to cover for mistakes that the developers made, complained to Leo that it was unfair that it was someone from development who got "to stay home and relax on the sofa."

Predictably, perhaps, new technology issues were also raised. If this was truly the way the company was going to move, then a major commitment of additional resources was going to be required: new hardware, software, and telecommunication equipment was going to be necessary. With this, it was clear that the IS division would have to figure out a way to provide technical support for specific teleworker problems that would be difficult to solve because they would not be able to go to the site. Leo, especially, had serious concerns about whether this level of investment in telecommuting was warranted at this time.

Results in the Rest of ABC

In HR, Bob was also receiving a lot of inquiries from ABC employees, not only in IS but also in the rest of the company who had heard about Glenn's arrangement and who were also interested in telecommuting. Bob knew that the question of whether telecommuting was going to be a company wide option would have to be answered fairly soon. Additionally, if the telecommuting option was opened up to other employees, then, this would have to be addressed at the company level. His research into what other companies were doing made it clear to Bob that ABC was getting into something much larger than anyone had imagined when they first discussed telework three months ago. For example, exactly how would people be compensated and reviewed? What technology would be supplied by the company? How would worker's compensation work if someone got hurt at home? What new training was going to be necessary and for which groups in the company? Would ABC develop these training courses or would they look for outside consultants who specialized in establishing a telecommuting environment? Bob had gathered some initial statistics that other companies had published regarding their experiences (see Table 2) to share with Kelly. But, he knew he would need to determine not only the tangible but also the intangible costs and benefits associated with telecommuting that ABC would incur if it moved ahead.

Table 2. Employer's costs and benefits per teleworker

Employer Costs per Teleworker		Employer Benefits per Teleworker	
Computer costs and associated hardware	$3,200	Cubicle Savings	$5,000
New phone lines (ISDN more expensive)	$100-$500	Absenteeism Decreased	10%
Home office furniture (desk, chair, lighting, supplies)	$1,000	Productivity Increases	15%-25%
Selection and Training	$175	Office space savings	1000 sq. ft.
Moving costs such as materials	$240	Turnover Decreased	5% of salary (costs for searching/ training)
Annual Costs (e.g. phone lines)	$1,000	Park Expenses Decreased Environmental savings due to less vehicle commuting Increased Competitiveness	

Source: JALA International, Inc., 1998; Fitzer, 1997

The Next Step

It was now three months since the IS division had made the decision to experiment with the telecommuting option. As she prepared for her meeting with her IS managers and Bob Loskins, Kelly Watson wondered what the next step was going to be. Given the reactions to the experiment, should they go ahead and make telecommuting more widely available? If they chose this option, what had to happen to ensure its success? Should they put it on hold and delay the decision? Or, should they abandon it as something that was not feasible in American Bank Corporation?

DISCUSSION: CURRENT CHALLENGES OF INTEGRATING WORK, HOME, AND TECHNOLOGY

The preceding case points to a number of issues that highlight the altered bounding of work and home created by a telecommuting arrangement and the influence of new technology. Obviously, Glenn is no longer physically present at work. Glenn's work arrives at the office electronically but he isn't there, thus creating some dynamics that require active attention. This separation or unbundling of the worker from principal work location raises issues not only with regard to immediate work relationships, namely Glenn's boss, co-workers, and subordinates, but also more peripheral relationships, those organizational employees who are acquainted with Glenn or simply with Glenn's new work arrangement. Additionally, Glenn's physical absence from corporate offices also has more long-term implications for his development opportunities and advancement. At the same time, the creation of a workspace in the home bundles together home and family life with work life. Boundaries that facilitated the creation of a distinct way of living that was non-work-obliged are no longer present, and their absence raises concerns for work, family, and leisure. All these issues will require attendance and active maintenance on the part of the teleworker. Finally, the technology that allows telecommuting to be viable surfaces new issues that both the organization and Glenn will have to address.

Getting Work Done From a Distance

All parties immediately feel the physical distance between and the separation of the teleworker from the organization involved. Absence from the office creates a void, and the managers, co-workers, and others left behind who are unfamiliar with telecommuting have no experiences from which they can draw to fill it. Consequently, it is easy for the ambiguity associated with what the teleworker might be doing away from the office to be resolved with images of activities that are non-work associated. Perceptions that Glenn is involved in non-work related activities (e.g., watching TV, or playing golf) are understandable, if inaccurate. Second, Glenn's physical separation from those he supervises requires revisiting how he enacts his role as a manager. The interpersonal, informational, and decisional functions that comprise the traditional managerial role typically include a sizable face-to-face component (Mintzberg, 1973). A new way of enacting these roles will be necessary. Third, the potential for deterioration in all work relationships must be acknowledged and proactively managed. And, there is, indeed, evidence to suggest that telecommuting does have a destabilizing effect on work relationships that is expressed approximately six months into the new arrangement. The good news is that a recovery and mutual adjustment seems to be possible after about a year (Reinsch, 1997).

These three sets of issues clearly point to the need for a "learning and adjustment" period for those organizational members touched by the new telecommuting work arrangement to discover and establish new ways of working. It is clear that training programs need to be offered for managers and teleworkers that address issues such as communications and time management. Additionally, as the case indicated, Glenn had a strong working relationship with his manager, Harriet, and his ability to continue to provide quality contributions to ABC under the new work arrangement was never in doubt. Nevertheless, the issue of ensuring a teleworker's actual and "visible" productivity is central to this new work arrangement.

As indicated earlier, learning how to supervise others' efforts in a distributed mode can be particularly difficult and would seem to require a careful and explicit consideration of both the supervisors' and their subordinates' preferred work styles. This has been a challenge for Glenn and for his employees. Prior to telecommuting, Glenn ensured the productivity and quality of his department's work through an informal management-by-walking-around style. In order to stay on top of his department's contributions and particularly the efforts of those employees who performed better with closer monitoring, Glenn had to find ways to formalize his supervisory activities.

The above discussions emphasize the importance of developing new ways to ensure the contributions of the teleworker and underscore the importance of paying close attention to the network of relationships in which any of us organizational members are embedded. Teleworkers have to make projecting their virtual presence and developing substitute communication channels to stay connected with their relational network a priority (Bjorner, 1997).

Issues of Development and Advancement

Before concluding our discussion of the impact of separation from the principal work site, it is important to raise the question, what happens to a teleworker's advancement opportunities and career path? Certainly the lack of visibility that we have already discussed may cause some problems if not actively managed. As one AT&T teleworker rather crassly put it, separation from the workplace means a loss of "suck up" opportunities (Hequet, 1996). It is important for the teleworker to take responsibility for charting his development path under this new work arrangement, for example by actively seeking and initiating new challenging work projects. However, because this is uncharted territory for most organizations, we simply do not know the answers to this. Will the teleworker have to return to the home office to maintain upward mobility or must the teleworker trade advancement for location flexibility?

Dealing with Confounded Boundaries Between Work and Home

Telecommuting muddies the boundaries between work and home, and the telecommuter can cycle continuously among personal, family and work related tasks. This arrangement provides a potential flexibility not only in where, but also, to a degree, when the work takes place. While some aspects of the work, most notably contact and coordination with office workers and clients, need to take place during corporate business hours, the independent tasks that make telecommuting a viable option for a given employee can be completed at any time. The 24 hours/7 day week, thus, potentially becomes the frame within which work can be accomplished.

This continuous time frame brings both positive and negative attributes for the teleworker. On the positive side is the flexibility to complete work during times when the teleworker is most productive. Additionally, this potential flexibility of where and when to work opens up opportunities for the teleworker to improve their relationship with family members by permitting a more active role in family life (Hill, 1995; Olsen, 1987). For example, being available to pick a child up from school or to coach a sports team whose practice would typically begin 45 minutes before the parent could get home from work. Thus telecommuting allows the teleworker to work at some tasks early or late in the day, saving times in between for family.

On the negative side, imposing appropriate time boundaries around work and separating and conserving the quality of home life, including relationships with the family are critical. Indeed, there is a real possibility that unless specific steps are taken to define specific work boundaries, for example, by establishing a daily work agenda, that one's work life may overwhelm the home, temporally (there is precious little time that work does not impinge on) and emotionally (family members falling prey to work related frustrations). As is true with regard to the workplace, a learning and adjustment period are also needed with regard to the home.

Technical Issues: Connecting the Teleworker

The technology makes it possible to work anywhere any time providing the flexibility increasingly prized by so many of us on how, when, and where we get our work done. In order for Glenn to telecommute, the technical environment had to be set-up. This requires a negotiation between the teleworker and the organization to fully understand what equipment is necessary. Having the proper technology is essential for successful telecommuting (Artz, 1996). At a minimum, the teleworker must have a telephone, with many teleworkers also possessing a computer and a modem. The Internet has provided a low cost means to connect the teleworker to the office. E-mail has become a primary tool for communication within organizations thus facilitating an additional means to keep the teleworker connected. One teleworker made the following comment, "You learn to live by e-mail. That was my water cooler" (Murphy, 1998). Other equipment that may be required include printer, fax machine, copier, pager, answering machine, mobile phone, and additional phone lines. Exactly who is responsible for these costs needs to be addressed. Also, the slow access to the organization's LAN (local area network) was a constant source of frustration for Glenn. It is important to recognize that the phone line has become an umbilical cord that provides the connection of the teleworker to the organization. When there are technical problems (and there will be), this can render the teleworker lifeless. Solving technical problems remotely are often more challenging and procedures need to be established to address these issues.

REFERENCES

Ancona, D., Kochan, T., Scully, M., Van Maanen, J., & Westney, D. E. (1996). Workforce management: Employment relationships in changing organizations. In *Managing for the future*. Cincinnati, OH: South-Western College Publishing.

Artz, T. (1996). *Technologies for enabling telecommuting*. Retrieved April 1, 1998, from http://www.cba.uga.edu/management/rwatson/tc96/papers/artz/artz/toc.html

Bjorner, S. (1997). Changing places: Building new workplace infrastructures. *Online, 21*(6), 8-9.

Business Week. (1994, October 17). The new world of work, 76.

Ehrlich, C. J. (1994). Creating an employer-employee relationship for the future. *Human Resource Management, 33*(3), 491-501.

Fitzer, M. M. (1997). Managing from afar: Performance and rewards in a telecommuting environment. *Compensation & Benefits Review, 29*(1), 65.

Hequet, M. (1996, August). Virtually working: Dispatches from the home front. *Training*, 29-35.

Hill, E. J. (1995). The perceived influence of mobile telework on aspects of work life and family life: An exploratory study. *Dissertation Abstracts International, 56*(10), 4161A. (University Microfilms No. DA9603489)

Hodson, N. (1997). *Teleworking ... A different way of doing things*. Retrieved April 1, 1998, from http://www.teleworker.com/seminar/index.html

ITD Telecommuting Task Force Report. (1997). Retrieved April 1, 1998, from http://www.itd.umich.edu/telecommuting/report/

JALA International, Inc. (1998). Home-based telecommuting cost-benefit analysis. Retrieved January 7, 1998, from http://www.jala.com/homecba.htm

Langhoff, J. (1998). *FAQs about telecommuting...* . Retrieved September 26, 1998, from http://www.hanghoff.com/faqs.html

Mintzberg, H. (1973). *The nature of managerial work.* New York: Harper & Row.

Murphy, K. (1998). Web fosters telecommuting boom, and many in the industry take part. *Internet World.* Retrieved February 9, 1998, from http://www.internet.com

Nilles, J. M. (1997). *Making telecommuting happen: A guide for telemanagers and telecommuters.* John Wiley & Sons.

Olsen, M. H. (1987). Telework: Practical experience and future prospects. In R. E. Kraut (Ed.), *Technology and the transformation of white collar work* (pp. 135-152). Hillsdale NJ: Lawrence Erlbaum Associates.

Piskurich, G. M. (1996). Making telecommuting. *Training & Development, 50,* 20+.

Reinsch, N. L., Jr. (1997). Relationships between telecommuting workers and their managers: An exploratory study. *The Journal of Business Communication, 34*(4), 343-369.

U.S. General Service Administration, Office of Workplace Initiatives. (1995). *Interim report: Federal Interagency Telecommuting Centers.*

Gigi G. Kelly is assistant professor of MIS at the College of William and Mary. She holds a BBA from James Madison University, an MBA from Old Dominion University, and a PhD from the University of Georgia. Dr. Kelly teaches in the areas of systems analysis and design, information resource management, and decision support systems. Her research interests focuses on the various effects of IT on the work environment, including telecommuting, virtual group development, and enterprise-wide systems implementations. She has extensive consulting experience in MIS and has published articles in Journal of MIS, Small Group Research, Computerworld, *various conference proceedings and book chapters.*

Karen Locke, PhD, is associate professor of business administration at the College of William and Mary. She joined the faculty there in 1989 after earning her PhD in organizational behavior from Case Western Reserve University. Her investigative work focuses on the written construction of scientific work in the organizational studies community, on the examination of emotionality in the workplace, and more recently on the impact of telecommuting on home and work. Karen's work has appeared in journals such as Academy of Management Journal, Organization Science, Journal of Management Inquiry, *and* Studies in Organization, Culture and Society. *Additionally, she has co-authored the book* Composing Qualitative Research *(Sage).*

This case was previously published in the *Annals of Cases on Information Technology Applicaitons and Management in Organizations,* Volume 1/1999, pp. 213-223, © 1999.

Chapter XIII

Application of an Object-Oriented Metasystem in University Information System Development

Petr C. Smolik, Brno University of Technology, Czech Republic

Thomas Hruška, Brno University of Technology, Czech Republic

EXECUTIVE SUMMARY

This case presents experience from design and implementation of a university information system at the Brno University of Technology. The newly built system is expected to provide the students and staff with better tools for communication within the university's independent faculties, departments, and central administration. An object-oriented metasystem approach was used by the IT Department to describe the services offered by the university information system and to generate needed program code for run-time operation of the system. This approach so far produced good results over the period of two years when the pilot project started, nevertheless there are some shortcomings that still have to be resolved.

BACKGROUND

This case comes from a large and very old organization. The organization was established by a decree of Kaiser Franz Joseph I in 1899, and, since then, it has grown to be the third largest state owned organization of its kind in the Czech lands. Currently with almost 2,000 employees and around 15,000 full-time students, the Brno University of Technology is an institution of higher education with eight independent faculties: Faculty of Civil Engineering, Mechanical Engineering, Electrical Engineering and Communication, Architecture, Business and Management, Fine Arts, Chemistry, and Information Technology — offered degrees range from bachelor's to master's to PhDs and MBAs.

During their rise, information technologies found increasingly their place in some parts of the school. Some computers were used for research in various departments, some found their place at the central Accounting and Personnel Department. The Internet took the school by storm and the more IT-oriented departments slowly started offering some content on the Web. Development of some department level or faculty level information systems started. These systems have been used either by the staff to report research and teaching activities or by students to obtain the class requirements. Apart from the central economic agenda, faculties started attempts to use software to organize the student agenda. Further, the university has been slowly forced to provide electronic reports to the Ministry of Education in order to get desired annual funding.

There was growing need to organize the information systems into one compound information system or into a set of flexibly communicating systems. As opposed to other universities in the area, where faculties are very tightly centrally controlled, the faculties at this university are very independent and the central control is low. Regarding information systems, this might be an advantage for the stronger and more technical faculties, since they have enough strength to bring up their own systems that meet their needs. Nonetheless, the smaller faculties become more dependent on the centrally-offered information technologies. Therefore, the decision to centralize or decentralize IT could not be made, and it is probable that both need to be supported at once.

So who is the one to work on solving these issues? Of course, there is a central IT Department that for many years dealt with these issues and took care of the unquestionably central agendas and computer networks. The IT Department is placed right under the school's top management, though it has no power to force the faculties to make any particular decisions about their IT. The IT Department is there purely to offer services both to the central administration and the diverse faculties.

SETTING THE STAGE

A few years ago in 1999, the university found itself in a situation that "just something had to be done with the university information systems." The amount of paperwork reporting had been increasing and the general acceptance of information technologies, as something available to do something about, grew. Three of the eight faculties managed to develop their own Web-based systems for their staff and students, and were able to provide electronic reports to the central administration. Student agenda was behind, since it was all paperwork, and, at the same time, some data were entered into the faculty instances of student agenda software, which, in turn, did not provide much output to anyone, especially not the students. This was a good time to think about the issues of centrality vs. decentrality of the future solution for the university information system.

There were several players. The IT Department, the one expected to do something about it, was the major player. The other players formed during the time. A Committee for University Information System was formed on the "right below top" management level and was the one to make the IT Department to act. Later, in order to get individual faculties involved more closely, system integrators were established at individual faculties. These integrators then started meeting with the IT Department at regular meetings. Thus, the cooperation was initiated with enough room for debate.

The committee decided that the IT Department should use XML technologies to enable faculties communicate data with the central information system. This was the time (Summer, 2000) when the IT Department decided with the committee to do a pilot project on the technologies forming the XML-able central information system. At that time, there was a great need for science and research management system, and because it was not overly complicated, it was selected to be the pilot project. It was good system to start with also because the system has many levels of checks and reporting all the way up through the school administration to the Ministry of Education.

Now, with the planned pilot project came the last player, the technology provider. A local IT company provided expertise in implementing XML based e-business systems and suggested how to go about implementing the whole university information system. The company presented a first draft version of the software to the IT Department and the cooperation on the pilot project started.

Requirements for the pilot project were the following:

- Web user access to central storage of science and research activities (HTML)
- User authentication and authorization for access to specific resources
- Interfaces for existing systems to submit and request data online (XML)
- Utilization of open technologies and relative database independence
- Further extensibility both on the side of the center and the different faculties

CASE DESCRIPTION

University information systems are specific in the scale of various data produced and processed by different groups of users. E-business systems used by companies to communicate their business data with their partners usually exchange just a few types of business documents, such as orders, stock quotes, or invoices. Even though these documents change a little over time, there is no dramatic need to keep adding new types of them to cover other business needs. Nevertheless, in the university environment with many diverse and changing activities, there is high demand for information system flexibility. New services for new groups of people need to be designed and redesigned continuously. Therefore, there is need for effective means to build user-to-system or system-to-system interfaces. An on-going growth of the information system has to be taken care of.

In order to manage a growing system, it might be useful to have a metasystem that defines and manipulates the underlying information system. Information systems usually contain one or several databases where data are stored. For each piece of data in any of the databases, the metasystem would need to have a description (metadata) (Tannenbaum, 2001). There exists an object-oriented mechanism that may be used to define the metadata and provide means to access the data based on the descriptions.

The information system that is being developed at the Brno University of Technology uses the concept and concept view definitions presented in this case. Concepts are used to describe data in object-oriented fashion and concept views are used to describe more complex views on interrelated data (Amble, 2001). Since the definitions are always accompanied with human readable navigation and support texts, it is possible to use them as metadata for generation of Web enabled services for human users or for system-to-system communication.

It is expected that using a metasystem to define, redefine, refine, and generate underlying system speeds implementation time and improves quality of the resulting expanding system. This is because it should enable the project team to focus on the services that need to be offered to users instead of continuously having to program the services at low level with many chances of mistake.

The IT Department decided to base the solution on open technologies. The main one of them is XML (Extensible Markup Language) used for platform-neutral representation of complex data (Smolik, 1999; Birbeck, 2001; Ray, 2001; Extensible Markup Language [XML], 2002; Learn XML, 2002). Since there was just a little experience with XML, the IT Department requested expertise from the local company that had already two years experience using XML in commerce applications. The company provided its Java, XML, and XSL based Web-application framework with extension that enables XSL style-sheets to be used for generation of the XSL style-sheets used to provide the system Web pages (McLaughlin, 2001; Tidwell, 2001; XSL Transformations [XSLT], 2002).

Since the system developed is always defined in metadata on the metasystem level, it is possible to generate application not only for the provided applications framework, used for the pilot, but for other available application frameworks or servers. That ensures independence of the resulting solution on a specific platform.

The whole business model is built on the provider company's model of information system development where there are two fundamental partners. One is an experienced technology provider but knows a little about the problem domain of the resulting application. And the other has very good know-how in the problem domain, but doesn't have much experience with the technologies. Either of the partners may, for specific projects, outsource the knowledge and work of the other. In the case of this pilot project, the IT Department contracted the provider to outsource the technologies, but itself it is taking care of defining the university information system know-how.

The following subsections provide closer look on the technologies used to satisfy all the project requirements.

Conceptual Model

This section presents the fundamental mechanism that has been used for object-oriented description of data for the pilot project and later for the whole university information system. It might get little more technical, but on the other hand, may serve as a good background for understanding the implementation side of the case.

The metasystem uses descriptions, called concepts, to provide static views of systems it defines and generates. The concept definitions were first employed in the G2 object-oriented database system. The G2 model of concepts provides the most complete picture of what could be implemented at the metasystem level to describe data of processed in systems. Even though the current version of the metasystem implements only part of the specification, it is presented here in the full scale.

Concept Definition

CDL (Concept Definition Language) is the specification language for the conceptual object modelling of G2 system applications. It exists in two syntax forms - the textual form and UML. The object model consists of *objects*. An object is an ordered tuple of property values. Each property has *name* and *data type*. *Concept* is the Cartesian product of the named property data types. For the time being, concept is equivalent to the well-known notion of *class*, but with expanded semantics. Each property can be parameterized, i.e., each property can have an arbitrary number of parameters. Each parameter must also have some *name* and *data type*.

An example could be a Person object. Each person can have a *list of names*, a *family name,* and a *birth date*. Concepts are used to define objects. The following concept defines the names Name, Family, and DateOfBirth as properties with their corresponding data types **String** and **Date**. The Name property is parameterized. The Order parameter is used to define the order of person names.

```
Concept Person
Properties
   Name(Order: Integer):   String
   Family:                 String
   DateOfBirth:            Date
End Concept
```

It is not stated whether the property is a date or an algorithm. The concept defines the interface of the object for a user communication. The implementation is defined elsewhere.

The basic set of elementary data types includes **Byte**, **Integer**, **Long**, **Single**, **Double**, **Currency**, **Date**, **String**, and **Boolean**.

Relationships and Collections

The model consists of objects, nevertheless objects do not exist separately. In opposition to the relational model, each object can define possible relationships to other objects. Two kinds of relationships are distinguished: *relationships 1:1* and *relationships 1:N*.

The 1:1 relationship definition expresses that one object, an *owner* of the relationship, is bound to another object, *member* of the relationship (for example, a person has usually one father). The 1:N relationship expresses that one owner is bound to a (possibly empty) set of members (a person and its children). The way the model expresses relationships is shown below, but first *collection* of objects need to be defined.

The natural purpose of objects is to establish collections. For example, some person has a collection of children or some document has a collection of lines. Further, it should be noticed that collections of objects could have properties themselves (i.e., an average age).

An object and its collection can be defined in one concept (here the notion of concept starts to differ from the one of a class).

```
Concept Person/People
Properties
   Name(Order: Integer):    String
   Family:                  String
   DateOfBirth:             Date
   Father:                  Person
   Children:                Persons
   AverageAge:              Integer [Where=Col]


End Concept
```

Because concepts define both objects and collections of objects, there is one name for a single object (concept name) and another name for the collection (concept collection name). In the example, *one person* is defined under the name Person and the *collection of people* under the name People. To be able to distinguish between properties of a single object and of its collection, the Where attribute may be used in the property definition. This attribute can have values Alone, Col, or (Col, Alone). It means that the property belongs to a single object, to a collection, or to both. Usually, most properties belong to a single object, and therefore the Where = Alone attribute is implicit.

An *extent* is a very important collection type of the model. It is a collection of all objects of the same concept. The database system maintains this collection automatically. Extents are the main entry points to object-oriented databases.

Relationships are defined as properties with a data type equal to either concept name or a concept collection name. If the concept name is used then the 1:1 relationship is established and if the concept collection name is used then the 1:N relationship is established.

In the above example, the relationship Father is 1:1 relationship to the object Person and the relationship Children is 1:N relationship to the collection of People.

Object-Oriented Data Access

The previous section shows how data in a system may be described on the metasystem level using the concept descriptions. This section will explore way how the data may be expressed as instances of concepts (objects) and how collections of interrelated objects may be expressed (Dodds, 2001). Further, the XML Data Access (XDA) query language will be shown. XDA was developed at the technology provider company and has been used on several e-commerce and e-business projects (Smolik, 2000). The system realized at the Brno University of Technology uses XDA to enable flexible data access to various types of databases. Moreover, XDA is complemented with a Web services access control methodology (WSAC) that allows limiting access to objects based on user and group rights to services and objects within those services. WSAC was also developed by the provider company and will not be explained within the scope of this case, nevertheless it is very useful for controlling access of many different groups of people to various parts of a university-wide information system.

Expressing Objects in XML

There are two ways how data in a database might be accessed. Either they could be accessed directly by application objects via an object interface, or they could be accessed externally through an XML interface.

Each object or a collection of objects of a certain concept may be expressed in XML exactly as defined in the concept definition (Carlson, 2001). The property values of the objects are enclosed in elements tagged with the corresponding property names. All the property value elements are then enclosed in an element tagged with the name of the corresponding concept. For example, a sample *object* of the Person concept may look like the following:

```
<Person oid="25456787">
  <Name>Drundun</Name>
  <Family>Hallway</Family>
  <DateOfBirth>1955-05-16</DateOfBirth>
</Person>
```

Similarly, individual objects may be wrapped into *collections of objects* using the concept collection name also defined in the concept definition. Then the individual person objects are enclosed in the corresponding collection element:

```
<People>
  <Person oid="25456787">
    <Name>Drundun</Name>
    <Family>Hallway</Family>
    <DateOfBirth>1955-05-16</DateOfBirth>
  </Person>
  <Person oid="25456788">
    <Name>Miranda</Name>
    <Family>Hallway</Family>
    <DateOfBirth>1959-11-07</DateOfBirth>
  </Person>
</People>
```

Concept Views

The simple objects presented so far could be also expressed as simple rows of a table as in known from result-sets in relational databases. Nevertheless, in the world of objects there need for much more than just this. It should be possible to express complex data structures corresponding to *views of related objects*. Concept definition allows properties to define relationships with other concepts. According to the Person concept, each person may have a father and several children.

That may be expressed in the following way:

```
<Person oid="25456787">
  <Father oid="12456477">
```

```
    <Name>Eduard</Name>
    <Family>Hallway</Family>
    <DateOfBirth>1926-01-21</DateOfBirth>
  </Father>
  <Children>
    <Child oid="84544587">
      <Name>Monica</Name>
      <Family>Hallway</Family>
      <DateOfBirth>1978-10-29</DateOfBirth>
    </Child>
  </Children>
</Person>
```

Since objects are often interrelated in complex way, mechanism is needed that would allow for accessing only the data corresponding to few relationships that are of interest. The view of related objects called a concept view is defined in a concept view definition. Definition for the above view might be the following:

```
<ConceptView name="PeopleWithFathersAndChildren">
  <Include concept="Person">
    <Include relationship="Father"/>
    <Include relationship="Children"/>
  </Include>
</ConceptView>
```

XML Data Access (XDA)

In order to communicate with an object-oriented database a query language is needed. There could be three types of queries. Queries that get data, update/create data, or delete data. Each query specifies which concept or concept view defines the data of interest. The get data queries further include constraints limiting the scope of the selected data. The update data queries include the data to being updated or inserted, and the delete data queries include the necessary object ids that identify the objects to be deleted.

For example to obtain collection of all Person objects available the following query may be used:

```
<GetData concept="Person"/>
```

To obtain collection of all the people, whose family name is "Hallway," the following query could be used:

```
<GetData concept="Person">
  <Where>
    <Eq property="Family" value="Hallway"/>
  </Where>
</GetData>
```

Concept view queries are built similarly. To obtain all people with their fathers and children the following might be an appropriate query:

```
<GetData conceptview="PeopleWithFathersAndChildren"/>
```

The concept view queries might also be constrain on any level of the included subtree as in the following example, which would provide all people whose child's name is Monica.

```
<GetData conceptview="PeopleWithFathersAndChildren">
  <Where>
    <Eq property="/Person/Child/Name" value="Monica"/>
  </Where>
</GetData>
```

Finally, here are examples of both an update data and delete data queries:

```
<UpdateData concept="Person" insert="yes">
  <Data>
    <Person oid="25456787">
      <DateOfBirth>1955-05-17</DateOfBirth>
    </Person>
  </Data>
</UpdateData>

<DeleteData concept="Person">
  <Data>
    <Person oid="25456787"/>
  <Data>
</DeleteData>
```

These were just simple examples of concept and concept view queries demonstrating the object access mechanism usable to access data in a database. This exact mechanism was used for the pilot project to provide XML-based access to data in various relational databases and has been used both as the XML interface for the faculties to the central system and as source of data from the databases to be visualized in the user interface that was built.

Web Services

Web services could be considered as either human or machine interfaces to information systems. Web services are accessible over the Internet where human users use browsers to view HTML pages or other systems communicate XML documents. XDA mentioned in the previous section serves as a Web services provider for other systems to access data in the central university system, and it also enables the business logic tier to access data in an object-oriented fashion.

Figure 1 shows general Web services provider architecture. On the right side, each Web service accesses data in various data sources, such as databases or directly

Figure 1. Web services provider architecture

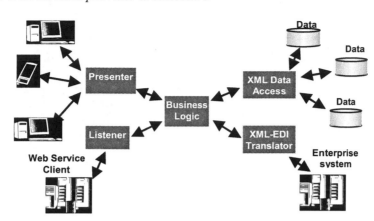

accessible enterprise information systems. On the left side, each Web service communicates with the outside world. It could present its services directly to users via the Presenter, which transforms the pure data into appropriate Web pages, and to other systems, it provides its services via the Listener, which listens for service requests and provides service replies. All communication with the outside world is based on the HTTP protocol, and the data is represented either in some visualization format for users (HTML, ...) or in XML for other systems. Any Web service may use other Web services to obtain necessary information or to have computations done. Web services are mostly seen as services provided by a machine to another machine, nevertheless, they could be also seen as services directly offered to human users, because the logic of the service is equivalent in both cases, only the visualization part is being added for the comfort of the user. And further, if there is a good service description then it means that the graphical user interface could be easily generated. Web services therefore might be considered as both machine and human accessible.

Web services enable communication of data among companies or independent parts of an organization (i.e., business entities) (Cerami, 2002; Oellermann, 2001; *A platform for Web services*, 2002). Each business entity may offer their Web services to other entities. Because the use of Web services has to be easy and connecting to new services needs to be simple, there will be standard ways to describe, offer, and find services. The most promising language for description of Web services is the XML-based Web Services Description Language (WSDL) jointly developed by Microsoft and IBM. Further, the Universal Description, Discovery and Integration (UDDI) specification describes Web services and programmatic interface that define a simple framework for describing any kind of service, and finally SOAP as the transfer wrapper (Graham, 2001; Universal Description, Discovery and Integration [UDDI], 2002).

As mentioned earlier, there are two generic types of a Web service. Ones that provide access to data and ones that do computations. Example of the first case could be sending and an order or receiving an invoice. Example of the second case could be requesting price for sending a package of specified size and weight from place A to place

B. Web services that access data could be defined in terms of concept views that were presented above. Web service could, therefore, be defined as a concept view enriched with documentation texts and other information related to level of access to objects within the service. This enriched description could be considered as the service description. Each service thus defines what view of interrelated objects is available within that service. For each concept in the service definition defines what operation (list, create, edit, delete) is allowed. The service may define implicit filters on accessible objects and other information needed for proper documentation or visualization. Further constraints on what objects could be accessed by what users within given service are put by a separate mechanism (Web services access control) mentioned earlier.

Metasystem

Previous sections discussed ways to represent data about a system (metadata). From the metasystem point of view there are several types of metadata:

- Metadata that define the data in the systems (concepts)
- Metadata that define views on data with their hierarchical relationships (concept views)
- Metadata that define Web services (service descriptions)

In a way, service description in its complete form may take form of a personal process. Personal process is then the activity that the user performs in order to take a full advantage of a service, for example, the service "Send purchase order." From the data perspective, an XML document containing proper order head and order items needs to be sent. For any user, preparing such a document is a process of entering order-head information and adding items. This personal process information would be also needed as part of the service description, so that suitable editor may be generated for each service.

In the future, attempts to add metadata that would define organizational processes should be made. Organizational process could be seen as the process of using different Web services by different users in order to reach an organizational goal. Usually, organizational processes are executed by individual users implicitly, and they have no explicit definition, or they are hard-wired in the system implementation. By making organizational processes explicit by representing them on the metasystem level, it could be expected that the continuous growth of the underlying system could be better controlled.

The current implementation of the metasystem, as shown in Figure 2, uses concept definition to define objects manipulated by the system, uses service descriptions (with concept views included) to define Web services with their corresponding personal processes. Using the Generator, metasystem generates several parts of a system. The XML interfaces to services are the full-blown Web service listeners presented in the previous section. The User interfaces to services are the user Presenters for each Web service. All the system parts are generated from a universal description to a specific system program code using the Generating styles. There are special styles for generation into different target environments. The current version of the metasystem generates XML service descriptions and XSL style sheets used by the provider company's

Figure 2. System generated by a metasystem

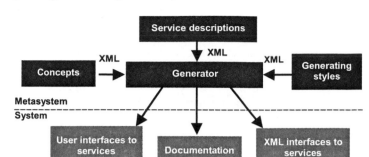

applications framework, but it is possible to add generating styles to generate applications in Java, JSP, ASP, PHP, WebSphere, WebLogic, or Oracle AS.

Based on both concept definitions and service definitions, the metasystem is able to generate not only proper interfaces, but also documentation. In development and maintenance of a system offering wide range of services and encompassing many different types of data, there is a high need for good documentation. For this reason, the concept definitions are further complemented with human readable descriptions of all properties and concepts themselves, so the complete information as what the data means is presentable.

CURRENT CHALLENGES/PROBLEMS FACING THE ORGANIZATION

The pilot project went relatively well. The technologies proved their usability and a science and research reporting system is in operation. The smaller faculties that have no information system in place use the generated central Web-based access to the university information system and the larger faculties are making attempts to communicate the required data from and to their systems via the XML interface. The system has been appended with other functional areas and slowly even some parts of the student agenda have been added.

Nevertheless, there are problems as with any information system that should serve such diverse needs and the needs of thousands of users within a university. The primary problem is the lack of people taking part on building the system resulting in slow movement in adding new functions and functional areas to the system, which results in groups of unsatisfied users.

There are two major reasons for the lack of IT personnel: insufficient funding and the difficulty to employ experienced developers at the university. IT professionals are very expensive, and it is almost impossible to employ ones in the university environment, since it poses political problem of them having higher salary than maybe the dean himself.

Furthermore, the university is not able to afford a regular IT project from its budget. So far, the project has been an inexpensive one since it employed only part-time consultant from the technology provider company, one internal IT Department employee (also not full time), and one system integrator that helped to define the central system metadata. Regarding this fact, the project has obtained almost miraculous results within about one year from the start. The two internal university staff have defined over a hundred of database tables and their descriptions in the metasystem, and defined the suitable Web services for user and system access to various areas of the data for different groups of users (including many other services than just the ones related to the pilot science and research). The part-time consultant ensured that the metasystem functioned properly and implemented some required improvements, which proved that the technology provider business model seems to work well.

Even though, the question how to go further in the university information system development stays unanswered. Apart from the funding problems, there are also other problems with the further development. The technologies employed are very advanced and seem to really speed up the development, but there are just few professionals that understand them, thus they are also very expensive. On the other hand, it is possible to employ bigger number of cheap inexperienced programmers to do the job in more simple technologies and by raw people force, which would probably result in much less elegant and harder to maintain solution. These questions will not be answered in this case and they are now open even for the case actors themselves.

REFERENCES

Amble, S. W. (2001). *The object primer*. Cambridge Univ Pr (Trd).

A platform for Web services. (2002). Retrieved January 30, 2002, from http://msdn.microsoft.com/library/techart/Websvcs_platform.htm

Birbeck, M. (2001). *Professional Xml (programmer to programmer): 2nd edition*. Wrox Press, Inc.

Carlson, D. (2001). *Modeling XML applications with UML: Practical e-Business applications*. Addison-Wesley.

Cerami, E. (2002). *Web services essentials (O'Reilly XML)*. O'Reilly & Associates.

Dodds, D., & Watt, A. (2001). *Professional XML meta data*. Wrox Press Inc.

Extensible Markup Language (XML). (2002). Retrieved January 30, 2002, from http://www.w3.org/TR/REC-xml

Graham, S. (2001). *Building Web services with Java: Making sense of XML, SOAP, WSDL and UDDI*. Sams.

Hruška, T. (1998). Multiple class objects. *ISM 98 Workshop Proceedings* (pp. 15-22).

Hruška, T., & Mácel, M. (1999). G2 — Component architecture of object-oriented database system. *ISM 99 Workshop Proceedings* (pp. 119-124).

Learn XML, XSL, XPath, SOAP, WSDL, SQL, HTML, Web building. (2002). Retrieved June 16, 2002, from http://www.w3schools.com/

McLaughlin, B. (2001). *Java & XML, 2nd edition: Solutions to real-world problems*. O'Reilly & Associates.

Oellermann, W. L., Jr. (2001). *Architecting Web services*. APress.

Ray, E. T. (2001). *Learning XML*. O'Reilly & Associates.

Smolik, P. (1999). The importance of Extensible Markup Language. *ISM 99 Workshop Proceedings*.

Smolik, P., & Tesacek, J. (2000). Data source independent XML data access. *ISM Conference 2000 Proceedings* (pp. 17-22).

Tannenbaum, A. (2001). *Metadata solutions: Using metamodels, repositories, XML, and enterprise portals to generate information on demand.* Addison Wesley Professional.

Tidwell, D. (2001). *XSLT.* O'Reilly & Associates.

Universal Description, Discovery and Integration (UDDI). (2002). Retrieved January 30, 2002, from http://www.uddi.org/

XSL Transformations (XSLT). (2002). Retrieved January 30, 2002, from http://www.w3.org/TR/xslt

This work has been done within the project CEZ: MSM 262200012 "Research in Information and Control Systems" and also supported by the Grant Agency of the Czech Republic grant "Environment for Development, Modeling and Application of Heterogeneous Systems," GA102/01/1485.

Petr C. Smolik, after completing his BS in computer science at the Valdosta State University, Valdosta, Georgia, continued at the Faculty of Information Technologies of the Brno University of Technology, Czech Republic. There he completed his MS and continued in the PhD program since 1998. During that time he divided his duties among teaching parts of the information systems and object modeling courses, and real-world implementation of new e-business solutions for local companies. He continues to put his research interests in metasystems for description, generation, and maintenance of information systems into real-world solutions not only at the Brno University of Technology but also at one of the largest local banks.

Tomas Hruška graduated from the Brno University of Technology, Czech Republic. Since 1978, he has worked in the Department of Computers (Faculty of Information Technology - FIT), Brno University of Technology. At present, he is the dean of FIT. From 1978-1983, he dealt with research in the area of compiler implementation for simulation languages. From 1983-1989, he concentrated on design and implementation of both general-purpose and problem oriented languages. Since 1987 he has participated in the project of C language compiler. Since 1990, he has worked on the implementation of information systems. He deals with an implementation of an object-oriented database systems as a tool for modern information systems design now.

This case was previously published in the *Annals of Cases on Information Technology*, Volume 5/2003, pp. 585-607, © 2003.

Chapter XIV

Integrating Information Technologies into Large Organizations

Gretchen L. Gottlich, NASA Langley Research Center, USA

John M. Meyer, NASA Langley Research Center, USA

Michael L. Nelson, NASA Langley Research Center, USA

David J. Bianco, Computer Sciences Corporation, USA

EXECUTIVE SUMMARY

NASA Langley Research Center's product is aerospace research information. To this end, Langley uses information technology tools in three distinct ways. First, information technology tools are used in the production of information via computation, analysis, data collection and reduction. Second, information technology tools assist in streamlining business processes,particularly those that are primarily communication based. By applying these information tools to administrative activities, Langley spends fewer resources on managing itself and can allocate more resources for research. Third, Langley uses information technology tools to disseminate its aerospace research information, resulting in faster turn around time from the laboratory to the end-customer. This chapter describes how information technology tools are currently cutting cost and adding value for NASA Langley internal and external customers. Three components from a larger strategic WWW framework are highlighted: Geographic Information Systems (GIS), Integrated Computing Environment (ICE), and LANTERN (Langley's intranet). Based on experiences with these and related projects at Langley, we suggest that there are four pillars of information technology project success: training; provision of useful services; access to enabling tools; and advertising and advocacy.

BACKGROUND

Established in 1917 as the first national civil aeronautics laboratory, Langley's mission is to be a world leader in pioneering aerospace science and innovative technology for U.S. aeronautical and space application. The center is dedicated to serving traditional aerospace customers and to transferring aerospace technology to non-traditional customers in response to changing national priorities.

More than half of Langley's effort is in aeronautics, improving today's aircraft and developing ideas and technology for future aircraft. The center's wind tunnels and other unique research facilities, testing techniques and computer modeling capabilities aid in the investigation of the full flight range — from general aviation and transport aircraft through hypersonic vehicle concepts.

The center manages a dynamic program in atmospheric sciences, investigating the origins, chemistry and transport mechanisms that govern the Earth's atmosphere. A key component of this study is to understand the impact of human activity on our planet. Langley is also contributing to the development of the Earth Observation System (EOS), a major part of the international Mission to Planet Earth.

To better reflect the needs of its customers the center has recently implemented a major reorganization of its management and operating structure consisting of: customer interface groups; a Research Group and an Internal Operations Group. Langley Research Center is a world class research laboratory which has a staff of well trained and highly productive scientists, engineers and support personnel, as shown in Tables 1 and 2 (Office of Public Affairs, 1995).

Other pertinent workforce facts include:

- Civil Service Employees 2,508 (Fiscal Year 1995)
- Contract Employees 1,975 (Fiscal Year 1995)
- Fiscal Year 1994 total procurements: US$525,000,000
- Fiscal Year 1995 Payroll: US$144,500,000 (includes all compensation)
- Total Program Year 1995 Budget: US$643,700,000

The NASA Langley Research Center occupies 787 acres of government-owned land and shares aircraft runways, utilities and some facilities with neighboring Langley Air Force Base. The center's more than 220 buildings represent an original investment of $687 million and have a replacement value of over US$2 billion. Langley's experimental facilities are: aerothermodynamic, subsonic, transonic, supersonic and hypersonic wind tunnels as well as scramjet engine tunnels. Langley's unique facilities include:

- Nation's only large flight Reynolds Number transonic tunnel
- Nation's only transonic dynamic loads/flutter tunnel
- Nation's only aerodynamic spin tunnel
- Nation's only high-Reynolds Number supersonic quiet tunnel
- Nationally unique aircraft landing loads and impact dynamics facility
- Highly specialized aero structures and materials research laboratories

Langley's 30 wind tunnels cover the entire speed range from 0 mph to nearly Mach 25. In addition to these unique facilities, Langley houses facilities for structures,

Table 1. Skill mix (1995 data)

Function	Number	% of Workforce
Scientific/Engineering	1,256	50.0%
Administrative	271	11.0%
Tech/Craft/Production	757	30.1%
Clerical	224	8.9%

Table 2. Formal education distribution (1995 data)

Degree	Number	% of Workforce
Doctoral	282	11.2%
Master	573	22.8%
Bachelor	647	25.8%
Associate	469	18.7%
Some college	260	10.4%
H.S. Diploma	268	10.7%
Other	9	0.4 %

materials and acoustics research, flight electronics, flight systems, simulators, simulation facility components, and a scientific and general purpose computing complex.

SETTING THE STAGE

NASA Langley Research Center's product is aerospace research information and its production can depicted by the data relation model in Figure 1. Due to the technical nature of Langley's work, use of some information technology tools at Langley was already common place. A campus-wide Transmission Control Protocol/Internet Protocol (TCP/IP) network, known as "LaRCNET" (Shoosmith, 1993), has been in place since 1986, and use of e-mail, USENET news, file sharing, and other electronic communications were wide spread.

Langley uses information technology tools in three distinct ways in the production cycle of aerospace information. First, information technology is used in a variety of methods during research: computation, analysis, data collection, data reduction, etc. (Wieseman, 1994). Second, Langley uses information technology tools to assist in streamlining its business processes, especially those that are communication based. By applying these information tools to administrative activities, Langley spends fewer resources managing itself and can allocate more resources for research. Lastly, Langley

Figure 1. Simple work relation model at Langley

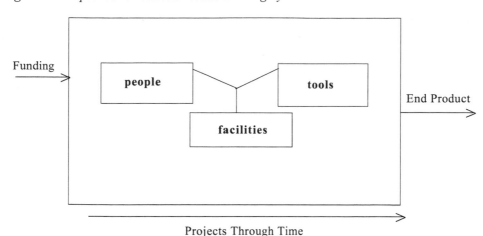

Projects Through Time

uses information technology tools to disseminate its information. Depending on the media format, security classification, and targeted customer, the end product can be a technical report, software and data sets, or general technology availability notices. Distribution is by traditional hard copy and Internet publishing methods. NASA Langley technical publications are currently distributed on the World Wide Web (WWW) by the Langley Technical Report Server (Nelson et. al., 1994) and the NASA Technical Report Server (NTRS) (Nelson, et al., 1995); some software via the Langley Software Server (LSS); and technologies available for licensing via the Technology Opportunity Showcase (TOPS) (Nelson & Bianco, 1994).

This work relation model for aerospace information requires: (1) *people* to do the work, (2) *tools* to do the job, (3) and *facilities* to house the tools and people. Managing the research and technology process involves allocating varying resources to achieve the optimum relationship between these unique data sets.

However, decreasing workforce and budgets, coupled with increasing workloads has forced Langley to investigate technologies with a non-linear return on investment. Based upon the success of earlier experiments of using WWW for electronic publishing, increasing WWW usage and focusing it on a few core functionalities was seen as a method to achieve the desired cost savings and efficiency gains.

PROJECT DESCRIPTION

Overview

The desire was to build upon the existing TCP/IP infrastructure and budding WWW user community to provide universal access to key services at Langley. The World Wide Web (Berners-Lee, 1992) is a TCP/IP based, wide area hypertext information system that is available for all popular computer platforms: Mac, PC, and Unix. In June 1993 Langley became the first center in the Agency to have a World Wide Web home page for distributing information internally and to the public. Growth of home pages at the center and throughout the world was rapid; Langley Web space currently consists of 17,000+ pages, and the entire Web is estimated at 50 million pages (Bray, 1996). The rapid growth at Langley, however, was due to the efforts of a few World Wide Web enthusiasts. In early 1995, NASA Langley dedicated 1.5 full-time equivalents to officially handle the integration and coordination of WWW services and facilities. Membership in the team is dynamic and is a grass roots effort, with new members constantly being recruited. The team consists of 13 additional members from across the center and has occasionally grown to 126 for events such as Internet Fair 2 (See Appendix A for Uniform Resource Locators [URLs]). Under this center-wide self-directed team approach the WWW Team provides vision, develops strategy, and procures resources to implement its mission. The WWW Team's mission is "To implement an integrated, easy to use, and cost efficient information environment."

There were a number of distributed WWW activities occurring simultaneously, but there were three that benefited from central coordination and provided significant savings: LANTERN, Langley's intranet; Cicero, a software distribution mechanism; and an online version of the Langley Master Plan, a facilities and physical plant reference guide. Each of these significant activities corresponds to a component in the work model in Figure 1: LANTERN is an asynchronous communications tool for people, Cicero provides point and click access to software tools, and the online Master Plan is a continuously updated record of the physical center and its facilities. Although the three projects did not run concurrently, they form the foundation of a holistic information technology strategic plan for Langley.

Technology Concerns

Technology was not a limiting factor in any of the projects. Aside from having a large pool of creative technical talent at Langley, most of the tools involved in the Master Plan, Cicero and LANTERN were developed wholly or in part by the Internet community at large or were commercial-off-the-shelf (COTS) tools. By leveraging the output of a world-wide development team, creating WWW services becomes largely systems integration, not systems development. Turn-key solutions do not exist freely on the Internet; but their components do. The following is a summary of the technical challenges for each of the project components, with the details appearing in the next section.

For the Master Plan project, the technical challenges were largely database integration, and converting legacy hard copy information to electronic format. Once the information had been extracted from closed, proprietary databases, or converted to electronic form, manipulation of the data is a well understood process.

Cicero required the innovative application of existing Unix tools, with some Perl script "glue" to hold it all together. The technology for Cicero has existed for some time, its the defined process available through a well-known, consistent WWW user interface that makes it a full product.

LANTERN required the least amount of additional technology of all. The very nature of an intranet dictates the reuse of wide area networking tools in a local area networking scenario (Auditore, 1995; Sprout, 1995). The challenge in establishing a successful intranet is the social and political groundwork, not the technology.

Management and Organizational Concerns

Management, organization, and logistical concerns are by far the biggest challenges to an information technology project. In fact, solving the "technical" part of a project is often the "easy" part. Information technology projects impact existing processes and must overcome both those that have a real or perceived interest in existing processes as well as the inertia of human nature. They key to mitigating the resistance to the projects is to involve all stake holders of the existing process early in the planning and development of the information technology project (Moreton, 1995).

In the case of the Master Plan, this involved working with those that operated and maintained hard copy drawings, maps, and various proprietary databases. By involving the current information holders early in the process, resistance to creating an online version was minimized.

Cicero illustrates a socio-management challenge on multiple levels. There were many people that had wanted Cicero-like functionality for some time, and eagerly embraced it as soon as it was available. Others were won over through several demonstrations, seminars, and referrals from other users. Even though Cicero is widely praised by all that use it, it has the side effect of serving as a catalyst for existing, unresolved debate in how to handle distributed systems support. So even though a product can meet customer expectations, its impact has to be understood within the larger organizational context.

LANTERN's challenge was similar to Cicero's, but on a larger scale. Not only was a new communications process introduced, but its impact on existing processes had to be clearly communicated to all involved. The LANTERN team accomplished this by first gaining the support of a senior member of management, then publicizing and then holding a number of informal design shop meetings that encouraged employees to contribute suggestions for process improvement at Langley. The suggestions that were within the scope of LANTERN were addressed, and explanation was given for why the remainder were outside the scope of LANTERN. In addition to the design shop meetings, feedback was encouraged via e-mail, hard copy, and telephone.

CURRENT STATUS OF THE PROJECT

This section introduces the history and current details about each major component of the WWW project at Langley.

Master Plan

A 1994 survey of the Internal Operations Group identified many problems associated with how institutional information at Langley was gathered, stored, updated, distributed and used. One of the most significant challenges found was that large volumes of data were held in many different formats. As many as 95 different management information systems were in use in the Internal Operations Group (Ball et al., 1994). The survey found that these systems were developed with a "variety of software packages on a variety of platforms." The study also noted, "Many are standalone and unique to functional organizations" and with regard to these systems, "There is a large amount of manual data entry." This diversity in the location and formats of data as well as the lack

Figure 2. Non-integrated databases for each component

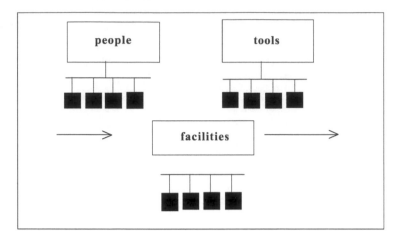

of integration of the various systems (as illustrated in Figure 2) makes complex queries (i.e., across multiple databases) very difficult and time consuming. One of these databases was the Master Plan, the architectural blue prints of each building and total center layout for Langley. The Master Plan is a benchmark when new construction or changes need to be done to any of the buildings or adjacent land area. The Master Plan includes data such as underground utilities, building dimensions, ground composition, title, and abstract data.

The Langley Master Plan had traditionally been produced by hand as a collection of sheets or "plates" which were duplicated and distributed in hard copy form using offset printing. Approximately 70 copies of these tabloid sized booklets consisting of 150 color plates each were distributed for each release. In 1979, the cost for producing the Master Plan was approximately US$500,000. Because of this costly and labor-intensive process, the document was only revised every 5-10 years. This update schedule was typically "crisis-driven," by events such as major building additions and other large construction projects. Because of this haphazard revision cycle, and because of the document being almost always out of date, the Master Plan was only used as a sort of "coffee table reference" to be marked up by planners wishing to make changes. Actual site planning was done with separate drawings for Preliminary and Critical Design reviews. The nature of the data in the document limited its usefulness as well. The plates, existing only as pictures on paper, could not contain or effectively reference other attribute data. This static format did not enable "what-if" planning or make it easy to visualize or analyze different options. This format also dictated that a completely separate database, existing totally on paper, be maintained for the Master Plan and individual building site plans, since the Master Plan was not drawn to sufficient levels of detail or accuracy to be used as an engineering working document. The level of effort required to produce the Master Plan increased dramatically in 1986 because of escalated requirements for documentation of environmental and historical information. In order to make the Master Plan more current, cheaper to maintain, and carry additional information, it was converted to CADD (Computer Aided Design/Drafting) electronic format beginning in 1989.

The development of this electronic version took place between 1989 and 1992 with a total cost (including printing and distributing 60 copies) of approximately US$300,000. The required number of complete copies distributed was reduced by making the electronic drawing files available over the Langley computer network. The level of detail and accuracy of these drawings allowed them to be used for engineering planning, both in the creation of proposed site plans and for discussion in design reviews. This alleviated the need for separate drawing databases and the extra maintenance associated with multiple copies.

The ability to associate these drawings with additional features such as environmental and historical data allowed other organizations outside the Master Planning office to use them for reference and analysis. This electronic Master Plan was demonstrated to a NASA-wide "lessons learned" conference as a model for the other centers to strive toward. Even though this migration to electronic files was a necessary and cost-effective step, the emergence of new technologies for handling spatial data would prove to eclipse this version with even greater savings and more useful data. This new technology, geographic information systems, would revolutionize the way planners at Langley thought about maps.

Begun around 1993 as a prototype of a future Master Plan, this transition to GIS has been accepted as the latest incarnation of this "document." Actually, Master Plan is really a database, tying together electronic drawing sets and tabular data into a seamless hierarchical system of information gathering, display, and analysis. Unprecedented accuracy is possible through the use of digital aerial photography and Global Positioning System (GPS) data, which can be readily assimilated into the system. Through the use of appropriate queries, the separate plates of the old Master Plan can be generated, but the real power of this system is to allow detailed analysis and scenario planning using any number of variables, data relationships, and constraints.

By integrating the graphical databases of underground utilities, building layouts, environmental zones, roads, etc., with tabular databases such as personnel information, space utilization, energy metering, and real property, the GIS Master Plan allows complex analysis and planning to be easily performed in a fraction of the time. This data can now be used for tasks which were all but impossible before. The system was recently used to settle a 20-year-old dispute over some remote property used by Langley for drop model testing. Using the aerial photos, GPS surveys, and property and tax maps digitized into the system, property boundaries were precisely located, and plots generated which convinced a federal judge that Langley was the legal owner of the property in question. During the threat of hurricane Felix in September of 1995, flood contours and building outlines were used to determine where to place sandbags, and what equipment to elevate and how high, to minimize water damage.

Not only has the Master Plan database become a much more useful planning and response tool, it has truly become the most current source of spatial information at the center. Because live links to various databases are used to gather information on demand, every query is answered with the latest data from every source. The GIS version of the Master Plan was developed between 1993 and 1996 at a cost of less than US$70,000. The decreasing cost of producing the Master Plan in various formats is shown in Figure 3. The ability to perform queries and view result over the LaRCNET network has drastically

Figure 3. Decreasing master plan cost

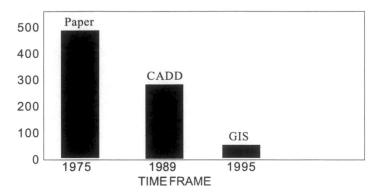

reduced the need for printed copies. Individual users can print a report as needed at their local printer, and almost no one needs a print of the entire database. Although network access has made the Master Plan available and useful to a much greater user community, some training and computer expertise is required to effectively use the system. That is changing rapidly through the development of World-Wide-Web-based tools and services.

Cicero

Historically, distributed computing at Langley has been a collection of small groups of computers, separately maintained and staffed by individual branches or projects, held together by a center-wide TCP/IP backbone. Little work sharing existed between the individual system administrators, especially those employed by different organizations. The Langley Computer Users' Committee (LCUC) was established to provide a unified voice for Langley users to provide their requirements to the computing center. Under its auspices, the Distributed Systems Subcommittee provided a way for system administrators to get to know each other and to cooperate in order to provide better service to their users.

The LCUC and its various subcommittees have proved enormously successful. LCUC-sponsored user's groups and initiatives have helped individual system administrators stay current with the important computing issues at Langley. In addition, they offer a repository of sources and binaries for popular software packages, which has been one of their most successful offerings. It represents a form of load-balancing among the system administrators. Unfortunately, the software is considered "unsupported." Anyone at Langley who wishes can contribute to the archive, but there is no way to determine if the software they configured and compiled will work on any other given system. The system administration methods at Langley vary widely, and software configured for one set of machines may not be suitable for another. Thus, the load-balancing provided by these software archives, while useful, is not quite as useful as it could be if common configuration differences were overcome.

To help overcome these problems, Langley formed the Integrated Computing Environment (ICE) team in late 1993. ICE's objectives are twofold: the first is to provide a common computing environment on each of the major platforms in use. By providing a common set of software tools across all platforms, users will feel comfortable no matter what hardware or OS is being used. The second objective is to ease the administrative burden on the maintainers of these systems. We accomplish this by serving as a clearinghouse for important tools and information for example, vendor security information and by providing optional standards for system installation and configuration.

In February 1994, the ICE team started to look seriously at the problem of providing stable, pre-configured versions of its Unix software for use by the Langley community. Learning from and building upon the success of the LCUC software archive, the ICE team offered a set of applications and software that is made available on each of the widely used platforms at Langley. The team projected that some system administrators would prefer to get their entire /usr/local partition from ICE through a method such as Rdist (Rdist is a Unix program to maintain identical copies of file over multiple hosts) that are executing that would serve to keep them tightly in sync with the latest supported versions of the software for their platform. The ICE team recognized, however, that this would not be the normal case. Most system administrators would prefer to retain the privilege of determining exactly what software is present on their systems rather than trust an automated method. Cicero was born out of this need to provide a solution that would address those concerns, while still providing an easy, semi-automated method of acquiring supported ICE software.

Using the WWW method, administrators can use WWW clients to browse the list of supported software for their particular hardware and operating system combination. Clicking on the software's name will download the package information file and cause ciceinst (Cicero install) to run on the local machine. Because the package files have digital signatures there is no possibility of accidentally installing malicious software. If the package is not signed correctly, ciceinst refuses to proceed with the installation.

LANTERN

NASA Langley Web activities prior to the summer of 1995 were largely constructed for external customers. However, a growing number of information resources that were not appropriate or applicable for external dissemination were becoming common. At that time, the WWW Team decided that a separate information space was necessary dedicated to the business of doing business inside Langley. WWW and computer networks were no longer the domain of the computational scientists, but all organizational users could now access information of interest to them on the internal Web pages (Sprout, 1995). LANTERN is Langley's intranet (LANgley inTERNal home page.) LANTERN's value is that an employee can access institutional knowledge and initiate processes from their desktop 24 hours a day seven days a week ("is Bob in this week?") and "storage" of information ("Where did I put that notice, I can't find anything on my desk?") is provided in a single well known location. In order to build an effective intranet the team also recognized that there were a number of activities that needed to be done to ensure successful integration of an intranet into Langley's work environment, most importantly training.

The WWW Team developed a two-hour training class taught by volunteers to give instruction to the entire center on the fundamentals of the Internet and WWW browser Navigation. During the first offering of classes 288 slots were filled in less than two hours and 265 people were place on a waiting list. The team then scheduled two one-day retreats. On May 23, 1995, the WWW Team met to develop the design requirements and work out the process methodology for gathering for customer requirements for the center's intranet. The three significant design goals that came out of the process were: it must be easy to use by all; it must be easy to maintain; and it must provide useful information that employees want.

To gather data an electronic message was sent to the entire center explaining what an intranet was, how it would streamline administrative processes, and how they could participate. The team also scheduled five two-hour round table discussions open to the entire center for requirements gathering. Equipped with the data from the first retreat, collected e-mail, and data gathered from the round table discussions, the team met two weeks later for an all day retreat to analyze and process all data into categories and develop the initial architecture for LANTERN.

The WWW Team felt that it needed to inform most potential users and so the Researcher News, the center's bi-weekly publication was asked to participate in the design process. Timed very closely with the first release of LANTERN, June 13, 1995, was a two-page article in the center's *Researcher News* explaining LANTERN, its scope, and how interested employees could participate.

SUCCESSES AND FAILURES

Master Plan

In 1995, Langley was beginning to use the Web as an itranet tool for many of its internal data services. It was apparent that the ideal way for the largest number of users on various computing platforms to use the Master Plan and other GIS data was to make them accessible through WWW browsers. The first such a service was the Langley Building and Room Locator. This interface presents a form to the user into which a building number and room number are entered, and maps of both the building and room are pulled from the GIS database and presented on the user's WWW browser.

This service has been made a part of the Langley electronic post office, also Web based, which allows a search by name, e-mail address, etc., and after presenting the address, organization, phone number, and other pertinent contact information, the user can access the Building and Room Locator to find that person's office. Since live database connections are used, and the maps and building drawings are generated on the fly, the system automatically shows the current room assignment, correct building, etc. This is a limited use of GIS on the Web, but it is a proof of concept for the eventual interface to the Master Plan. Users will be able to perform complex queries on multiple data sets and generate specific graphics that show the relationships between those data sets. Information such as energy use by building, maintenance costs per organization, population density within buildings, and even combinations of those data will be available, forming the core of a decision support system for downsizing, building closure planning, consolidation of similar spaces, personnel relocation planning, etc.

With sufficient "hooks" into many databases, the types of information available are nearly unlimited. Eventually, updates to the database will be possible using advances in WWW technology and digital signature and authentication techniques. This will allow anyone to maintain their data with little training. Because of this universal accessibility, the data will converge to greater accuracy, since each user will be interested in correcting the particular information associated with their facility or function. It is anticipated that a significant number of possible queries will be available over the WWW before 1997. Hopefully, with easy access to the Master Plan, hard copy distributions will be not be needed again. The Master Plan will evolve into a complete, detailed virtual model of Langley that is always up to date, available to everyone, and sophisticated enough to answer any question about the people and infrastructure of the center.

Cicero

Cicero has been very well received by the Langley community. There was, however, one unanticipated objection to providing this service. During an informal presentation of Cicero to one of the LCUC subcommittees, it was pointed out that we had perhaps made software "too easy" to install. System administration methods vary widely at Langley. In many places, users have root access in order to perform certain tasks relating to their research (e.g., loading a real-time sensor device driver). If Cicero provided an easy to use "Click to Install" method of updating system binaries, would the users not be tempted to install any piece of software that looked interesting to them? Potentially, this could cause such problems as filling up available disk space or catching other users by surprise when a piece of software is suddenly upgraded to a newer, slightly different, version.

This is, unfortunately, a symptom of a more serious problem: users who have root access yet cannot be trusted to use it wisely. For Cicero, a compromise was available. Cicero uses the syslog facility to log a message each time a software package is installed. While this does not prevent users with root access from installing software on their own, it does provide a useful audit trail.

The system administration community gives Cicero high acclaim because saves significant time and frees them to do other work without worrying about the chore of software installation. In the 10 hours it takes to build and install X Window System, they can help users, re-configure workstations, etc. Users also are pleased with this concept, since it means they receive much more personal attention from their system administrator. Package installation systems like Cicero are time savers for large sites with distributed system administration tasks. Langley has already seen significant cost savings. Packages that used to take three hours for an experienced system administrator to install now takes two minutes.

LANTERN

LANTERN quickly integrated into Langley culture. Through the outreach efforts, people took ownership in LANTERN and enthusiastically contributed to the services available, a sample of which is in Table 3.

In November, 1995 the weekly newsletter ceased hard copy circulation and is available only through LANTERN, at an annual cost savings of US$25,000. In addition, the editor of the newsletter has reduced preparation time of each issue from 1.5 days to two hours. The center is also transitioning all small purchases to a WWW transaction based system. This service will become available in April 1996 and the estimated annual cost saving is US$100,000. Significant cost reductions are being realized from the elimination of redundant paper processing with the center's Management Instructions Manuals. Secretaries no longer have to spend time replacing outdated sheets in the

Table 3. Twenty of the 60+ LANTERN Services

1.	Forms (81 currently used Center-wide)
2.	Health Services
3.	Holiday and Payroll dates
4.	How Do I ? (by subject, computers, procurement, printing, publishing, scheduling, reservation)
5.	Idle Property Screening List
6.	Instructions for putting a paper on Langley Technical Report Server (LTRS)
7.	Intranet feedback form
8.	Inventory Database
9.	Job Order Numbers
10.	Langley Activities Association (Intramural Activities)
11.	Langley Child Development Center
12.	Langley Federal Credit Union
13.	larc.announce (newsgroup which archives all messages sent Center-wide)
14.	LaRC Street Map
15.	LaRC This Week (Weekly Newsletter)
16.	LaRC Townmeeting Question Form
17.	Institutional Studies Library
18.	Recycling Information
19.	Services Directory
20.	Software License Information

Management Instruction Manual binders with updated versions. This is significant in that there is at least 50 copies of the Management Instructions Manuals across the center and each need to be kept up to date. On average a secretary spends an hour a week on this activity. Keeping all 50 sets up to date used to consume nearly a collective person-year! Also eliminated are costs incurred from making the initial copies and mailing them to all offices that have manuals.

New communications tools create new communications patterns (Evard, 1993). The most obvious communications shift caused by WWW and LANTERN is that people are comfortable with passing pointers to information collections, instead of trying to pass entire data object(s) through e-mail, hard copy, conversation, etc. This lessens the effort required for certain communication tasks and encourages more data sharing. There has been a tremendous growth of sharing of information in the past year. Many databases where information was solely the domain of a particular organization are now available to the entire center. For example, there is an equipment database that tracks who owns what equipment and where it is located. Previously an employee had to call the inventory office or and fill out a form to request an inventory of equipment. Now it is possible to query the equipment database in 12 different ways and get the data online in a matter of seconds.

Another time saving device is the center's electronic post office which started out as an X.500 server then subsequently was ported to the Web with necessary common gateway interface (CGI) scripts to incorporate the building and room locator service (Master Plan data). Currently, the electronic post office has 10,000+ accesses weekly. It is also noteworthy that the center telephone directory is printed once every year and half at a cost of US$15,000 and is out of date the moment its printed. However, the electronic post office receives weekly downloads from the personnel database and automatically does its updates at virtually negligible cost. Its these types of efficiencies that is maximizing information sharing among employees and increases work productivity while cutting costs.

Not exactly a failure, but a situation to be aware of is the cost of success. LANTERN has been successful, but it also raises expectations in the user community. Users, encouraged by the usefulness of the initial intranet, expect that anything can be done for everyone. Setting priorities and curbing expectations are difficult management tasks that occur in the wake of a successful project.

EPILOGUE AND LESSONS LEARNED

Through the Master Plan, Cicero, LANTERN, and the other related WWW projects, we have found that there are four pillars that are key to information technology project success (Figure 4):

- *Training* — How to use emerging information technology tools
- *Providing Useful Services* — Services Langley's staff needs and wants
- *Enabling Tools* — Providing information access tools to the customers
- *Advertising and Advocacy* — NASA's products and services around the nation and internally

By applying resources equally across each pillar, the Master Plan, Cicero and LANTERN have been successful individually, and have provided the foundation for a large strategic information interchange infrastructure for the center. Table 4 shows major activities for each project component and it is measured impact for people at Langley and the external customers.

The WWW, GIS, and ICE project teams have already been introduced. TAG and K-12 refer to the technology transfer and educational outreach project teams, respectively.

Figure 4. Four pillars of information technology success

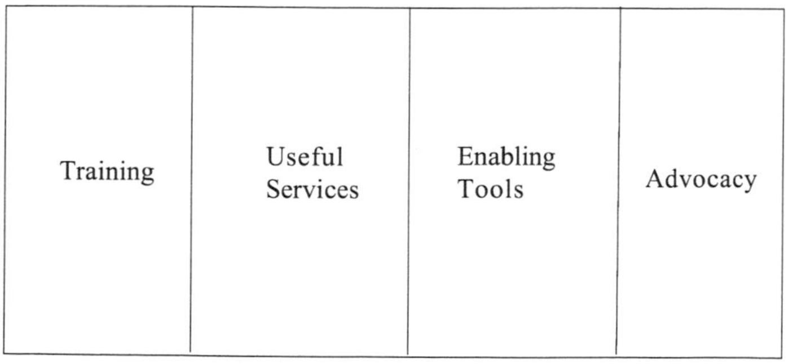

Training	Useful Services	Enabling Tools	Advocacy

Table 4. Activities and the impact chart (March 1-December 1, 1995)

Area	Activity	Participating Team	Number of People Reached
Training	WWW browser training classes	WWW, GIS	750
	3 WWW reference sheets	WWW	1,500
	WWW browser training video	WWW	500
Services	NASA Technical Report Server	WWW	100,000
	Langley Software Server	WWW, TAG	1,000
	Technology Opportunity Showcase online	WWW, TAG	15,000
	LANTERN - Langley's intranet	ICE, GIS , WWW	5,000
	WWW caching server	ICE	5,000
	Building & Room Locator	GIS	15,000
Enabling Tools	Netscape Site License	WWW	2,500
	MacHTTP Site License	WWW	300
Advocacy	TOPS '95	WWW, ICE, GIS, TAG	2,500
	Oshkosh '95	WWW, K-12, TAG	10,000
	Internet '95	WWW, GIS, ICE	400
	Internet Fair 2	WWW, GIS, ICE	2,000

Specific lessons we would give to others considering integrating information technology into their business environment include the following.

Get a strong senior management sponsor with resources.

WWW efforts have the highest probability of success when they begin as grass roots efforts, develop a small number of success stories, and then sell the concept to a sympathetic member of senior management. It is inevitable that WWW successes will be viewed as threats (real or imagined) by other members of the organizations, and a sponsor offers a greater level of protection.

Find someone who is a WWW evangelist with leadership skills.

Until a WWW effort is institutionalized, success is dependent on generally a single person who can both produce useful WWW services with limited resources while enduring changing perceptions of WWW technologies, from "when will this silver bullet fix my problem" to "WWW is a waste of time."

Train everyone how to use your tools and services.

Just as nearly all employees are both consumers and producers of information in traditional media formats, the same holds true for electronic format. It is everyone's responsibility to both use and contribute to an organizations information space.

ADVERTISE, ADVERTISE, ADVERTISE and then ADVERTISE some more.

Advertisement and outreach is a never ending effort. The resources required will go down when critical mass is required, but it is never "done." A danger to watch out for is that the technical organizations that generally assist in the development of WWW resources are generally unskilled in the advocacy of such resources, especially to non-technical customers.

Find and GET the best technical support you can find.

Services that are down, unavailable, or stale can do more harm than good. Enlist the best technical people possible, but ensure they have the commitment necessary for information maintenance.

World Wide Web has been the catalyst for many Information Technology projects at NASA Langley. The Master Plan, Cicero, and LANTERN are selected because they support each segment of the NASA Langley Work Relation Model; Facilities, Tools, and People respectively. For providing useful services, the focus for the future is to integrate more Master Plan data, push for wider application of Cicero, and transition more hard copy administrative data to LANTERN. Langley's intranet continues to grow, with even more procurement and tracking services slated for WWW interfaces. For training, in March 1996 the center began HTML authoring classes. For enabling tools, an Agency-wide site license for Netscape has been purchased. Advertising and outreach continues at every possible center event and at a personal, one-on-one advocacy and education. How do you know you have succeeded? When phrases such as "Well we can just put it up on LANTERN" are commonly overheard in the cafeteria.

QUESTIONS FOR DISCUSSION

1. The focus of this paper has been mostly applicable to large organizations. How would the issues, resolution, and services presented here be applicable to small organizations? What would be the same? What would not be applicable? What new challenges would exist that large organizations do not face?

2. WWW and related technologies are often applied to integrating a number of legacy database systems. If an organization could start with a clean slate, what would be an appropriate Information System architecture and what role would WWW play in the new system?

3. The Master Plan and related geographic information systems showed a clear path of evolution from hard copy to initial WWW interface. Discuss how a traditional management information system would make such an evolution. What additional security or privacy issues are introduced by the switch to management from geographic data?

4. The NASA Langley WWW Team proposed the "Four Pillars of Information Technology Project Success" which are: Training, Developing useful services, Providing enabling tools, and Advertising available resources. Discuss this four-pronged model. Are there additional pillars? Could the model be simplified to fewer pillars? Are there cases where the four pillars could be fully satisfied, but a project could fail?

5.   The concept was introduced of a grass roots team or project transitioning to official status with senior management approval and sponsorship. What is the best time for this transition to take place? What are the dangers of it occurring too early? Too late? What (if any) about this concept is applicable only to Information Technology projects?

REFERENCES

Auditore, P. (1995, Summer-Fall). Weaving a web inside the corporation: The future of collaborative computing. *Computer Technology Review*, 14-17.

Ball, W. B., Meyer, J. M., Gage, R. L., & Waravdekar, J. W. (1995). GIS Business Plan. NASA Langley Research Center. Retrieved from http://gis-www.larc.nasa.gov/bplan/

Berners-Lee, T., Calliau, R., Groff, J. F., & Pollermand, B. (1992). World-Wide Web: The information universe. *Electronic Networking: Research, Applications, and Policy, 2*(1), 52-58.

Bray, T. (1996). Measuring the Web. *World Wide Web Journal, 1*(3), 141-154.

Evard, R. (1993, November 1-5). Collaborative networked communication: MUDs as system tools. *Proceedings of the 7th System Administration Conference (LISA '93)*, Monterey, California (pp. 1-8).

Moreton, R. (1995). Transforming the organization: The contribution of the information systems function. *Journal of Strategic Information Systems, 4*, 149-164.

Nelson, M. L., & Bianco, D. J. (1994, October 18-20). The World Wide Web and technology transfer at NASA Langley Research Center. *Proceedings of the Second International World Wide Web Conference*, Chicago, Illinois (pp. 701-710).

Nelson, M. L., Gottlich, G. L., & Bianco, D. J. (1994). *World Wide Web implementation of the Langley Technical Report Server* (NASA TM-109162).

Nelson, M. L., Gottlich, G. L., Bianco, D. J., Paulson, S. S., Binkley, R. L., Kellogg, Y. D., Beaumont, C. J., Schmunk, R. B., Kurtz, M. J., Accomazzi, A., & Syed, O. (1995). The NASA Technical Report Server. *Internet Research: Electronic Networking Applications and Policy, 5*(2), 25-36.

Office of Public Affairs, NASA Langley. (1995). NASA Langley Fact Sheet. Retrieved from http://www.larc.nasa.gov/org/pao/PAIS/Langley.html

Shoosmith, J. (1993). *Introduction to the LaRC Central Scientific Computing Compex* (NASA TM-104092).

Sprout, A. (1995). The Internet inside your company. *Fortune, 132*(11), 161-168.

Wieseman, C. D. (Ed.). (1994). *The role of computers in research and development at NASA Langley Research Center* (NASA CP-10159).

Appendix A. URLs of listed resources

Resource	Uniform Resource Locator
Building and Room Locator	http://gis-www.larc.nasa.gov/cgi-bin/locator
Cicero	http://ice-www.larc.nasa.gov/ICE/doc/Cicero/
Integrated Computing Environment	http://ice-www.larc.nasa.gov/ICE/
Internet Fair 2	http://www.larc.nasa.gov/if2/
Langley Computer Users Committee	http://cabsparc.larc.nasa.gov/LCUC/lcuc.html
Langley Home Page	http://www.larc.nasa.gov/
LANTERN	(restricted access)
Master Plan	http://gis-www.larc.nasa.gov/
Post Office	http://post.larc.nasa.gov/cgi-bin/whois.pl
WWW Training	http://boardwalk.larc.nasa.gov/isd-www/training.html

Gretchen L. Gottlich's principal contributions at NASA included digital library work and World Wide Web team management.

John M. Meyer has been active in numerous areas at NASA Langley, including CADD, design of real-time video and data acquisition systems, and geographic information systems.

Michael L. Nelson works for NASA Langley Research Center's Information Systems Division, contributing to a variety of projects and he is the Webmaster for NASA Langley Research Center.

David J. Bianco began working for Computer Sciences Corporation in November of 1993. He has assisted NASA Langley's Information Systems and Services Division in many areas.

This case was previously published in J. Liebowitz & M. Khosrow-Pour (Eds.), *Cases on Information Technology Management in Modern Organizations*, pp.209-224, © 1997.

Chapter XV

Globe Telecom:
Succeeding in the Philippine Telecommunications Economy

Ryan C. LaBrie, Arizona State University, USA

Ajay S. Vinzé, Arizona State University, USA

EXECUTIVE SUMMARY

This case examines the role and implications of deregulation in the telecommunications sector on an IT-based services organization in the Philippines. Reports from international lending institutions suggest that investments in the telecommunications sector typically produce up to a 30-fold impact on the economy. Predictions like these have caused several of the emerging economies throughout the world to deregulate their telecommunications infrastructure in an attempt to leverage this economic potential. This case study specifically examines the actions of Globe Telecom from just prior to the 1993 Philippine deregulation through the present. Globe has continued to succeed despite the competition against the Philippine Long Distance Telephone Company, which at one time controlled over 90% of the telephone lines in the Philippines. Globe has been able to do this through strategic partnerships, mergers, and acquisitions. Furthermore, Globe has developed into a leading wireless provider by its effective use of modern information technology.

SETTING THE STAGE

Consider Fe Reyes. The resident of Quezon City, Manila's biggest residential district, waited nearly three decades for the nation's monopoly telephone service, Philippine Long Distance Telephone Co., to reach her doorstep. But last year, thanks to the 1993 deregulation that allowed rival companies to start offering phone service, she got a new company to install a line in just three days.[1]

The telecommunications sector in the Philippines was deregulated in 1993. Prior to the deregulation, the government-sponsored Philippine Long Distance Telephone Company (PLDT) handled the infrastructure and services requirements related to telecommunications. For most practical purposes, PLDT was commonly viewed as an operational arm of the government's Department of Transportation and Communications. Since the deregulation of 1993, over 150 new telecommunications infrastructure providers have been formed. Five players have now emerged as the leading keepers of telecommunications for the Philippines. This change has had a significant impact for the Philippines and for the Southeast Asian region in general. This new environment raises a variety of economic and technological issues that organizations need to recognize as they operate in the Philippines. With its geographical compositions of over 7,100 islands, the Philippines provides some unique challenges for information technologies and telecommunications. This case examines the current status of investments in the Philippines telecommunications infrastructure and their implications. Using a single representative organization-Globe Telecom-financial, competitive, regulatory, and technology pressures and opportunities are examined in light of a recently deregulated telecommunications sector. Using Globe Telecom as a focus organization, this case includes a macro perspective and provides insights and information that illustrate the impacts from a national and regional (Southeast Asia) perspective.

The pervasive utilization of information technology throughout the telecommunications sector inherently makes it ideally suited to study. Furthermore, economically speaking, the international investment banking sector has suggested that investments in the telecommunications sector typically produce a 30-fold return on investment for a host nation's economy. At a macro level, telecommunications can be viewed as an indicator of a country's development status. At an organizational level, telecommunications can be a source of competitive advantage (Clemons & McFarlan, 1986).

Understanding the Philippines

The Philippines unique geographical composition makes it an excellent case for a telecommunications study. Composed of over 7,100 islands, the Philippines is located in Southeast Asia off the coasts of China, Vietnam, and Malaysia, between the South China Sea and the Philippine Sea (see Exhibit 1). The nation encompasses an area of approximately 300,000 sq. km., comparable to the size of Arizona. There are roughly 80 million inhabitants of the Philippines, and approximately 11 million of those are located in metro Manila. Quezon City, within metro Manila, is the seat of the country's capital, while Makati is metro Manila's financial district. The Philippines has two official languages: Filipino and English. In fact, the Philippines is the third largest English-speaking country in the world, maintaining a 95% literacy rate. The Philippines gained their independences from the United States in 1946. Since that time, they have slowly moved toward a democracy, finally ratifying their new Constitution on February 2, 1987.

The Philippines is a member of the United Nations and the Association of South East Asian Nations (ASEAN). ASEAN plays a key role in the region and is comprised of the following countries: Brunei, Cambodia, Indonesia, Laos, Malaysia, Myanmar, Philippines, Singapore, Thailand, and Vietnam. The functional goals of ASEAN is to accelerate the economic growth, social progress and cultural development in the region through joint endeavors in the spirit of equality and partnership in order to strengthen the foundation for a prosperous and peaceful community of Southeast Asian nations. They also aim to promote regional peace and stability through abiding respect for justice and the rule of law in the relationship among countries in the region and adherence to the principles of the United Nations Charter. The ASEAN region has a population of approximately 500 million, a total area of 4.5 million square kilometers, a combined gross domestic product of US$737 billion, and a total trade of US$720 billion.[2]

In 1998, the Philippine economy — a mixture of agriculture, light industry, and supporting services — deteriorated as a result of spillover from the Asian financial crisis and poor weather conditions. Growth fell from 5% in 1997 to approximately -0.5% in 1998, but since has recovered to roughly 3% in 1999 and 3.6% in 2000. The government has promised to continue its economic reforms to help the Philippines match the pace of development in the newly industrialized countries of East Asia. This strategy includes improving infrastructure, overhauling the tax system to bolster government revenues, moving toward further deregulation and privatization of the economy, and increasing trade integration with the region.[3] In 2000, the inflation rate was estimated to be 5%, the unemployment rate 10%, national debt US$52 billion, and GDP US$76.7 billion.

The monetary unit of the Philippines is the Philippine Peso. Over the past few years, the Philippine pesos per U.S. dollar has devaluated quite dramatically due in large part to the economic phenomenon known as the "Asian Flu." This economic downturn was widespread and lasted throughout much of the late 1990s. Figure 1 shows the Philippine Peso dropping approximately 50% in five years, from almost US$0.04 to just under US$0.02 in value. During the first two years of the new millennium, the Philippine Peso has stopped its decline and has stabilized against the U.S. dollar.

The Philippine government has gone through a number of changes in the recent years. In January 2001, President Estrada was found unable to rule by the Philippine Supreme Court due to the large number of resignations in key cabinet positions. Vice President Gloria Macapagal-Arroyo assumed the presidency for the remainder of the term. The next presidential elections will be held May 2004. Prior to the Estrada presidency, other presidential reigns included Ferdinand Marcos (1965-1986), Corazon Aquinos (1986-1992), and Fidel Ramos (1992-1998).

History of Telecommunications in the Philippines

Until 1993, the Philippine telecommunications sector was completely dominated by a single, privately-owned company. Philippine Long Distance Telephone Company (PLDT) provided 95% of all telephone service in the Philippines. Their record of poor service and even worse investment left the nation with just 1.1 phone lines per 100 residents, and a backlog of over 800,000 requests with as much as a five-year wait. Consider the following story reported in the *Asian Wall Street Journal*.

Figure 1. The "Asian Flu" effect and the devaluation of the Philippine Peso (Source: National Telecommunications Commission, Republic of the Philippines)

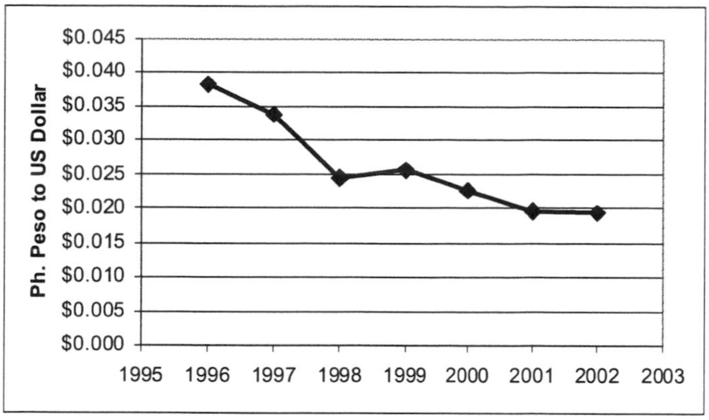

Bella Tan had just given birth to her first-born son when she and her husband applied for a phone line. Her son is now 17 years old. A daughter, now 11, has been added to the family. The phone line still hasn't arrived.[4]

In 1990, investment in the Philippines telecommunications sector was approximately 1% of GDP-about one-fourth of other Asian countries (World Bank, 2000).

In 1993, President Ramos signed two executive orders (Executive Order 59 and 109), in an attempt to spur competition in this sector. Executive Order 59 (see Exhibit 2) required PLDT to interoperate with other carriers, forcing them to share in the lucrative long distances market and to provide access to its subscribers. Executive Order 109 (see Exhibit 3) awarded local exchange licenses to other operators. In exchange for offering cellular or international gateways, the government also required the installation of landlines by those operators. For each license granted, a cellular company was required to build 400,000 landlines by 1998. Similarly, for each international gateway operator, 300,000 local lines were required to be added. The target of 4,000,000 new phone lines set under Executive Order 109 were met and even exceeded; with PLDT's contribution of over 1,250,000 lines, the total count of lines exceeded 5,250,000. Table 1 shows the number of lines committed and installed under the Basic Telephone program.

Table 2 shows the current status of telephone line distribution per region, including the actual number of lines installed and number of lines subscribed to. This table gives two teledensity numbers, one of capacity and one of subscription. While 9.05 is a substantial increase over the 1.1 teledensity previously provided by PLDT, and meets the goals the Philippine government set for the telecommunications sector for the year 2000, it is rather misleading in that a large number of those lines go unused. It is suspected that a large number of those lines run into business/office buildings and are not fully utilized, whereas those living in rural areas, and, in some cases, many who live in metro areas still go without. Investigating the actual subscription rates tells a whole different story. Subscribed teledensity shows that only four individuals out of every 100 have a

Table 1. Telephone line commitments and installations under the basic telephone program (Source: National Telecommunications Commission, Republic of the Philippines)

Carrier	Total lines required under Executive Order 109	Total lines committed under revised rollout	Cumulative lines as of 1998
Digitel	300,000	337,932	337,932
Globe Telecom	700,000	705,205	705,205
ICC/Banyantel	300,000	341,410	341,410
Islacom	700,000	701,330	701,330
Philcom	300,000	305,706	305,706
PILTEL	400,000	417,858	417,858
PLDT	0	1,254,372	1,254,372
PT&T	300,000	300,000	300,000
SMART	700,000	700,310	700,310
ETPI	300,000	300,497	200,050
All Carriers	**4,000,000**	**5,364,620**	**5,264,173**

Table 2. Teledensity in the Philippines by region (Source: National Telecommunications Commission, Republic of the Philippines)

REGION	TELEPHONE LINES	SUBSCRIBERS	POPULATION	TELEDENSITY	
				TELELINES	SUBSCRIBED
I	256,828	104,712	4,140,531	6.20	2.53
II	41,246	29,948	2,812,589	1.47	1.06
III	513,626	222,915	7,686,845	6.68	2.90
IV	1,086,030	470,817	11,301,272	9.61	4.17
V	136,465	61,047	4,755,820	2.87	1.28
VI	331,576	151,315	6,324,098	5.24	2.39
VII	484,968	182,278	5,539,177	8.76	3.29
VIII	100,468	48,272	3,743,895	2.68	1.29
IX	160,537	26,641	3,152,009	5.09	0.85
X	188,827	76,510	4,441,739	4.25	1.72
XI	366,971	126,168	5,749,821	6.38	2.19
XII	76,245	26,139	2,660,270	2.87	0.98
NCR	3,025,164	1,481,269	10,405,479	29.07	14.24
CAR	88,052	44,592	1,400,490	6.29	3.18
ARMM	48,959	8,764	2,206,106	2.22	0.40
TOTAL	**6,905,962**	**3,061,387**	**76,320,141**	**9.05**	**4.01**

telephone line, as compared to nearly 11 out of every 100 people for the rest of Asia and just over 17 per 100 people for the entire world's average.

A Tale of Two Regions

The telecommunications sector in the Philippines is really a tale of two regions, metro Manila and the rest of the Philippines. NCR is the National Capital Region, which includes metro Manila. Outside of NCR and Region IV, no other region even remotely nears the Philippine national teledensity average. With those two exceptions all other regions fall significantly below the national teledensity average.

Table 3. Philippine telecommunications industry structure (Source: National Telecommunications Commission, Republic of the Philippines)

TELECOM SERVICE	1997	1998	1999	2000
Local Exchange Carrier Service	76	76	76	77
Cellular Mobile Telephone Service	5	5	5	5
Paging Service	15	15	15	15
Public Trunk Repeater Service	10	10	10	10
International Gateway Facility	11	11	11	11
Satellite Service	3	3	3	3
International Record Carrier	5	5	5	5
Domestic Record Carrier	6	6	6	6
Very Small Aperture Terminal	4	4	4	5
Public Coastal Station	12	12	12	12
Radiotelephone	5	5	5	5
Value-Added Service	47	70	106	156

Telecommunications is more than just "plain old telephone service" (POTS). Like any country, the Philippines telecommunications industry is a mixture of a number of different services, some of which have remained relatively constant over that past several years, and some that have grown rapidly. Table 3 breaks down the Philippine telecommunication industry per the governmental recognized categories.

BACKGROUND[5]

In 1993, Globe Telecom was one of the first two companies granted licenses under Executive Order 109. At this time, Globe began its long uphill crusade against the Goliath Philippines Long Distance Telephone Company. Prior to 1993, Globe Telecom was an international data carrier allowing it to offer telephone and telegram services. Globe Telecom traces its roots back to the 1920s, descending directly from Dollaradio, a ship-to-shore radio and telegraph company later renamed Globe Wireless Limited (GWL). Globe is also heir to Philippine Press Wireless, Inc. (PREWI), founded to advocate independence in the Commonwealth era, and Clavecilla Radio System (CRS). A merger between GWL, PREWI, and Mackay Radio Telegraph established Globe-Mackay Cable and Radio Corporation (GMCR) in 1930. When GMCR sold 60% of its stocks to then Ayala and Co. in 1974, it had already been strengthened by this colorful history of partnerships. In the 1990s, GMCR and CRS merged to form GMCR, Inc., later renamed Globe Telecom, a leading telecommunications company offering domestic and international services.[6]

In 1993, Globe Telecom partnered Singapore Telecom International (SingTel). This partnership gave SingTel a 40% ownership in Globe Telecom (maximum allowed by Philippine law) and gave Globe Telecom the capital and expertise to grow in the burgeoning Philippine telecommunication environment.

In March 2001, Globe Telecom acquired Islacom as a wholly owned subsidiary. This acquisition brought many benefits to Globe Telecom including a new global partner-Deutsche Telekom, access to additional frequency spectrum, enlarged geographic landline access, and economies of scale cost savings capabilities. With this acquisition, the current capital structure of Globe Telecom is shown in Figure 2.

Figure 2. Ownership structure of Globe Telecom (Source: Philippine SEC 17-Q 1Q 2002 Report, http://www.globe.com.ph)

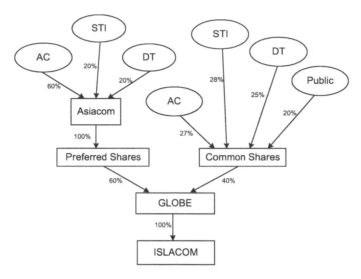

Where AC is the Ayala Corporation, STI is Singapore Telecommunication International, DT is Deutsche Telekom, and Public is the remaining shares available to public investors.

Globe is now a full-service telecommunications company offering cellular mobile telephone system (CMTS), fixed telephone and international communications services, International Private Leased (IPL) lines, Internet access, VSAT (Very Small Aperture Terminal) service, Inter-Exchange Carrier service, Frame Relay, Value-added Network Services (VANS) and other domestic data communications services.

Like any modern corporation, Globe Telecom is driven by a vision. Globe's vision draws upon its mission and value statements. It stresses their desire to be viewed as a solutions company that is best in its class. The full text of Globe Telecom's vision, mission and values statements can be found in the appendix (see Exhibit 4). Globe's mission statement supports their vision by suggesting what is important to them. Their goal is to improve the quality of life to the following identified key stakeholders: customers, shareholders, employees, community, and government. While many companies create vision and mission statements, few make them publicly available. Globe has made these principal statements as well as their core value statements available via their Internet Web site. Their value statements help to keep them focused on a few set of core competencies that include commitments to customers, employees, integrity, excellence, teamwork, society, and the environment.

The managing structure of Globe Telecom is congruent with any modern corporation. They are lead by a board of directors and a balanced executive management team (see Exhibit 5). Globe's achievements in recent years have prompted the company, during its 2001 annual stockholders meeting, to request that the Board of Directors be increased from the present 11 to a total of 15. Growth in boards is typically due to successful growth

Figure 3. Globe Telecom revenue composition for 2001 (Source: Globe Telecom Shareholders' Reports)

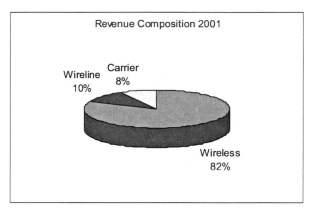

of the company; as a company grows, diversifies, and enters new markets additional board member are usually brought on for their insights. More recently, in the December 27, 2001-January 3, 2002 issue of *Far East Economic Review*, Globe Telecom made its debut appearance in the top ten best businesses, landing at number six in both leadership and quality.

Globe's primary revenue generator is their wireless product division. Wireless sales made up 82% of its 2001 revenue, followed by their wireline division and their carrier service. Figure 3 shows the breakdown of Globe Telecom's revenues between its three major divisions.

In 1999, Globe Telecom quadrupled their wireless subscription base. In 2000, Globe continued rapid wireless growth and nearly tripled its numbers of subscribers to more than 2.5 million users. This rapid pace continued as Globe nearly doubled its wireless subscription in 2001. Globe Telecom's wireless subscription base for the previous six years is shown in Figure 4.

Figure 5 shows that, despite the Asian Flu, Globe Telecom has been able to double its revenue consistently for the past five years, largely due to its growth in its wireless offerings. This increase in revenue has led to a share value increase of 27% for fiscal year 2000, even as the Philippine Stock Exchange Index declined by 30%.

CASE DESCRIPTION

The evolving story of Globe Telecom within the context of the telecommunications sector of the Philippines needs be examined at multiple levels. At a macro level, the story of Globe Telecom is analogous to the biblical story of David versus Goliath. Compared to PLDT, Globe Telecom is a rather small firm; however, it has firmly established itself as the number two telecommunications firm in the Philippines. At an organizational level, Globe Telecom has improved its position through strategic partnerships, mergers, and acquisitions. Finally, at a micro level, Globe has continued to use advances in information

Figure 4. Globe Telecom Wireless Subscribers, 1996-2001 (Source: Globe Telecom 2001 Shareholders' Reports and SEC Filings)

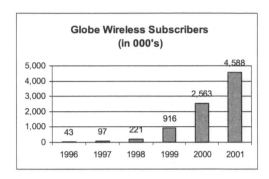

Figure 5. Globe Telecom revenue, 1997-2000 (Source: Globe Telecom Shareholders' Reports)

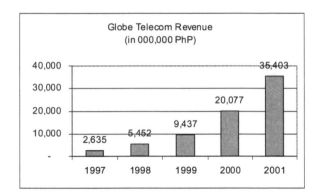

technology to provide innovations such as 100% digital wireless offerings, text messaging, and a host of other value-added offerings that you would come to expect from any modern telecommunications firm. Each of these levels is discussed in greater detail in the next section.

Deregulation Provides Opportunity for the Small Firm

In 1993, Globe Telecom was basically a long distance provider. It was in the initial phases of developing a cellular business, but it had no subscribers. The Philippine Long Distance Telephone Company (PLDT) completely monopolized the wireline market, controlling over 90% of all telephone service in the Philippines. Globe Telecom understood that it could not complete in the telecommunications market head to head with PLDT if its strategy was to provide additional wireline telephone service for the Filipino people. It had to differentiate itself, and it did that by providing an all-digital network infrastructure, thus laying the foundation for high quality wireline, wireless cellular, and wireless text messaging offering for its customers. Fast forward to 2000, while PLDT still maintains an 80% hold on the wireline market Globe Telecom has a majority of the digital wireless offering, garnering a 48% market share.

Partnerships, Mergers, and Acquisitions

Being a relatively small firm, Globe Telecom needed to find regional and global partners in order to survive the telecommunications shakedown of the past decade. As noted previously, prior to 1993, PLDT was the dominant telecommunications firm in the Philippines. With the liberalization of the telecommunications sector in 1993 by President Ramos, a host of smaller firms, Globe Telecom included, each attempted to carve out their own piece of the telecommunications pie. The Philippines went from a single player to 11 additional players within the span of four short years. As the telecommunications sector matures in the Philippines, consolidation is taking place, leading to just a handful of major players left competing. Most analysts now agree that Globe has firmly established itself as the number two telecommunications firm in the Philippines. It can be argued that this ranking is due in part to Globe's choices in strategic partnering. Very early on, the Ayala Corporation recognized that it needed a strong regional partner with expertise in world-leading telecommunications operations. Singapore Telecommunications International (SingTel) was the perfect partner to provide this leadership and established Globe as a serious contender against PLDT. Not to remain stagnate during the consolidation period beginning in the very late 1990s and the early part of 2000s, Globe Telecom acquired Islacom in 2001; this acquisition not only brought increased market share within the Philippines, but added a global partner in Deutsche Telekom. This acquisition could not have been timelier in the respect that just one year earlier PLDT had acquired Smart Communications, the second leading wireless communications firm in the Philippines to add to its previous wireless subsidiary, Pilipino Telephone Company (PilTel).

Wireless: The Enabling Technology of the Philippines

The Philippines telecommunications industry is a classical example of an emerging economy that has taken advantage of what has been described as a "leapfrog effect" in technology diffusion. The leapfrog effect is defined as when an old technology is largely bypassed in favor of a newer technology in a developing nation. Historically, it has been quite a challenge to lay landlines in the Philippines. Even after the liberalization of the telecommunications sector in 1993, landline subscription has only risen from 1.1% teledensity to 4% teledensity. Teledensity is defined as the number of subscribed phone lines per 100 inhabitants. While this is nearly a 400% increase in less than a decade, it still woefully trails the rest of the world's teledensity average of 16% and Europe's, Oceania's, and the America's teledensity averages of 35-40%. Figure 6 shows ASEAN national landline teledensity for the prior four years, as well as graphically displaying the Philippine landline teledensity against the ASEAN, Asian, and World landline teledensity averages.

Figure 6 shows that the Philippines clearly trails the ASEAN regional average. Furthermore, it shows that the Philippines has less than half of the number of telephone lines as the rest of Asia and less than one fourth the average number of subscribed lines per inhabitant as the rest of the world.

Figure 6. Landline teledensity (Source: International Telecommunications Union, 2002[7])

COUNTRY	1998	1999	2000	2001
Philippines	3.42	3.88	4.00	4.02
Brunei	24.68	24.59	24.52	24.52
Cambodia	0.21	0.23	0.24	0.25
Indonesia	2.70	2.91	3.14	3.70
Laos	0.55	0.66	0.75	0.93
Malaysia	20.16	20.30	19.92	19.91
Myanmar	0.52	0.55	0.56	0.58
Singapore	45.99	48.20	48.45	47.17
Thailand	8.49	8.70	9.23	9.39
Vietnam	2.25	2.68	3.19	3.76
ASEAN	4.25	4.50	4.75	5.11
Asia	7.46	8.33	9.55	10.84
World	14.38	15.18	16.17	17.19

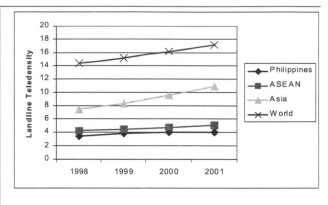

From Lagging to Leading

However, examining the cellular teledensity numbers quite a different story unfolds; one in which the Philippines, as a nation, is a leader rather than a laggard. From 1999 to 2000, cellular growth went from approximately 2.9 million to 6.5 million or an increase of 3.6 million. This is compared to landline growth from approximately 2.9 million to 3.1 million for a total increase of a mere 169,000. Figure 7 graphically depicts this rapid pace of cellular subscription growth rate in the Philippines.

This phenomenal growth has easily outpaced the other ASEAN nations, the rest of Asia, and nearly matches world as shown in Figure 8.

Another way to examine this data is to compare cellular subscribers to traditional landline subscribers as a percentage of usage (see Figure 9). Comparing landlines against cellular lines, it is shown that in 1999 cellular use was equal to landline use in the Philippines and by 2000 it had doubled landline use. Philippine cellular service increased from 41% to 77% in four years (and at the time 41% was an extremely high cellular to landline percentage). Figure 9 illustrates these comparisons.

Here it is interesting to note that as a percentage of total phone subscribers the Philippines lead all the regional and world averages. This demonstrates how the leapfrog effect has taken place in the telecommunications sector of the Philippines. All of this shows that Filipinos who choose to communicate have a greater opportunity with wireless options than with traditional landlines.

The Philippine Text Messaging Craze

Another phenomenon that is worthy of mentioning is the Philippines leads the world in text messaging usage. A recent *Asian Wall Street Journal*[8] article reports that text messaging is in the neighborhood of 70 million messages a day as compared with an estimated 30 million messages a day in all of Europe. Other reports note that text messaging may be hitting 100 million messages a day in the Philippines. To illustrate the power of text messaging in the Philippines, the same article examines how the mobile text

Figure 7. Cellular subscribers in the Philippines (Source: National Telecommunications Commission, Republic of the Philippines)

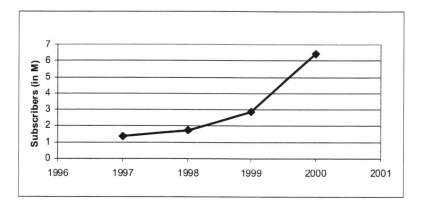

Figure 8. Cellular teledensity (Source: International Telecommunications Union, 2002)

COUNTRY	1998	1999	2000	2001
Philippines	2.38	3.83	8.44	13.70
Brunei	15.60	20.52	28.94	28.94
Cambodia	0.54	0.73	1.00	1.66
Indonesia	0.52	1.06	1.73	2.47
Laos	0.12	0.23	0.23	.52
Malaysia	10.11	13.70	21.32	29.95
Myanmar	0.02	0.03	0.06	0.03
Singapore	28.32	41.88	68.38	69.20
Thailand	3.33	3.90	5.04	11.87
Vietnam	0.29	0.42	0.99	1.54
ASEAN	1.68	2.45	4.20	6.58
Asia	3.08	4.53	6.59	9.25
World	5.40	8.21	12.15	15.48

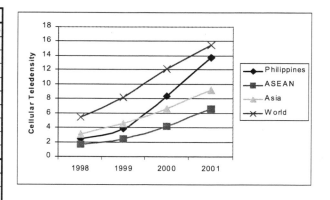

messaging network was utilized to mobilize hundreds of thousands of demonstrators to protest against former President Estrada. These protests eventually led to the ousting of Mr. Estrada in favor of the more telecommunications friendly Ms. Arroyo, the Philippines current president.

CURRENT CHALLENGES

The Philippines still struggles with getting landlines to many of its residents. This is a challenge not only for Globe Telecom, but also for the Philippine telecommunications

Figure 9. Cellular service as a percentage of total telephone percentage (Source: International Telecommunications Union, 2002)

COUNTRY	1998	1999	2000	2001
Philippines	*41.0*	*49.6*	*67.8*	*77.3*
Brunei	38.7	45.5	54.1	54.1
Cambodia	71.7	76.3	80.9	87.0
Indonesia	16.1	26.8	35.5	40.0
Laos	18.5	25.6	23.7	36.0
Malaysia	33.4	40.3	51.7	60.1
Myanmar	3.6	4.4	9.9	4.7
Singapore	38.1	46.5	58.5	59.5
Thailand	28.2	31	35.3	55.8
Vietnam	11.3	13.5	23.7	29.1
ASEAN	*21.5*	*28.5*	*38.0*	*44.6*
Asia	*29.2*	*35.2*	*40.8*	*46.0*
World	*27.3*	*35.1*	*42.9*	*47.4*

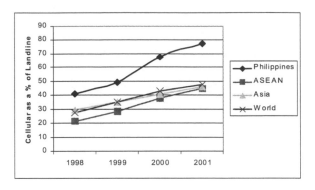

sector as a whole. It is going to take additional work by all of the telecommunications firms and the Philippine government to address this problem adequately.

The Philippines are still in the midst of an economic slow down that hit all of Asia during the decade of the 1990s. This Asian Flu has devalued the Philippine Peso by over 50% in the course of five years. The unemployment rate still hovers around 10%, and the average income is only US$3,800 per year. This sort of economic environment is challenging for any company to exist in, let alone succeed in. It is interesting to note that even the recent September 11, 2001, World Trade Center tragedy made it into Globe Telecoms third-quarter 2001 report as a negative economic effect.

Another serious challenge Globe Telecom is facing is that of customer retention. As in the United States and much of the rest of the world, the cellular telecommunications firms are providing strong incentives to switch from competitors. As Globe Telecom seeks to maintain, and even grow, in this particular market, it needs to seek out new and innovative ways of maintaining customer loyalty and continue to market aggressively for new subscribers, either as competition converts or first-time cellular users. A recent story noted a major blow to Globe's cellular subscriber base when a long time contract with the Philippine government was not renewed. The new contract was awarded to Smart Communications, which as you might have guessed, is a wholly owned subsidiary of PLDT.

Related to both the hard economic times and customer retention, Globe Telecom faces an equally tough challenge in maintaining a quality revenue stream. Pricing structures in this competitive market are leading to narrower margins. While providing free text messaging is a huge benefit to its customers and a direct pressure to its competitors, it is not conducive to providing value to the shareholders. Globe Telecom needs to continue finding appropriate ways to maintain and grow its revenue stream. This will require a combination of new value-added product offerings as well as growing its subscription base.

Globe Telecom and other telecommunications companies within the Philippines need to continue various partnerships in building additional high quality, cost-effective networks across the country. They need to lessen their dependence on the PLDT network where they must share revenue. Seamless interconnection is necessary for subscribers from both sides; however, less reliance on PLDT will increase profits and enhance customer satisfaction.

CONCLUSION

Development-social and economic-without telecommunications is not possible; but neither is telecommunications in a country without development. It will take more than a few telephone wires to break out of this vicious circle. (Ure, 1995, p.65)

While the Philippines have made significant progress since 1993, the country has a long way to go to be considered on par with the rest of the industrialized world with respects to telecommunications. The challenges faced in the Philippines are so large that no single telecommunications firm can solely alleviate them by themselves. It will take continuing commitments from the Philippine government, Philippine telecommunications companies like Globe Telecom, and a host of international partners to provide adequate telecommunications to the Philippine people.

Globe Telecom has demonstrated that an innovative firm can compete and survive in a industry that at one time was monopolistically dominated by a single player. Attempting to compete with PLDT directly in landlines would have caused almost certain failure for Globe. However, by focusing on a differentiated strategy offering long distance, landline, and wireless communications, Globe was able to compete successfully in several different areas of telecommunications. This diversity has lead Globe Telecom to become a leader in digital wireless communications in the Philippines.

Strong alliances with Singapore Telecommunications International, and more recently Deutsche Telekom, have allowed Globe to remain competitive and boost its recognition as a worldwide telecommunications company. With the strategic acquisition of Islacom, Globe was able to expand its holdings in the Philippines, acquiring additional bandwidth and subscribers, all while maintaining profitability.

It can be argued that Globe Telecom has been successful due in part to their aggressive use of information technology. While the telecommunications industry in general is highly dependent on technology, Globe Telecom has captured and maintained competitive advantage through their dedication early on to a 100% digital infrastructure. This decision has enabled them to provide superior wireless communications quality and services. Furthermore, it has allowed Globe to lead the Philippines in text messaging, a communication format so vital in the Philippines that they lead the world in its usage. As shown in Exhibit 6 Globe Telecom continues to reap rewards due to the innovational offerings it is able to provide to its customers based on its advanced digital informational technology infrastructure.

This case sheds light on the difficulties of telecommunications in the Philippines. From the lack of infrastructure-four lines per 100 inhabitants, to the geographical challenges of an island nation with some 7,100 islands. Despite these and a variety of other challenges, Globe Telecom has grown into one of the leading telecommunications providers in the Philippines. Strategic decisions by Globe Telecom have resulted in their recognized leadership in customer-service quality. By being the first to develop a 100% digital wireless network, they were able to lead the way in text messaging, used more in the Philippines than anywhere else in the world. Their commitment to innovation through technology provides another example of how Globe Telecom uses its resources to maintain a competitive advantage. Globe Telecom is now a leading business in the Philippines, and a worldwide example of how a telecommunications firm can succeed in an emerging economy.

REFERENCES

Clemons, E. K., & McFarlan, F. W. (1986, July-August). Telecom: Hook up or lose out. *Harvard Business Review*, 91-97.

Clifford, M. (1994, October 6). Talk is cheap. *Far Eastern Economic Review*, 78-80.

Hookway, J. (2001, December 27). Review 200/Philippines: Bon appetit. *Far Eastern Economic Review*.

Liden, J., & Reyes, R. (1996, April 15). Asian infrastructure: Telecommunications — Philippines: Aggressive marketing pushes prices lower. *Asian Wall Street Journal*, p. S10.

Lopez, L. (2001, March 8). Manila picks up reform pace — New government pledges to enhance telecommunications industry competition — Popular mobile networks support people power, attract foreign interest. *Asian Wall Street Journal*, p. N2.

National Telecommunications Commission, Republic of the Philippines. (n.d.). Retreived May 31, 2002, from http://www.ntc.gov.ph/

Reyes, R., & Liden, J. (1997, June 10). Asian infrastructure: Telecommunications — Philippines: A vast and speedy deregulation has made this country a test case — And a basket case. *Asian Wall Street Journal*, p. S3.

Riedinger, J. (1994). The Philippines in 1993: Halting steps toward liberalization. *Asian Survey, 34*(2), 139-146.

Smith, P. L., & Staple, G. (1994). *Telecommunications sector reform in Asia: Toward a new pragmatism*. Washington, DC: The World Bank.

Ure, J. (1995). *Telecommunications in Asia: Policy, planning and development*. Hong Kong: Hong Kong University Press.

Weiss, E. (1994). Privatization and growth in South East Asia. *Telecommunications, 28*(5), 95-101.

World Bank. (2000). *Private solutions for infrastructure: Opportunities for the Philippines*. Washington, DC: The World Bank.

World factbook. (2001). Retrieved from http://www.cia.gov/cia/publications/factbook/index.html

ENDNOTES

1   *Asian Wall Street Journal*, June 10, 1997
2   ASEAN Objectives, found at http://www.aseansec.org
3   Information obtained from the *World Factbook 2001* at http://www.cia.gov

4   *Asian Wall Street Journal*, April 15, 1996
5   All data used in this case came from publicly available sources and the management
    team of Globe Telecom has not been consulted for the analysis or conclusions
    stated in this case study.
6   http://www.globe.com.ph
7   http://www.itu.org
8   *Asian Wall Street Journal*, March 8, 2001

APPENDIX

Exhibit 1. Map of the Philippines and its Surroundings

Exhibit 2. Executive Order No. 59

PRESCRIBING THE POLICY GUIDELINES FOR COMPULSORY INTERCONNEC-TION OF AUTHORIZED PUBLIC TELECOMMUNICATIONS CARRIERS IN ORDER TO CREATE A UNIVERSALLY ACCESSIBLE AND FULLY INTEGRATED NATIONWIDE TELECOMMUNICATIONS NETWORK AND THEREBY ENCOURAGE GREATER PRI-VATE SECTOR INVESTMENT IN TELECOMMUNICATIONS

WHEREAS, in recognition of the vital role of communications in nation-building, it has become the objective of government to promote advancement in the field of telecommunications and the expansion of telecommunications services and facilities in all areas of the Philippines;

WHEREAS, there is a need to enhance effective competition in the telecommunications industry in order to promote the State policy of providing the environment for the emergence of communications structures suitable to the balanced flow of information into, out of, and across the country;

WHEREAS, there is a need to maximize the use of telecommunications facilities available and to encourage investment in telecommunications infrastructure by service providers duly authorized by the National Telecommunications Commission (NTC);

WHEREAS, there is a need to ensure that all users of the public telecommunications service have access to all other users of the service wherever they may be within the Philippines at an acceptable standard of service and at reasonable cost;

WHEREAS, the much needed advancement in the field of telecommunications and expansion of telecommunications services and facilities will be promoted by the effective interconnection of public telecommunications carriers or service operators;

WHEREAS, the Supreme Court of the Philippines, in the case of Philippine Long Distance Telephone Co. v. The National Telecommunications Commission [G.R. No. 88404, 18 October 1990, 190 SCRA 717, 734], categorically declared that "Rep. Act No. 6849, or the Municipal Telephone Act of 1989, approved on 8 February 1990, mandates interconnection providing as it does that 'all domestic telecommunications carriers or utilities . . . shall be interconnected to the public switch telephone network.'";

WHEREAS, under Executive Order No. 546 dated 23 July 1979, as amended, the NTC has the power, as the public interest may require, "to encourage a larger and more effective use of communications facilities, and to maintain effective competition among private entities whenever the NTC finds it reasonably feasible"; and

WHEREAS, there is a need to prescribe the consolidated policy guidelines to implement Rep. Act No. 6849 and Executive Order No. 546, as amended.

NOW, THEREFORE, I, FIDEL V. RAMOS, President of the Republic of the Philippines, by virtue of the powers vested in me by law, do hereby order:

Section 1. The NTC shall expedite the interconnection of all NTC authorized public telecommunications carriers into a universally accessible and fully integrated nationwide telecommunications network for the benefit of the public.

Section 2. Interconnection between NTC authorized public telecommunications carriers shall be compulsory. Interconnection shall mean the linkage, by wire, radio, satellite or other means, of tow or more existing telecommunications carriers or operators with one another for the purpose of allowing or enabling the subscribers of one carrier or operator to access or reach the subscribers of the other carriers or operators.

Section 3. Interconnection shall be established and maintained at such point or points of connections, preferably at the local exchanges level and at the junction side of trunk exchanges as are required within a reasonable time frame and shall be for sufficient capacity and in sufficient number to enable messages conveyed or to be conveyed to conveniently meet all reasonable traffic demands for conveyance of messages between the system of the parties involved in the interconnection.

Section 4. Interconnection shall permit the customer of either party freedom of choice on whose system the customer wishes his call to be routed regardless which system provides the exchange line connecting to the local exchange. Such a choice may be done initially through the use of distinct carrier access code assigned to the relevant connectable system and ultimately, as the local exchange providers upgrade to stored-program-controlled (SPC) exchanges, comparatively efficient interconnect (CEI) or equal access pre-programmed option.

Section 5. Interconnection shall be mandatory with regard to connecting other telecommunications services such as but not limited to value-added services of radio paging, trunking radio, store and forward systems of facsimile or messaging (voice or data), packet switching and circuit data switching (including the conveyance of messages which have been or are to be transmitted or received at such points of connection), information and other services as the NTC may determine to be in the interest of the public and in the attainment of the objective of a universally accessible, fully integrated nationwide telecommunications network.

Section 6. Interconnection shall be negotiated and effected through bilateral negotiations between the parties involved subject to certain technical/operational and traffic settlement rules to be promulgated by the NTC; Provided, that if the parties fail to reach an agreement within ninety (90) days from date of notice to the NTC and the other party of the request to negotiate, the NTC shall, on application of any of the parties involved, determine the terms and conditions that the parties have not agreed upon but which appear to the NTC to be reasonably necessary to effect a workable and equitable interconnection and traffic settlement.

Section 7. Interconnection among public communications carriers shall be effected in such a manner that permits rerouting of calls from an international gateway operator which is rendered inoperative, whether in whole or in part, in the event of strikes, lock-outs, disasters, calamities and similar causes, to another international gateway operator not so affected. A public telecommunications carrier shall be allowed such permits to operate an international gateway as may be necessary to service its own network requirements; Provided, that its subsidiaries shall not be given a permit to operate another international gateway.

Section 8. In prescribing the applicable technical/operational and traffic settlement rules, the NTC shall consider the following:

8.1 The technical/operational rules should conform with the relevant recommendations of the Consultative Committee on International Telegraph and Telephone (CCITT) and the International Telecommunications Union (ITU).

8.2 For traffic settlement rules:

(a) Either meet-on-the-air and/or midpoint circuit interconnection between parties;

(b) For local exchange point of interconnection, settlement shall be on the basis of volume of traffic on the local connection based on per minute with day and night rate differential. In case of store and forward services for facsimile, data and voice mail, settlement shall be on the basis of equivalent monthly trunk line charges as generally charged by the local exchange carrier (LEC) to its customer owning their own PABX;

(c) For junction exchange point of interconnection, settlement shall be on the basis of volume of traffic carrier over:

(i) short haul connection not exceeding 150 kilometers; and

(ii) long haul connection exceeding 150 kilometers.

Similarly, a per minute rate shall be evolved with day and night differential. The determination of the per minute rate is based on the principle of recognizing recovery of the toll related cost and fair return of the investment of the facilities employed in making the toll call exchange between the systems.

(d) Subsidies which shall be approved on the basis of the sound public policy shall be allowed in two (2) ways:

(i) for operator assisted calls - an operator surcharge kept by the system that employs the operator; and

(ii) access charge - the principle of access charge is an assistance to the unprofitable rural telephone development, remote pay stations, etc., thereby assuring the universal service obligation of the PSTN operators. The introduction of the access charge may result in a charge that will be passed on to the subscribers of the PSTN.

Section 9. Interconnection shall at all times satisfy the requirements of effective competition and shall be effected in a non-discriminatory manner.

Section 10. The Points of Connection (PC) between public telecommunications carriers shall be defined by the NTC, and the apportionment of costs and division of revenues resulting from interconnection of telecommunications networks shall be approved or prescribed by the NTC.

Section 12. Interconnection and revenue-sharing agreements approved or prescribed by the NTC may be revoked, revised, or amended as the NTC deems fit in the interest of the public service.

Section 13. In the implementation of this Executive Order, the NTC may, after due notice and hearing, impose the following penalties in case of violation of any of the provisions hereof:

13.1. Imposition of such administrative fines, penalties and sanctions as may be allowed or prescribed by existing laws;

13.2. Suspension of further action on all pending and future applications for permits, licenses or authorizations of the violating carrier or operator and in which particular case, the NTC shall be exempted from compliance with the provisions of Executive Order No. 26 dated 7 October 1992 on the period for the disposition of cases or matters pending before it;

13.3. With the approval of the President, directive to the appropriate government financial or lending institutions to withhold the releases on any loan or credit accommodation which the violating carrier or operator may have with them;

13.4. Disqualification of the employees, officers or directors of the violating carrier or operator from being employed in any enterprise or entity under the supervision of the NTC; and

13.5. In appropriate cases, suspension of the authorized rates for any service or services of the violating carrier or operator without disruption of its services to the public.

Section 14. The NTC is directed to promulgate the implementing rules to this Executive Order within ninety (90) days from the date of effectivity hereof.

Section 15. All executive orders, administrative orders, and other issuance inconsistent herewith are hereby repealed, modified or amended accordingly.

Section 16. This Executive Order shall take effect immediately.

DONE in the City of Manila, this 24th day of February in the year of Our Lord, Nineteen Hundred and Ninety-Three.

(Sgd.) FIDEL V. RAMOS
By the President
(Sgd.) ANTONIO T. CARPIO
Chief Presidential Legal Counsel

Exhibit 3. Executive Order No. 109

POLICY TO IMPROVE THE PROVISION OF LOCAL EXCHANGE CARRIER SERVICE

WHEREAS, local exchange service is fundamental to the goal of providing universal access to basic and other telecommunications services;

WHEREAS, during the development phase, cost-based pricing of services such as national and international long distance and other telecommunications services may be employed to generate funds which my then be used to subsidize the local exchange service;

WHEREAS, while the telecommunications sector as a whole is profitable, the profits mainly come from the toll services particularly from the international long distance service; and

WHEREAS, there is a need to promulgate new policy directives to meet the targets of Government through the National Telecommunications Development Plan (NTDP) of the Department of Transportation and Communications (DOTC), specifically: (1) to ensure the orderly development of the telecommunications sector through the provision of service to all areas of the country, (2) to satisfy the unserviced demand for telephones and (3) to provide healthy competition among authorized service providers.

NOW, THEREFORE, I, FIDEL V. RAMOS, President of the Republic of the Philippines, by virtue of the powers vested in me by law do hereby order:

Section 1. Definition of Terms. The following definitions shall apply within the context of this policy:

(a) Basic Telecommunications Service - refers to local exchange residence and business telephone service and telegraph service without additional features;

(b) Cost-based pricing - refers to a system of pricing in which the actual cost of providing service establishes the basic charge to which a fixed mark-up is added to collect a standard charge to all users without discrimination;

(c) Local Exchange Carrier Service - refers to a telecommunications service, primarily but not limited to voice-to-voice service, within a contiguous geographic area furnished to individual

Subscribers under a common local exchange rate schedule;

(d) Value-based pricing - also known as value of service pricing refers to a system of pricing where cost of .service establishes the minimum charge and a variable mark-up is added to collect revenue from those who value .the service more highly; and

(e) Universal Access - refers to the availability of reliable and affordable telecommunications service in both urban and rural areas of the country.

Section 2 Objective. The objective of this policy is to improve the provision of local exchange service in unserved and underserved areas as defined by the National

Telecommunications Commission (NTC), thus promoting universal access to basic telecommunications service.

Section 3. General Policy. The Government shall pursue the policy of democratization in the ownership and operation of telecommunication facilities and services.

Section 4. Cross-Subsidy. Until universal access to basic telecommunications service is achieved, and such service is priced to reflect actual costs, local exchange service - shall continue to be cross-subsidized by other telecommunications services within the same company.

Section 5. Service- Packaging. Authorized international gateway operators shall be required to provide local exchange service in unserved and underserved areas, including Metro Manila, within three (3) years from the grant of an authority from the NTC, under the following guidelines:

(a) Authorized gateway operators shall provide a minimum of three hundred (300) local exchange lines per international switch termination;

(b) At least one (1) rural exchange line shall be provided for every ten (10) urban local exchange lines installed;

(c) The establishment of Public Calling Offices at the rural barangay level shall be given an appropriate credit by the NTC towards the obligation to provide local exchange service.

The above figures are derived from the following factors: number of exchange lines, number of international switch-terminations, traffic, grade of service and demand;

(d) No permit for an international gateway facility shall be granted an applicant unless there is a clear showing that it can establish the necessary foreign correspondenceships; and

(e) Carriers already providing local exchange service in accordance with Section (a), (b) and (c) shall be authorized to operate an international gateway subject to applicable laws.

Section 6. Subsidiary. The subsidiaries of a public telecommunication carrier operating an authorized international gateway shall not be allowed to operate another gateway in accordance with Executive Order No. 59 (1993).

For this purpose, a telecommunications company shall be considered as a subsidiary if any or all of the following conditions exist:

(a) The two companies share the services of key operating and management personnel;

(b) The shareholdings of one company, together with the shareholdings of its stockholders, in the other company aggregate more than fifty percent (50%) of the outstanding capital stock of the letter company; or

(c) One company and its stockholders have a combined exposure in the other company in the form of loans, advances, or asset-lease equivalent to more than fifty percent (50%) of the capital accounts of the other company.

Section 7. Cellular Mobile Telephone System. Authorized international gateway operator may also be authorized to provide Cellular Mobile Telephone System (CMTS) service and other non-basic telecommunications service which are possible source of subsidy for local exchange carrier service.

Section 8. Non-Basic Services. Authorized providers of other non-basic telecommunications service which are possible sources of subsidy shall be required to provide local exchange carrier service in accordance with guidelines, rules and regulations prescribed by the NTC.

Section 9. Duration of Services. The obligation to provide local exchange carrier service shall remain in force for as long as the service providers described in Sections 5, 7 and 8 hold their authorizations to provide their respective non-basic services.

Section 10. Other Requirements. The foregoing provisions shall be without prejudice to the other requirements for the grant of franchises and Certificates of Public Convenience and Necessity.

Section 11. Interconnection Requirement. All telecommunications service networks shall be interconnected in a non-discriminatory manner in accordance with Executive order No. 59 (1993) and its implementing guidelines.

Section 12. Financial Reporting Requirements. The internal subsidy flows shall be made explicit in the financial reporting system of the telecommunications service providers.

Section 13. Policy Implementation. The NTC is hereby directed to promulgate the guidelines, rules and regulations to implement this Executive Order within (30) thirty days from the effective date of this Executive Order.

Section 14. Violations. Any violation of the Executive Order shall be subject to the same penalties provided for in Section 13 of Executive Order No. 59 (1993).

Section 15. Transitory Provisions. Existing telecommunications servicee providers described in Section 5, 7 and 8 shall have a period of five (5) years to comply with the above requirements to provide local exchange service.

Section 16. Pending Applications. Telecommunications service providers with existing and pending applications for International Gateway Facility, Cellular Mobile System (CMTS) and other Value Added Services (VAS) providers need not revise their applications with the NTC. However, upon issuance of the Provisional Authority of CPCN, as the case may be, they shall be given a period of three (3) months within which to submit and file the necessary applications for local exchange service in accordance with the provisions hereof.

Section 17. Repealing Clause. All executive orders, administrative orders and other Executive issuance inconsistent herewith are hereby repealed, modified or amended accord.

Section 18. Effectivity. This Executive Order shall take effect immediately.

DONE in the City of Manila, this 12th day of July in the year of the Lord, Nineteen Hundred and Ninety-Three.

(sgd) FIDEL V. RAMOS
By the President:
TEOFISTO T. GUINGONA, JR.
Executive Secretary

Exhibit 4. Globe Telecom's Business Vision, Mission, and Value Statements

Vision

The pursuit of our mission is guided by the company's vision and actualizes our corporate values:

Globe Telecom provides more than just lines. We advance the quality of life of individuals and organizations by delivering the SOLUTIONS to their communications-based needs. We provide quality and personalized service that exceeds our customers' needs and expectations. We are driven by a culture of excellence and innovation, enabled by best-in-market talent and superior operating effectiveness and flexibility. WE ARE THE COMPANY OF CHOICE BECAUSE, IN WHAT WE PROVIDE, WE ARE THE BEST.

Mission

Our mission is to advance the quality of life by delivering the best solutions to the communications-based needs of our subscribing publics. We take lead of the industry as the service provider of choice. We secure our competitive edge by packaging solutions enhanced by pioneering innovations in service delivery, customer care, and the best appropriate technologies. We acknowledge the importance of our key stakeholders. In fulfilling our mission, we create value for:

- Customers: Customer satisfaction is the key to our success. We help individuals improve their way of life and organizations do their business better.
- Shareholders: Our business is sustained by the commitment of Ayala Corporation and Singapore Telecom International. We take pride and build on the value our shareholders provide. In return, we maximize the value of their investments.
- Employees: Our human resources are our most valuable assets. We provide gainful employment that promotes the dignity of work and professional growth and thus attract and retain best-in-market talent.
- Community: Community support is vital. We will act as responsible citizens in the communities in which we operate.
- Government: We are the partners of government in nation building. We support and participate in the formation of policies, programs and actions that promote fair competition, judicious regulation and economic prosperity.

Values

These values are the anchor of our corporate existence:

- Customer Commitment — A steadfast pledge to provide only the best to the customer
- Excellence — The relentless pursuit of outstanding performance
- Integrity —A faithful adherence to the highest ethical standards

- Primacy and Worth of the Individual — Respect for every employee as a unique individual, a professional in his own right, and with his own sense of dignity and self-worth
- Teamwork — The collective drive to achieve the company's vision and mission and uphold the company's values
- Commitment to Society and the Environment — A responsibility to uplift the quality of people's lives and protect the environment

Exhibit 5. Globe Telecom Leadership

Board of Directors	Executive Management Team
Jaime Agusto Zobel de Ayala II *Chairman*	Gerardo C. Ablaza, Jr. *President & Chief Executive Officer*
Lee Shin Koi *Co-Vice Chairman*	Edward Ying *Chief Operating Adviser*
Delfin L. Lazaro *Co-Vice Chairman*	Gil B. Genio *Islacom Chief Operating Officer & Senior Vice President*
Renato O. Marzan *Corporate Secretary*	Manuel R. De los Santos *Senior Vice President – Wireless Data*
Gerardo C. Ablaza, Jr.	Delfin C. Gonzalez, Jr. *Senior Vice President & Chief Financial Officer*
Fernando Zobel de Ayala	Oscar L. Contreras, Jr. *Senior Vice President – Human Resources*
Lucas Chow	Rodolfo A. Salalima *Senior Vice President – Corporate & Regulatory Affairs*
Rufino Luis T. Manotok	Rafael L. Llave *Vice President – Logistics & Management Services*
Mark Anthony N. Javier	Lizanne C. Uychaco *Vice President – Retail Operations & Centers Management*
Tay Chek Khoon	Rodell A. Garcia *Vice President & Chief Information Officer*
Edward Ying	Rebecca V. Ramirez *Vice President – Internal Audit*
	Emmanuel A. Aligada *Vice President – Customer Service*
	Joaquin L. Teng, Jr. *Vice President – Fixed Network Business*
	John W. Young *Vice President – Carrier Business*

Exhibit 6. Key Historical Advances in Globe Telecom, 1994-2002

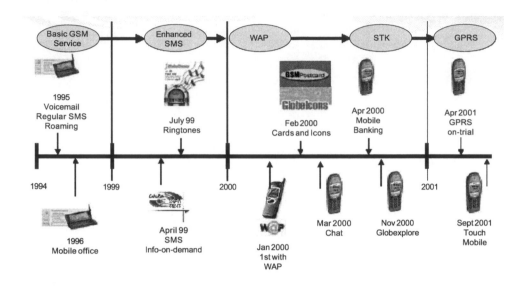

Source: *Globe Telecom Product Development Unit*

Ryan C. LaBrie is currently completing his doctoral studies in information systems at Arizona State University, USA. Prior to starting his doctoral program Mr. LaBrie worked at the Microsoft Corporation for ten years, most recently as a program manager in the Enterprise Knowledge Management organization. He has been involved with instructing at the university and corporate levels in the U.S. and internationally including Australia, France, Indonesia, Japan, Malaysia, Singapore, Thailand, and the UK. His teaching and research interests include: international information technology issues, database and data warehousing, and information ethics. Mr. LaBrie holds an MS in information systems and a BS in computer science from Seattle Pacific University.

Ajay S. Vinzé is the Davis distinguished professor of information management at Arizona State University, USA. He received his PhD in MIS from the University of Arizona. Dr. Vinze's research, teaching and consulting interests focus on both IS strategy and technology issues. He has worked on topics related to decision support and business intelligence, computer supported collaborative work and applications of artificial intelligence technology for business problem solving. His publications have appeared in many of the leading MIS journals. Before joining the academic environment, he was an IT consultant based in Southeast Asia. He is presently active with the business community in the U.S. with organizations like NASA, IBM, Motorola and internationally in Argentina, Australia, India, Mexico, New Zealand, Peru, Philippines, and Russia.

This case was previously published in the *Annals of Cases on Information Technology*, Volume 5/2003, pp. 333-357, © 2003.

Chapter XVI

Implementing a Wide-Area Network at a Naval Air Station:

A Stakeholder Analysis[1]

Susan Page Hocevar, Naval Postgraduate School, USA

Barry A. Frew, Naval Postgraduate School, USA

LCDR Virginia Callaghan Bayer, United States Navy, USA

EXECUTIVE SUMMARY

The Naval Air Systems Team is an organization wishing to capitalize on the benefits derived from connecting geographic stakeholders using wide-area network technologies. The introduction of common e-mail, file transfer, and directory services among these organizations is envisioned as a significant enabler to improve the quality of their aggregate product. At the same time this organization has decided to transform itself from a traditional functionally hierarchic organization to a competency based organization. The new model introduces a modified matrix organization consisting of integrated program teams at 22 geographically separate sites in the United States. This case study illustrates the use of a non-traditional approach to determine the requirements for the Naval Air Systems Team Wide-Area Network (NAVWAN). It is considered to be non-traditional because the case data enable the use of stakeholder analysis and SWOT (strengths, weaknesses, opportunities, threats) assessments to determine the requirements instead of asking functional proponents about function and data requirements. This is an action planning case. The case objective is to apply these methodologies and an understanding of organizational change to developing an action plan recommendation for implementation of a wide-area network.

BACKGROUND

The Naval Air Systems Team (NAST) is the component of the United States Department of Defense that is responsible for delivering aircraft and related systems to be operated, based, and supported at sea. To that end, this organization employs 42,000 civilians and 4,500 military personnel[2] at commands and bases throughout the country. Examples of products provided by this organization include air anti–submarine warfare mission systems; aircraft and related systems for aircraft carriers; maritime and air launched and strike weapons systems; and training in the operation and maintenance of these systems.

In April 1992, NAST, then headed by Vice Admiral (VADM) William C. Bowes, initiated a significant organizational restructuring as part of a large-scale change effort to enhance organizational effectiveness. The structure changed from that of a traditional functional hierarchy to a Competency Aligned Organization (CAO) which is a modified matrix organization that established dedicated Integrated Program Teams located at 22 different sites across the country. These teams are comprised of personnel from relevant functional competencies and coordinate activities that often span multiple command locations. A wide-area network (WAN) was identified as a critical infrastructure requirement for the success of these teams.

VADM Bowes became the champion for the implementation of a Naval Air System Team Wide-Area Network (NAVWAN) system. He viewed this infrastructure upgrade as critical to the success of the Competency Aligned Organization. He established a Demonstration-Validation team to perform the systems analysis, design, and implementation of the NAVWAN. This team identified several prototype implementation sites to be used to both validate the functionality of the NAVWAN and provide data to support a full system implementation.

As part of this effort, the Validation Team sponsored a research effort to conduct a stakeholder analysis at one of the prototype implementation sites. This analysis was designed as an alternative to the traditional design phase for a new information system implementation. The Department of the Navy has traditionally used a waterfall method to design and implement new information system technologies. This method begins with a requirements analysis and is followed by design, coding, testing, and maintenance. The focus of the case study presented in this chapter is on the requirements analysis phase. In the requirements analysis phase, the traditional waterfall method would focus on identifying specific types of data the system would need to be able to manipulate and on the business functions being performed.

The data gathered and presented in this case are intended to provide an alternative methodology for requirements analysis and implementation planning. These data were derived from interviews with representatives of each of the critical stakeholders at this implementation site. This case changes the traditional requirements analysis focus from that of data and function for the waterfall method to that of broader stakeholder issues in the application being developed.

This is an action planning case. The data presented provide information that can be used to develop a set of recommendations to be presented to the Validation Team and ultimately to VADM Bowes regarding the NAVWAN, including: planning strategies; user requirements; implementation strategies and schedules; resource allocation; training strategies and schedules; and maintenance strategies. The task of the reader, at the close of this case, will be to generate the recommendations supported by this stakeholder analysis.

SETTING THE STAGE: NAVWAN PROTOTYPE IMPLEMENTATION SITE

The Miramar Naval Air Station (NAS) is located in Southern California just north of the city of San Diego. The Naval Air Station chain of command is similar to other air stations and includes a Commanding Officer, Executive Officer, Supply Officer, Aviation Intermediate Maintenance Officer, Administrative Officer, Security Officer, and a Staff Civil Engineer. At this base, there are also several tenant activities. Tenant activities have their own chain of command, but they are located at the NAS and they rely on the NAS to provide infrastructure support including supply, facilities maintenance, and administrative services. NAS Miramar and the tenant activities who are participating in the Naval Air Systems Team (NAST) are potential users of the NAVWAN. (An Organization Chart is presented in the Appendix.)

Stakeholders

A stakeholder is defined as an individual or group who can affect or is affected by the achievement of a given mission, objective, or change strategy (Freeman, 1984; Roberts & King, 1989). From the perspective of site-level implementation, NAST and the NAVWAN Demonstration-Validation Team represent external stakeholders. Internal stakeholders include the departments within NAS Miramar and the tenant activities. Each stakeholder and the general mission of their organization is presented below. While the Department of the Navy uses "alphabet soup" acronyms, a translated title for each of these stakeholders is presented in quotes and these will be used in the presentation of data in the case. The official organization title and acronym are also included:

(a) *Validation Team.* The NAVWAN Demonstration-Validation Team was established by VADM Bowes. This team is led by the Program Manager responsible for the NAVWAN implementation at Headquarters, Naval Air Systems Command [NAVAIRSYSCOM].

(b) The *Maintenance Office* is responsible for all aspects of naval aviation maintenance and administration programs [Naval Aviation Maintenance Office: NAMO].

(c) The *Maintenance Depot* provides intermediate level aviation maintenance support for Pacific Fleet aviation activities. Intermediate level maintenance reflects a more complex maintenance action that is beyond the capability of squadron and station organizations [Naval Aviation Depot, NAS North Island: NADEP].

(d) *Pacific Region Aviation Command* promulgates policy and asset management direction to all Pacific Fleet aviation activities [Commander, Naval Air Force, U.S. Pacific Fleet: CNAP].

(e) The *SUPPLY Department* is responsible for all logistic and supply support required by the squadrons and tenant activities of NAS Miramar.

(f)   The *Intermediate Maintenance Department* is responsible for all intermediate maintenance services required by the squadrons at NAS Miramar [Aviation Intermediate Maintenance Department, NAS Miramar: AIMD].

(g)   The *ADMIN Department* is responsible for all postal and administrative services required by personnel of NAS Miramar and its tenant activities.

(h)   The *Civil Engineering Department* is responsible for all facilities maintenance, construction, hazardous waste management, and environmental conservation required by NAS Miramar and its tenant activities.

(i)   *Personnel* is the activity responsible for all personnel and disbursing functions for NAS Miramar and tenant activity personnel [Personnel Support Activity Detachment, NAS Miramar: PSD].

(j)   *Aviation Wing* are two aviation type wings that are responsible for administration of all operational, maintenance, and administrative support for the squadrons located at NAS Miramar. There is one stakeholder representative for both commands [Commander Fighter Wing Pacific: COMFITWING; and Airborne Early Warning Wing Pacific: COMAEWWING].

(k)   The *Engineering Support Facility* is the activity responsible for providing technical support and maintenance training to organizational activities throughout the Pacific Fleet [Naval Aviation Engineering Support Unit: NAESU].

(l)   The *Information Systems Department* is responsible for providing information technology support to NAS Miramar and tenant activities [Information Systems Support Office, NAS Miramar: ISSO].

Status Quo

All NAS Miramar offices are currently connected to an Ethernet local area network (LAN). Each stakeholder has implemented LANs in an autonomous way and the systems used at stakeholder activities are not well integrated. These stakeholders have diverse missions and responsibilities. The NAVWAN is intended to be a conduit for communication at the local, metropolitan, and global levels. This improved communication will better support the information requirements of the Competency Aligned Organization. Also important is the improved integrated communication capabilities offered to the larger Navy organization.

Although their missions differ, many of the above listed stakeholders in the NAVWAN have similar information technology requirements. Because of the NAVWAN open systems architecture, stakeholders do not expect it to significantly constrain their mission accomplishment. In fact, all the stakeholders anticipated significant benefits from interconnectivity with an array of Department of Defense, federal, academic, and civilian organizations.

The technologies needed to support the NAVWAN are available. The only technology oriented concerns, therefore, are whether the correct combination of technologies can be determined, acquired, implemented and maintained within this culture and resource environment.

PROJECT/CASE DESCRIPTION

For purposes of gathering the data for this case, representatives of each of the stakeholder groups were individually interviewed in regard to the NAVWAN implementation.[3] Analyzing the stakeholders' perceptions regarding NAVWAN provides information valuable to the identification of requirements, the implementation design process, as well as predictions of the ultimate effectiveness of NAVWAN. The information outlined below highlights themes that emerged from the stakeholder interviews regarding NAVWAN implementation.

Overall, the stakeholders see benefits to NAVWAN that would initially include: broader more integrated electronic mail, file transfer, and directory services. Because of these benefits, the stakeholders are strongly motivated to employ the NAVWAN. They anticipate it will significantly increase user productivity over the long run. However, there are also barriers to the implementation of the NAVWAN at the prototype site. Stakeholders identified potential barriers and, in some cases, offered solutions. Knowledge of the benefits, barriers and recommendations identified by stakeholders at this site offers the Validation Team important information as they proceed with planning full system implementation.

Stakeholder Functional Requirements

Many of the stakeholder information technology functional requirements are common to all the stakeholders. Office automation, including word processing and file transfer, is required by all stakeholders. Each stakeholder currently has word processing capability; however, the variety of vendors and versions supporting this capability pose a challenge for documentation management among the activities. Communication tools, such as message traffic management, bulletin boards, database management, and electronic mail applications, are other common requirements for all stakeholders.

Several stakeholders require access to a database management system (DBMS). This access must include a structured query capability for ad hoc queries. Several stakeholders require decision support tools including spreadsheets and graphic presentation. A few stakeholders require access to three dimensional graphics and access to technical CAD/CAM drawings. A few stakeholders require additional bandwidth in support of televideo conferencing. Some of these requirements are currently not being met and others are being supported by multiple LANs, electronic mail packages, modem connections, postal service, and voice mail.

Stakeholder Knowledge of Planned Capabilities of NAVWAN

At the time the interviews were conducted, 9 of the 12 stakeholders had a clear understanding of the initial capabilities of the NAVWAN. Three stakeholder groups had direct representation on the Validation Team led by the NAVWAN program manager: Maintenance Office, Maintenance Depot, and Pacific Region Aviation Command. In addition, five of the remaining stakeholder groups had attended an orientation brief presented by the Validation Team. This briefing provided extensive information about the capabilities of the NAVWAN and the impending implementation at NAS Miramar.

The three stakeholders least familiar with NAVWAN were the Civil Engineering Department, Personnel, and the Engineering Support Facility. These groups were aware of the NAVWAN implementation plan but were not familiar with its specific capabilities. As information technology managers, these stakeholders had a clear understanding of wide-area connectivity, but they were not certain of the exact functionality the NAVWAN will provide.

Stakeholder Perceptions of NAVWAN Benefits

"Increased communication is the single most important thing to come out of the NAVWAN implementation. Next would be potential cost cutting from reduced phone use, and third is the potential Internet access." This quote from one stakeholder representative summarizes the most common response regarding potential NAVWAN benefits. Other stakeholders offered different ways in which increased communication through the WAN would offer benefits. The Engineering Support Facility representative stated the NAVWAN would "improve coordination with the headquarters and the customers at the squadron level." Several stakeholders explained the value in terms of quicker response time: "It will mean less delay in communication, quicker responses and approvals of work." The Pacific Region Aviation Command added:

Communicating with all the wings at one time, vice sending individual correspondence or phoning them, will increase the speed, and there will be more communication. I believe this will help to prevent problems that occur when people are not informed.

Access to data was another benefit with multiple dimensions supported by several of the stakeholders. First, access to common databases will provide real-time information and reduce errors. According to the Engineering Support Facility representative, this should mean "less finger pointing and improved accountability of the information." The Maintenance Depot representative supported the benefit of improved integrity of data in the databases for logistics and aviation maintenance purposes. A related benefit, according to the Engineering Support Facility, would be improved, real-time technical advice that can contribute to such activities as the Integrated Logistics System Maintenance Training conferences, program management reviews, or investigations.

Cost cutting benefits were supported by several stakeholders who acknowledged savings due to decreased phone use, decreased travel requirements, and the potential for consolidation of support staffs. Regarding the latter, the Maintenance Depot representative commented, "Hopefully the NAVWAN will right-size the systems support staffs by consolidating everyone. There's too much duplication of effort by all the Automated Information System (AIS) staffs." Specifically cited were five information system staffs supporting organizations all closely located. This consolidation was seen as having an additional benefit, "The Navy needs economies of scale in purchasing hardware and software that we can't achieve unless we consolidate."

Elimination of paperwork, and progress toward a paperless Navy were benefits cited by the Personnel Activity representative: "We need to get rid of all the file cabinets." The Personnel Activity's goal is for all personnel records to be computerized and to electronically transfer records between commands. From their perspective, "The number one job of [Personnel] is to provide customer support. With WAN capability we will be able to improve customer service by responding more quickly."

The Intermediate Maintenance Department supported the value of NAVWAN to increased flexibility and responsiveness. He also plans to use NAVWAN to edit and endorse administrative paperwork. He identified an additional benefit by commenting, "It will solve the message traffic problems [on the Defense Message System (DMS)]; we'll be able to stop faxing people messages to make sure they got the stuff sent on DMS." While the Civil Engineer agreed NAVWAN would be an improvement over DMS, he added that the NAVWAN cannot replace DMS because it is not a secure network.

The assessment of the benefits of the NAVWAN from the perspective of the Validation Team emphasized "tangible cost reduction through circuit consolidation, reduced maintenance and support efforts, and improved configuration management." He also stated that "buying more capacity at a better price through economies of scale" would also reduce costs for all the NAVWAN users.

Stakeholder Perceptions of Barriers to Implementation

None of the stakeholders believed the implementation of the NAVWAN would significantly constrain the accomplishment of their functional requirements. Many stakeholders stated that most of their functional requirements are already met by other systems. As a result of the existing systems, the Aviation Wing representative explained that "point to point communication will help a lot, but you will have some duplication of effort until everyone is connected to the NAVWAN." Others felt the duplication may be more than transitional. Specifically, five stakeholder groups did not think elimination of these other systems would occur too easily because users are more familiar with the current system, and they believe existing systems are sufficient. These groups were the Information Systems Department, Maintenance Depot, Maintenance Office, ADMIN Department, and SUPPLY Department.

The issue of network security was raised by three stakeholders: Information Systems Department, ADMIN Department and the Aviation Wings. Each expressed concern that the potential for abuse and security risks would constrain the achievement of functional requirements using NAVWAN. They pointed out people could not use it for anything that is classified or of a sensitive nature. This would include a significant amount of message traffic. "People will still send messages [using DMS] for the important stuff, because that's the standard way of covering your six." ADMIN expressed a similar concern: "The need to protect sensitive information, such as investigations and HIV positive personnel management issues, still exists." The Maintenance Depot representative explored the security topic a bit further. "We are evaluating firewalls and filters, not necessarily to keep people from going out, but to keep other people from coming in."

While most stakeholder representatives stated NAVWAN use will increase productivity, they also expressed a concern that abuse of the system could be a productivity drain. "The temptation to surf all day on the Internet may be too great for some people."

Another concern about abuse of the NAVWAN was voiced by ADMIN who cautioned that people may use the NAVWAN to circumvent the chain of command. Implementing NAVWAN raises "power issues... not just communication issues." By implementing the NAVWAN, "the chain of command loses some power over their subordinates." The Intermediate Maintenance Department representative supported

ADMIN stating, "they [senior officers in the chain of command] don't want you talking to anyone outside the chain of command."

A Validation Team representative confirmed the potential challenge that increased communication capability poses to the chain of command. "E-mail is a democratizing agent in the organization." If an airman wants to send the commanding officer e-mail, he can do so without asking his supervisor or division officer. In addition, the effective utilization of the NAVWAN may not be fully supported by commanding officers (COs) and executive officers (XOs) because "the CO and XO want to see it and touch it or edit it before you send anything... This slows down the process and defeats the purpose."

A different source of resistance to implementation was voiced by the Aviation Wing representative who had experienced a problem with the earlier implementation of the Local Area Network (LAN):

People really resisted the network at the start because they are not computer literate and they didn't want to change ... they felt it would make their job harder, but they get over it once they understand ... Many of the potential users still have computer phobia and feel the network is too hard, but then they realize that they have the power to reach out, touch a button, and go across the world.

Nearly every stakeholder agreed money is the single biggest barrier to the NAVWAN implementation. "We just don't have the money we need to support the LAN, let alone the NAVWAN" was the initial reaction of the Aviation Wing representative. Money also limits the technical capabilities of the type wings. This concern was reinforced by the Personnel Activity representative who stated their biggest barrier is "definitely money for the hardware, software, and training."

A Validation Team representative disagrees the greatest barrier is money. He says "it is not so much money itself, but the ability to direct the money; especially where personnel are concerned." From his perspective as the program manager, the development effort is heavily dependent on civil service personnel. He gets frustrated because "we can't hire and fire people as we would like, and we don't have the right mix of people. Eighty percent of the IT people are not involved and the 20% who are, don't have the time to dedicate to it."

The problem of dedicating time and manpower was confirmed by several stakeholder representatives. In the Information Systems Department, there are only four people and they are always putting out fires in other places. "We simply don't have the time or people to conduct the necessary training and perform the network administration, maintenance, and support." The Maintenance Depot representative agrees there are limited technical personnel and sufficient manpower resources dedicated to the NAVWAN project. The representative from the Pacific Region Aviation Command adds that because there are so few people, "the barrier to implementation becomes the schedule and workload priorities of the implementation team... There are conflicting priorities for the people working on it, and they have difficulty knowing where the resources are going to go."

A related constraint identified by several stakeholders is the inadequacy of resources. For example, the LAN at the Engineering Support Facility has limited size and capacity, with only five terminals. This limits their ability to communicate with all the regional offices as well as with technical representatives when they are traveling. The

Maintenance Office representative expressed a concern regarding a different resource constraint. His concern is that the Validation Team responsible for implementing the NAVWAN does not have direct authority over people within the different stakeholder groups involved in the implementation. Because there are so many commands working together on this project, each with their own primary tasks, schedules, and resource priorities, NAVWAN may not get the attention it needs on critical issues of infrastructure coordination. He cited personal experience with this problem:

I have personally hooked up sites ahead of schedule because I needed them to be online and the infrastructure coordination problems could not be resolved. We had to do a lot of negotiating to get the support we need because the key people don't always work for us.

The representatives from both the Maintenance Depot and the Intermediate Maintenance Department concur. According to the former, "The rice bowls are difficult to overcome. People don't want to admit it, but they want to preserve their islands of communication." He believes people want to "put their own bridges into the NAVWAN, but they want to perpetuate isolated applications."

The Validation Team is also concerned about the political constraints against standardization.

Hopefully we will migrate to one standard because we now have 22 commands doing different things. Each program or project has their own network; each activity has a stovepipe. They are integrated vertically but not horizontally. Just because we have the infrastructure now doesn't mean they will be horizontally integrated. The infrastructure is not sufficient to solve the interoperability problems.

The Validation Team has not been able to overcome these interoperability problems at all the sites. Because they want to take advantage of the existing architecture, they cannot standardize everything to one system. They continue to have problems migrating the existing systems at the 22 NAVAIR activities to the four protocols selected for NAVWAN implementation. "It is technically feasible, but there may be some performance issues that we'll need to measure and correct" is the summary of the Validation Team leader.

RECOMMENDATIONS FOR NAVWAN IMPLEMENTATION

The data presented above provide the reader information for developing recommendations regarding NAVWAN implementation. The reader's responsibility is to act in the role of the case researcher, review and analyze the data, and prepare a set of recommendations with appropriate justification. The primary audience for these recommendations is the NAVWAN Program Manager and the Demonstration/Validation Team. They, in turn, will forward the recommendations to VADM Bowes and the Naval Air Systems Team.

It is predicted that an organization that deliberately manages its relationships with critical stakeholders will be more effective in mission accomplishment (Freeman, 1984; Roberts & King, 1989). In the case of any significant new technology implementation (the NAVWAN in this instance), failure to successfully manage stakeholder relationships can cause significant resistance to the planned change and thus cause an organization to fail to achieve its implementation goals.

The data presented on stakeholder perceptions of the advantages and potential barriers to implementation of the NAVWAN can be used to inform action planning. Analysis of the data can utilize strategic planning techniques such as the SWOT methodology that identifies strengths, weaknesses, opportunities and threats related to the planned implementation. Clearly, any change effort must examine both the benefits and barriers; and the success of the change effort will be influenced by the extent to which benefits are maximized and barriers are minimized. Research on organizational change has identified several categories of resistance to change. These sources of resistance can be either at the individual or the organizational level (Carrell, Jennings, & Heavrin, 1997). Individual sources of resistance include: fear of the unknown; threatened loss of self-interest; mistrust of change advocates; different perceptions of the value of the change initiative; preference for the status quo; concern regarding skill capabilities required to be successful. Organizational sources of resistance can include: structural inertia; bureaucratic inertia; organizational culture and norms; threatened power; threatened expertise; resource allocation.

The tasks and questions listed below apply the principles of stakeholder management by focusing attention on specific stakeholder interests and concerns regarding NAVWAN design and implementation. Additional information on stakeholder management can be found in the references listed at the end of this chapter.

Stakeholder Analysis Tasks

1. Identify and map the critical stakeholders.
2. Conduct a stakeholder audit by assessing the stakeholders' interests and concerns regarding the implementation of the NAVWAN.

   a. Summarize the benefits reported for NAVWAN.
   b. Summarize the concerns or impediments to implementation.

      i. Identify the individual-level sources of resistance.
      ii. Identify the organizational-level sources of resistance.

3. Identify ways in which each stakeholder is likely to support or resist the planned implementation and the extent to which they influence implementation success.
4. Use the results of the audit to develop implementation strategies that capitalize on stakeholders' interests and address or resolve their concerns. One option here is to use the SWOT analysis technique to formulate specific action recommendations.

   a. Identify strategies that capitalize on Strengths and Opportunities.
   b. Identify strategies that resolve or minimize the risk of Weaknesses and Threats (sources of resistance).

QUESTIONS FOR DISCUSSION

1. How does stakeholder analysis compare with the traditional approach to information technology design and implementation planning? What are the advantages and limitations of stakeholder analysis?
2. What is the likelihood that stakeholders will support or resist the planned implementation? How do stakeholder analysis and SWOT diagnosis inform this question?
3. What are the sources of resistance (individual and organizational) that must be managed in this case?
4. What action steps can be taken to optimize the successful implementation of the NAVWAN given stakeholder interests and concerns?
5. How is the risk position of the project impacted by the availability of SWOT and stakeholder analysis data? Risk should be considered with respect to functionality and with respect to end-user acceptance and use.

REFERENCES

Carrell, M. R., Jennings, D. F., & Heavrin, C. (1997). *Fundamentals of organizational behavior*. Upper Saddle, NJ: Prentice Hall.

Freeman, R. E. (1984). *Strategic management: A stakeholder approach*. Boston: Pitman Publishing.

Freeman, R. E., & Gilbert, D. R. (1987). Managing stakeholder relationships. In S. P. Sethi, & C. M. Falbe (Eds.), *Business and society: Dimensions of conflict and cooperation* (pp. 397-423). Lexington, MA: Lexington Books.

Roberts, N. C., & King, P. J. (1989, Winter). The stakeholder audit goes public. *Organizational Dynamics*, 63-79.

ENDNOTES

[1]   This case is based on: Bayer, V.C. (1995). *Analysis of Naval Air Systems Command wide-area network prototype implementation*. Masters thesis, Naval Postgraduate School, Monterey, CA.

[2]   Data from Fiscal Year 1994 report of NAST.

[3]   The responses are only attributable to the representative and may not reflect the opinions of the chain of command. Any mention of the stakeholder command should be attributed only to the stakeholder representative and not the organization.

APPENDIX

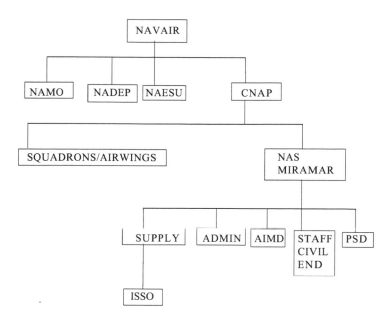

Susan Page Hocevar is currently an assistant professor in the Department of Systems Management at the Naval Postgraduate School in Monterey, CA. , where she teaches courses in organization and management to military officers earning a graduate degree. Her research and publications are in the areas of organization change, high involvement organizations, organizational reward systems, self-managed teams, organizational culture, quality of work life, and corporate social responsibility.

Barry Frew is an associate professor of information systems, Naval Postgraduate School (NPS). Before joining the NPS faculty in 1984, he spent several years as an information resource management practitioner involved with software development, technology acquisition and integration, and computer operations.

Lieutenant Commander Virginia Callaghan Bayer is currently serving on active duty in the United States Navy and pursuing a career subspecialty in corporate information systems management.

This case was previously published in J. Liebowitz & M. Khosrow-Pour (Eds.), *Cases on Information Technology Management in Modern Organizations*, pp. 72-82, © 1997.

Chapter XVII

Information Technology & FDA Compliance in the Pharmaceutical Industry

Raymond Papp, University of Tampa, USA

EXECUTIVE SUMMARY

Given the recent profitability of and demand for pharmaceuticals, from prescription antibiotics and analgesics like Ciproflaxin™ and OxyContin™ and men's health drugs such as Viagra™ and Vardenafil™ to over-the-counter Senokot™ laxatives and Betadine™ antiseptics, the rush to develop and market new pharmaceuticals has never been greater. The current process is complex and it often takes several years for a drug to reach the market due to the myriad of Food and Drug Administration (FDA) guidelines. Furthermore, the recent FDA guidelines mandating that all New Drug Applications (NDA) be submitted in electronic (paperless) format by the end of 2002 is a catalyst for change in the pharmaceutical industry (FDA proposes first requirement for electronic submission, 2002; New Drug Application [NDA], 2001). Bayer Pharmaceutical, like its competitors Purdue Pharma and Boots Healthcare, has begun to take steps to assure that its use of information technology will allow it to not only meet FDA guidelines, but achieve its corporate goals of improved efficiency and reduced operating costs.

BACKGROUND

The company has a long history, having been founded by Friedrich Bayer and Johann Friedrich Weskott in 1863 in Wuppertal, Germany. From its meager beginnings as a dyestuffs factory, Bayer has grown into a multi-billion dollar international chemical and health care company. Expansion took place rapidly for Bayer. In 1865, Bayer and

Weskott entered the coal tar dye business in the United States and began exporting intermediates. Further growth was achieved in 1876 with the opening of another dyestuffs factory in Moscow with the descendents of Bayer establishing the joint stock company Farbenfabriken vorm. Friedr. Bayer & Company. Additional factories soon opened in France and in 1884, under the guidance of chemist Carl Duisberg, Bayer scientists began receiving recognition for their pioneering discoveries. With the establishment of the Pharmaceutical Department in 1888, the stage was set for the most famous and historical discovery yet for Bayer. Dr. Felix Hoffman first synthesized acetylsalicylic acid in a chemically pure and stable form in 1897. Aspirin was registered as a trademark two years later in 1899; it is still the world's most popular over-the-counter pain medication. In 1925, Farbenfabriken vorm. Friedr. Bayer & Company merged with another company and became I.G. Farbenindustrie AG, which was later seized and broken up following the Second World War. Farbenfabriken Bayer AG was re-established in 1951, then changed its name to Bayer AG in 1972. The company remains Bayer AG; it reacquired the rights to the Bayer logo and trademark from Sterling Pharmaceuticals in 1986. Today, Bayer AG is ranked as the 117th largest company in the world with revenues topping $30 billion (see the Appendix). With headquarters in Leverkusen, Germany and with about 350 affiliated companies worldwide, the Bayer Group is represented on every continent. Bayer AG's business organization includes healthcare, agriculture, chemicals and polymers. Within the healthcare segment, the North American Pharmaceutical Division headquartered in Pittsburgh, Pennsylvania, accounts for more than $10 billion in annual revenues. The division has also recently achieved many business milestones, including $1 billion in annual sales for its antibiotic Ciproflaxin™ in 1999 and 2000 and a growth rate of 23% in 2000, which easily outpaces the prescription drug industry as a whole (BAYER AG Homepage, 2002). Bayer's highly recognizable trademark logo will unify the individual Bayer divisions as the company will migrate to a new corporate structure on January 1, 2003, when Bayer will become a management holding company with four legally independent operating subsidiaries (*A new Bayer — A new Bayer cross*, 2002).

SETTING THE STAGE

To better understand the scope of the changes Bayer must undergo to comply with the FDA's New Drug Application (NDA) process (*FDA proposes first requirement for electronic submissions*, 2002), a background in the FDA's role is important. The next section provides an overview of the process pharmaceutical firms must follow and the need to meet these guidelines.

New Drug Application Process

The Center for Drug Evaluation and Research (CDER) is a government agency whose job is to evaluate new drugs before they can be sold to the public. The CDER focuses on prescription and over-the-counter drugs, both brand name and generic, to ensure that they work correctly and that the health benefits outweigh the known risks. The information is also made available to doctors and patients to provide them with the information they need to use these medicines wisely.

The regulation and control of new drugs in the United States has been based on the New Drug Application (NDA). Each new drug has been subject to a NDA approval before it is allowed into the U.S. commercial market. Any data gathered during animal studies and human clinical trials become part of the NDA. (*About CDER*, 2002)

The Food & Drug Administration (FDA) has evolved considerably since its founding in 1938. When the Food, Drug, and Cosmetic Act (FD&C Act) was passed in 1938, NDAs were only required to contain information about an investigational drug's safety. In 1962, the Kefauver-Harris Amendments to the FD&C Act required NDAs to contain evidence that a new drug was effective for its intended use, and that the established benefits of the drug outweighed its known risks. In 1985, the FDA completed a comprehensive revision of the regulations pertaining to NDAs, commonly called the NDA Rewrite, whereby the modified content requirements restructured the way in which information and data are organized and presented in the NDA to expedite FDA reviews (*Questions about CDER*, 2002; *Benefit vs. risk: How FDA approves new drugs*, 2002).

Pipeline Products and NDAs

There are many challenges to the FDA mandate. Continued growth and improved efficiency are strategic goals for Bayer. Efficiency remains one of the key factors to continued success and the Pharmaceutical Division has recently implemented many restructuring and cost saving measures to improve not only operating costs, but also the adeptness with which processes are executed. For example, by implementing SAP, an enterprise-wide software solution company-wide, Bayer is trying to improve efficiency by integrating many of its existing systems. The SAP project is a long-term undertaking for Bayer and the complex technological and organizational issues are daunting. Another way Bayer is using information technology is with paper-free workflows (*Bayer continues expansion of e-commerce*, 2001).

Bayer's New Drug Application (NDA) submission project involves such a paperless plan and the expected benefits of this system include faster submission of NDAs. Slated as a replacement for the outdated and overburdened manual document management system, the updated system will allow for the timely submission of new drugs since the extremely competitive nature of the pharmaceutical industry necessitate faster and more efficient ways of getting FDA approval for these "cash cow" drugs.

To maintain its position within the competitive pharmaceutical industry, it is important to consistently have new development products in the pipeline. Pipeline products are used as criteria for judging the overall status of a pharmaceutical company as much as currently marketed products. Bayer feels they have maintained the desired growth trend through partnerships with high-tech drug discovery companies such as Millennium pharmaceutical (*Genomic technology pays off: Bayer and Millennium expand research program*, 2001) and most recently, with Aventis CropScience in its crop science division (*Bayer acquires Aventis CropScience*, 2001). Such partnerships will assist Bayer in reaching its goals for target compounds that may eventually become successful new drugs.

Together with the $1.1 billion in research and development expenditures and 3,880 pharmaceutical employees currently working on research and development for Bayer Corporation, the company is dedicated to building a productive pipeline and maintaining the growth trend of the past. In the near future, Bayer Pharmaceutical is planning on having a better alternative to Viagra™ all while continuing its promising work on early stage compounds that are being designed to combat cancer, asthma, osteoporosis, stroke and metabolic diseases (*More blockbusters in the pipeline*, 2001).

To its credit, Bayer Pharmaceutical has many successful and valuable FDA approved products on the market. These include Adalat™ and Baycol™ in the cardiovascular segment, Cipro™ and Avelox™ in the infectious disease segment, Glucobay™ for metabolic disorders, Nimotop™ for diseases of the central nervous system, and Kogenate™ to treat hemophilia. Combined, these products helped produce over $10 billion in revenues (see the Appendix) for Bayer Corporation. Incidentally, the recent bio-terrorist threats in late 2001 have seen the demand for one drug in particular, Cipro™ rise to unprecedented levels. The demand for similar drugs has never been greater and anything that can be done to speed up the NDA process is paramount. For example, as this case was being written, Bayer was in the process of obtaining FDA approval for Vardenafil™ a Viagra™type drug with fewer side effects, smaller doses and quicker response time (*Bayer files applications for Vardenafil in United States and Mexico*, 2001). This drug may be the biggest development for Bayer since Ciproflaxin™ and Aspirin™. Thus, timely filing of the NDA is critical to Bayer to keep pace in the competitive pharmaceuticals industry.

Further complicating the issue is the use of this new electronic document submission system for the first time. While the benefits for management are unquestionable, implementing a system for the first time while compiling a top priority NDA is difficult. The reason for this risk is not strictly monetary, although it is estimated that for every day the NDA is delayed, it will cost the company millions in revenues. Strict version control and validation guidelines have been set by the Food and Drug Administration, and the manual method did not meet these requirements (New Drug Application [NDA] Process, 2001).

CASE DESCRIPTION

Due to the Food and Drug Administration's mandate that all New Drug Applications be submitted in the electronic or paperless format (*FDA proposes first requirement for electronic submissions*, 2002), many information technology changes are taking place within Bayer Pharmaceutical. While most of the expenditures for the pharmaceutical industry are in the form of scientific research and development, many companies are now being forced to focus more on information technology. Bayer Pharmaceutical is not unique and its neglected document management system is now of prime concern and focus. In fact, one of its major competitors, Purdue Pharmaceutical, has already begun using such a system for the submission of NDAs. According to Martin Zak, Director of Information Technology:

Our goal at Purdue Pharma is to become one of the top ten pharmaceutical companies within the next ten years. One of the keys to achieving this goal is by using Documentum 4i eBusiness Platform to manage the content of our company's strategic assets. With Documentum 4i, we will be able to streamline the approval process and speed products to market without sacrificing quality or safety. In this way, Documentum not only helps us but also helps the patients we ultimately serve. (*Purdue Pharma L.P.: Speeding time to market with Documentum 4i*, 2001)

Bayer faces many of the same challenges, yet there are different systems and procedures for each department within the company. Unfortunately, this autonomy is not productive and makes complying with government regulations virtually impossible, reduced efficiency and expanded cost notwithstanding. There are also multiple systems, many badly outdated, currently in use. Separate applications for storing documents electronically, archiving documents, compiling information for submissions, interacting with foreign databases, and retrieving statistical analysis all require valuable resources to maintain. Furthermore, most of these systems do not communicate well with one another or other departments such as statistics or project management.

Thus, Bayer's critical business challenges for the short term are to find new ways to implement its processes in light of the electronic NDAs. Its competitors, such as Purdue Pharma and Boots Healthcare, have already begun this process. They propose to use technology to:

- Foster high growth within a competitive marketplace and accelerate time to market for new products while adhering to stringent Food and Drug Administration (FDA) deadlines and compliance requirements necessary for approval.
- Virtually assemble content, securely and efficiently control the flow of content, authorize and verify recipients, and track changes involved for submissions to, and compliance with, regulatory agencies. A missed paperwork deadline can mean a six-month delay in the approval process and impede time to market.
- Manage multiple forms of content from content owners at multiple sites throughout the world and publish to an internal Web site.
- Meet regulatory compliance for manufacturing, standard operating procedures, and training documents. Be able to produce documents on demand, decreasing potential future audits (*Purdue Pharma L.P.: Speeding time to market with Documentum 4i*, 2001).

By using a validated system for document management such as Documentum (like Purdue Pharma and Boots Healthcare have done), Bayer will be able to comply with the new FDA regulations, something not possible with the manual legacy system. Without superior technical support for many of the legacy systems, downtime will increase. All these factors lead to a forced decision to update. While there has been talk of a new document management system for some time, only now is it becoming reality.

The new system is currently undergoing many new challenges. The validation process and changeover are taking place at the same time. Instead of having a system in place and then using it for submission purposes, Bayer is developing standard operating procedures and using these new procedures at the same time. Furthermore, the legacy system is not being phased out. In fact, it has been the central archive for so many years, yet it cannot properly communicate with the new system. While the large-scale use of this legacy system will eventually be discontinued, it is not certain if it will ever disappear completely. What adds to the problem is that the old system was designed and built in-house for specific functions. The long-term goal is to integrate the old and new systems to take advantage of the current technology's speed and power (*Supporting pharmaceutical research and development*, 2001).

SUMMARY

With the onset of electronic submissions to the Food and Drug Administration and the need for paperless workflows to meet management goals of increased efficiency, many new information technology systems are being implemented at Bayer Pharmaceutical. This case will focus on the legacy system and the benefits the proposed electronic system will bring.

Complex Paper Trails

In the past, the compilation of NDAs was a difficult and complex undertaking. Copiers needed to be constantly maintained and even a truck rented to deliver the volumes of paper to the FDA. While it is still a complex process, the Regulatory Affairs department of Bayer Pharmaceutical handles all FDA contact for submitted drugs. Liaisons are designated for each drug and have complete access to reviewers. They are responsible for handling all review requests and maintaining FDA regulations for each of their assigned drugs. It is also Regulatory Affairs' responsibility to submit all documents to the FDA, including NDAs. This is done through hard copies that trickle down to the corresponding people in charge of each drug. For example, an investigator in the field must be approved by an Investigational Review Board in order to start administering a proposed drug to human patients. This is usually done in the latter phase of the testing procedure. Once the review board approves the investigator for the protocol in which they wish to take part, the paper work is sent to various departments for signatures by assigned directors. The paper copy is then sent to the liaison in Regulatory Affairs who must then make sure all procedures have been followed to comply with regulations. The liaison submits the proper copy to the FDA while another copy is archived. The document is given an archival number, scanned into the electronic archival database as an image, and filed with millions of other documents. Every department that sees this document, however, keeps a copy. Many of these copies are stored in different and unrelated systems. The archival system in Regulatory Affairs is a legacy database that was developed almost a decade ago and stores millions of links to hard copies archived in the files. The medical groups (departments) store their copy of this document in a newer relational database. This system assigns numbers to the document in a similar way to the Regulatory Affairs' database, but works much better for electronic distribution. The problem with the database is that if changes need to be made, the current document becomes useless. It must be printed for any revisions and version control is almost non-existent. Then it must be created again, given another code, and sent to Regulatory Affairs for updating and storage in their database. This is just one example of how the current workflow is inefficient. While distribution can be done electronically with the system, it is far from paperless in terms of revisions or version control. Finally, a hard copy must be entered in the Regulatory Affairs' database since it is the final storage of all documents.

Compilation for large submissions, such as NDAs, is also done by Regulatory Affairs. It is handled by the submission group of the department and guided by a project leader. Yet again, different systems are used. The electronic images are stored by archive in an image database. This system stores every page of every document as a single image. Therefore, if a document is 5,000 pages long, it consists of 5,000 images, which are not accepted by the FDA. Thus, the submission group must compile all necessary documents for every submission. This was formerly done (and for some types of FDA correspondence is still done) with paper copies. However, for electronic submissions, special steps must be taken. Submission must use yet another system to convert and compile all needed documents. This is usually done by manually scanning each final version into PDF format, but can also be done by converting the image files to PDF. The FDA requires the PDF format for the final published document. The main drawback to converting the image files to PDF is that the validated conversion only allows 50 images (pages) to be converted at once. The sections for each of the submissions are then stored until they are ready to be published. Once this is done, the electronic medium of choice (digital tape if the submission totals more than five full compact disks) is selected; the relevant data is stored, and sent to the FDA (*Supporting pharmaceutical research and development*, 2001; *Shortening time to market through innovation of regulatory approval process*, 2002).

Documentum and NDAs

Like its competitors, Purdue Pharmaceutical and Boots Healthcare, the implementation of Documentum, a validated system that allows for version control and audit functions, has been a complex and critical undertaking for Bayer Pharmaceutical. It is especially important in the integration of information technology and corporate goals. These goals include bringing compounds to market faster, improving Bayer's competitive advantage through time and cost savings, flexibility and management of content, global collaboration and improving the efficiency of the administrative aspects of the business. As stated earlier, the current systems used do not interact nor allow for the paperless workflow that is key to this desired efficiency. Documentum is a tool that handles all content management needs and allows compliance with the increasing demands of FDA requirements via its open standards-based architecture that allows easy integration of content management with its SAP applications. In addition, Documentum supports XML (eXtensible Markup Language), allowing disparate systems to be linked together and facilitates the movement of information between them in an automated fashion. With the complete rollout of Documentum scheduled for early 2003, Bayer should realize greater efficiency. Documentum uses linked documents allowing the reader to view a reference to specific page by clicking on the link rather than manually finding the correct volume and page on which the reference is printed (*Documentum in the Pharmaceuticals industry*, 2002).

The use of Adobe Acrobat and PDF files makes the process quick and easy (*U.S. Food & Drug Administration speeds review of new drug applications with Adobe Acrobat and PDF*, 2002). The only problem occurs when the document is changed and the links and table of contents need to be updated. Thus, a paperless flow of information and a single powerful, integrated tool which replaces many independent systems is the direction that the Regulatory community needs to head in order to achieve success and increased efficiency. This is especially important since Bayer has only done one other electronic NDA, that being Avelox™ last year (NDA Approvals for Calendar Year 2001, 2001).

Since Documentum is a completely validated system, it meets the FDA's stringent auditing requirements. It offers version control and security attributes that current systems simply do not have. It will be possible to electronically log those who view and/ or access the document, something the FDA requires be done. It will allow for an electronic environment throughout the entire Pharmaceutical division. For example, if the scenario described above were to occur in a Documentum environment, efficiency would be greatly increased. The investigator in the field would simply create his application using a standard template similar to the many paper variations used today. The Investigational Review Board could electronically approve this document. The research associate at the investigator's site could then place the document in Documentum, which would automatically give it an archive number. The document would then be routed automatically per standard operating procedure guidelines to the proper people. Electronic review and signatures would then be applied to the document. Simultaneously to this, the FDA liaison in Regulatory Affairs would be able to monitor the progress and view any revision until the point where the approved, completed document would be finalized by the liaison and readied for publishing. It is a streamlined, efficient process that will increase productivity and improve results. A list of everyone that revised or viewed the document can be generated electronically, replacing the cumbersome logbooks of the existing system (*Shortening time to market through innovation of regulatory approval process*, 2002; also see Appendix B).

Logistical Challenges

There are many challenges in the achievement of this transformation. First, training all the users and making them comfortable with the new system is a large undertaking, especially within a company of Bayer's size. There are also many logistical problems and geographic concerns. It has taken the German-based organization quite some time to realize this fact, but it has finally taken action, not only by the American-based Bayer Corporation, but also by the entire health sciences division of Bayer AG. Getting the pharmaceutical counterpart in Europe to agree to this new system will also be a challenge, since Germany is currently much farther behind in electronic submissions. A global submissions group could alleviate any shortcomings the German counterparts are experiencing (*Bayer's high-tech pharmaceutical research platform*, 2001).

With the company-wide top priority being the timely submission of NDAs in a complete, validated and electronic format, the most current challenge facing information systems is implementation on the move. Since there is no time for trial and error, precise forward-thinking decisions must be made. There is still much to be learned because any unknown problems will have to be dealt with as they arise. Bayer's profitability and long-term survival is at stake.

REFERENCES

A new Bayer — A new Bayer cross. (2002). Retrived from http://194.231.35.65/en/ bayerwelt/bayerkreuz/kreuzneu.php?id=0390302

About CDER. (2002). Retrieved from http://www.fda.gov/cder/about

Bayer acquires CropScience. (2001). Retrieved October 2, 2001, from http:// www.press.bayer.com/news/news.nsf/ID/NT0000E0B6

BAYER AG Homepage. (2002). Retrieved from http://www.bayer.com

Bayer continues expansion of e-commerce. (2001). Retrieved from http://press.bayer.com/ News/news.nsf/ID/NT000079C2

Bayer files applications for Vardenafil in United States and Mexico. (2001). Retrieved from http://news.bayer.com/news/news.nsf/ID/01-0281

Bayer's high-tech pharmaceutical research platform. (2001). Retrieved from http:// press.bayer.com/News/news.nsf/ID/NT00007936

Benefit vs. risk: How FDA approves new drugs. (2002). Retrieved from http://www.fda.gov/ fdac/special/newdrug/benefits.html

Documentum in the pharmaceuticals industry. (2002). Retrieved from http:// www.documentum.com/products/industry/pharmaceuticals.html

FDA proposes first requirement for electronic submissions. (2001). Retrieved from http:/ /www.hhs.gov/news/press/2002pres/20020501b.html

Genomic technology pays off: Bayer and Millennium expand research program. (2001). Retrieved from http://news.bayer.com/news/news.nsf/ID/01-0272

Managing strategic assets and speeding time to market. (2002). Retrieved from http:/ /www.documentum.com/products/customer/purdue.htm

More blockbusters in the pipeline. (2001). Retrieved from http://bayer.com/ geschaeftsbericht2000//eng/index.html

NDA Approvals for Calendar Year 2001. (2001). Retrieved from http://www.fda.gov/cder/ rdmt/ndaaps01cy.htm

New Drug Application (NDA). (2001). Retrieved from http://www.fda.gov/cder/foi/nda/

New Drug Application (NDA) Process. (2001). Retrieved from http://www.fda.gov/cder/ regulatory/applications/nda.htm

Purdue Pharma LP: Speeding time to market with Documentum 4i. (2001). Retrieved from http://www.documentum.com/products/customer/BP_purdue.html

Questions about CDER. (2002). Retrieved from http://www.fda.gov/cder/about/faq/ default.htm

Shortening time to market through innovation of regulatory approval process. (2002). Retrieved from http://www.documentum.com/products/customer/ boots_healthcare.htm

Supporting pharmaceutical research and development. (2001). Retrieved from http:// press.bayer.com/News/news.nsf/1D/NT0000795A

U.S. Food & Drug Administration speeds review of new drug applications with Adobe Acrobat and PDF. (2002). Retrieved from http://www.adobe.com/aboutadobe/ pressroom/pressreleases/199910/19991006fda.html

APPENDIX A

Bayer Key Data

Bayer Group		2000	1999	Change in %
Sales	• million	**30,971**	27,320	+ 13.4
Operating result	• million	**3,287**	3,357	- 2.1
Income before income taxes	• million	**2,990**	2,836	+ 5.4
Net income	• million	**1,816**	2,002	- 9.3
Gross cash flow	• million	**4,164**	3,192	+ 30.5
Stockholders' equity	• million	**16,377**	15,182	+ 7.9
Total assets	• million	**36,451**	31,279	+ 16.5
Capital expenditures	• million	**2,647**	2,632	+ 0.6
Employees	at year end	**122,100**	120,400	+ 1.4
Personnel expenses	• million	**7,735**	7,549	+ 2.5
Research and development expenses	• million	**2,393**	2,252	+ 6.3

Bayer AG		2000	1999	Change in %
Total dividend payment	• million	**1,022**	949	+ 7.7
Dividend per share	•	**1.40**	1.30	+ 7.7
Tax credit	•	**0.45**	0.08	-

Source: http://www.bayer.com/geschaeftsbericht2000//eng/index.html

Ten-Year Financial Summary

Bayer Group (• million)	1996	1997	1998	1999	2000
Net sales	24,853	28,124	28,062	27,320	30,971
Sales outside Germany	82,2%	83,9%	83,6%	84,3%	85,6%
Sales of foreign consolidated companies	65,4%	67,0%	67,5%	68,3%	69,0%
Operating result	2,306	3,077	3,155	3,357	3,287
Income before income taxes	2,282	2,611	2,728	2,836	2,990
Income after taxes	1,405	1,509	1,615	2,018	1,842
Noncurrent assets	**10,689**	**12,230**	**13,981**	**15,614**	**20,344**
Intangible assets	729	1,051	1,909	2,213	4,843
Property, plant and equipment	8,974	10,307	10,970	11,986	13,345
Investment	986	872	1,102	1,415	2,156

(continued on next page)

Bayer Group (• million)	1996	1997	1998	1999	2000
Current assets	**14,593**	**15,467**	**15,396**	**15,665**	**16,107**
Inventories	5,144	5,424	5,781	4,992	6,095
Receivables	7,028	7,588	7,894	7,533	9,308
Liquid assets	2,421	2,455	1,721	3,140	704
Stockholders´ equity	**10,765**	**12,232**	**12,779**	**15,182**	**16,377**
Capital stock of Bayer AG	1,851	1,867	1,867	1,870	1,870
Capital reserves and retained earnings	7,287	8,638	9,087	11,134	12,454
Net income	1,393	1,504	1,614	2,002	1,816
Minority stockholders´ interest	234	223	211	176	237
Liabilities	**14,517**	**15,465**	**16,598**	**16,097**	**20,074**
Provisions	7,057	7,275	7,271	6,714	7,163
Other liabilities	7,460	8,190	9,327	9,383	12,911
Total assets	**25,282**	**27,697**	**29,377**	**31,279**	**36,451**
Proportion of total assets					
Noncurrent assets	42.3%	44.2%	47.6%	49.9%	55.8%
Current assets	57.7%	55.8%	52.4%	50.1%	44.2%
Stockholders´ equity	42.6%	44.2%	43.5%	48.5%	44.9%
Liabilities	57.4%	55.8%	56.5%	51.5%	55.1%
Financial obligations	3,520	3,896	4,730	4,466	6,665
- Long-term	1,615	2,150	2,404	2,359	2,803
- Short term	1,905	1,746	2,326	2,107	3,862
Interest income (expense) - net	(44)	(157)	(179)	(196)	(311)
Noncurrent assets financed by stockholders´ equity	100.7%	100.0%	91.4%	97.2%	80.5%
Noncurrent assets and inventories financed by stockholders´ equity and long-term liabilities	114.9%	115.5%	106.1%	112.3%	93.9%
Return on sales	9.3%	11.0%	12.6%	11.2%	11.1%
Return on stockholders´ equity	14.0%	13.1%	12.9%	14.4%	11.7%

Source: http://www.bayer.com/geschaeftsbericht2000//eng/index.html

APPENDIX B

Benefits of Documentum

- **Competitive advantage through time and cost savings.** By streamlining the production and exchange of content related to regulatory submission processes, Documentum 4i TM enables to optimize its approval process. All documents containing information on product safety, manufacturing batch records, and marketing collateral can be compiled virtually and then sent electronically to the FDA.
- **Unprecedented content flexibility.** Documentum 4i allows easy maintenance and creation of new content in numerous forms-graphics, PDF, Microsoft Word, spreadsheet, etc. — and the ability to deliver it in a consistent format to multiple destinations.
- **Management of tremendous volume of content.** With Documentum 4i, all company information-from training records to manufacturing batch records to standard operating procedures for each employee — can be reviewed instantly. A document only needs to be created once — without replication or duplication of information.
- **Global collaboration.** By using content management workflow and version control capabilities to virtually assemble documents for different regulatory agencies in different countries, it becomes possible to speed time to market and quickly identify any potential geographic differences in a medication's public perception.
- **Ease of integration with other software.** The open, standards-based architecture of Documentum 4i allows easy integration of content management with its SAP applications. In addition, Documentum 4i strongly supports XML (eXtensible Markup Language), allowing disparate systems to be joined together and allows information to move between them in an automated fashion.

Adapted from: Managing strategic assets and speeding time to market *(2002)*

Raymond Papp is an associate professor in the Sykes College of Business at the University of Tampa, USA. Dr. Papp completed his doctorate in information management at Stevens Institute of Technology. His research interests include strategic alignment, IT for competitive advantage, distance learning, and pedagogical issues in IT. His book Strategic Information Technology: Opportunities for Competitive Advantage *(Idea Group Publishing, 2001) highlights the use of information systems to achieve competitive advantage and contains numerous cases on strategic information systems. He has published in* Annals of Cases on Information Technology, Communications of the AIS, *and* Industrial Management and Data Systems, *as well as presented at numerous national and international conferences.*

This case was previously published in the *Annals of Cases on Information Technology*, Volume 5/ 2003, pp. 262-273, © 2003.

Chapter XVIII

Norwel Equipment Co. Limited Partnership (L.P.) Internet Upgrade

Kenneth R. Walsh, Louisianna State University, USA

EXECUTIVE SUMMARY

Norwel Equipment Co. Limited Partnership (L.P.) is a Louisiana business retailer of construction equipment specializing in John Deere heavy-equipment and has secured exclusive John Deere rights for most of the State of Louisiana. Founded in 1972, Norwel is the sixth largest John Deere construction equipment dealer in the United States. This case illustrates business and technology issues facing Norwel. In mid-1999, the October 1ˢᵗ deadline for John Deere's requirement to communicate by e-mail was approaching and the response time of the Norwel's primary computers system, an AS/400, was increasing to the point where users were not satisfied with performance. Also users were requesting new computing services such as e-mail, document sharing, and Internet access. For example, the Parts Operations Manger suggested selling parts online and the Manager of the Used Equipment Division suggest supporting the sales staff through Internet connections. Managing Partner, Richard Hevey decided an upgrade to the networks and a connection to the Internet were needed. He is faced with both short-term and long-term decisions about Norwel's infrastructure.

BACKGROUND

Norwel Equipment is a Louisiana business retailer of construction equipment specializing in John Deere heavy-equipment. Founded in 1972, Norwel is the sixth largest John Deere construction equipment dealer in the United States. Although they sell new equipment, a significant amount of their business is in support, service, and used equipment. Used equipment sales and service cover all makes of construction equipment. Norwel values its partnership with Deere as its primary supplier of a high quality product and promises to "exceed customer expectations" in all of their business lines. In addition to John Deere equipment, Norwel has expanded in complimentary lines and has agreements with The Charles Machine Works (Ditch Witch), Ingersoll-Rand, Barko Hydraulics, Morbark, and Broce and Tramac. Norwel is headquartered in Baton Rouge, Louisiana with three offices in Baton Rouge and eight others throughout the state, and employs about 120 people statewide.

Founded in Shreveport, Norwel's name was derived from its location in Northwest Louisiana, but through expansion and acquisition now serves the entire state. Norwel grew rapidly expanding both its range of services and covered territories. Figure 1 lists some of Norwel's major expansion milestones. The central parts warehouse, established in 1997, allowed Norwel "to establish FASTPART" and "PARTS AT NIGHT", two unique parts distribution systems that provide the best parts service anywhere" (Norwel, 2000, p. 2). Norwel has a considerable used and refurbished parts business. Its Alexandria location is a major site for disassembling equipment and refurbishing parts. It has one of the most comprehensive used parts equipment inventories in the country with customers placing orders from all over the U.S. and abroad. Today, "Norwel's revenue is derived from seven areas of activity: (1) new equipment, (2) used equipment, (3) parts (new and used), (4) service labor, (5) rentals, (6) Industrial Hydraulics, and (7) Ditch Witch" (Roher, 2000, p. 170). Norwel's organizational structure is shown in Figure 2. Norwel is a partnership whose general partner owns Empire, another construction equipment company based in California. The two companies are managed independently as they primarily serve local markets.

In late 1999, Roher Capital Group developed an analysis of the construction equipment industry and Norwel. The following are excerpts from the Roher report:

The Construction Equipment Industry generates over $23 billion annually in domestic new equipment shipments. The market is supplied by many firms but is dominated primarily by several major manufacturers including Caterpillar, John Deere, Case Equipment, and Domatsu Dresser. Equipment is sold to the end-user or to rental companies through many firms who operate exclusive dealerships representing these and other manufacturers. John Deere currently utilizes a construction, equipment dealer network of 87 owner groups who operate approximately 461 locations throughout North America. According to John Deere, the average John Deere construction

Figure 1. Expansion summary (summarized from Norwel, 2000)

- 1972: Founded, 9/1 in Shreveport
- 1973: Acquired dealerships in Alexandria and Monroe
- 1979: Acquired Lake Charles John Deere market and opened new facility
- 1983: Acquired Baton Rouge and the area north of Lake Ponchatrain John Deere markets
- 1985: Added New Orleans market
- 1988: Added Used Part Division in Alexandria
- 1993: Acquired Ditch Witch dealership for Southern Louisiana
- 1996: Acquired Lafayette John Deere market, completing acquisition of state wide rights
- 1997: Acquired a hydraulic component repair and machine shop and leased a 15,000 sq. ft. central parts warehouse
- 1998: Began selling Morbark chippers and opened a new 3,000 sq. ft. engine and component remanufacturing center in Alexandria

Figure 2. Norwel management organizational chart

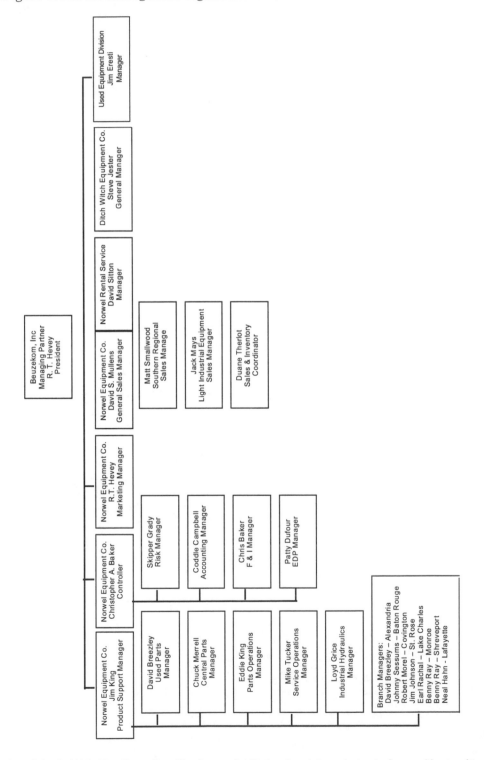

equipment dealer sells approximately 185 new units per year in an AOR [area of responsibility] that has a market potential of 1,040 units for an approximate 18% market share. Including direct sales by John Deere to national accounts, for which dealers receive selling commissions, the average John Deere Construction Equipment market share in North America is 23.5%. ... The average John Deere dealer generates approximately $42 million in revenues, $2.4 million in EBITDA [earnings before interest, taxes, depreciation and amortization] and maintains an asset base of approximately $26 million.

The industry is influenced by several key economic factors including new housing starts, public, and commercial construction spending, both for new projects and refurbishment/ infrastructure activities, and depending on the region, timberland development and demand for forestry products.

Demand for construction equipment has been consistently strong following the recessionary environment in 1991 and has closely tracked the growth of the U.S. Economy during this period.

Equipment sales in the industry are expected to hold steady well into the 21ˢᵗ century due to a continuation of the robust U.S. construction economy and economic recoveries in other parts of the world. Most industry experts point to a continued high level of employment in the U.S. with consumer confidence remaining at sustainable level, regardless of various international difficulties occurring. Federal funding for transportation and water-related infrastructure construction is expected to provide integral support for longer-term commitments than previously were possible. Demographics and high levels of expenditures in both public and private, construction market sectors are reinforcing federal funding commitments, bolstered by various state matching programs. This, together with strong single family housing construction and high levels of residential improvement activities, are expected to continue in the near future...

On the global front, demand for construction services and equipment continues to grow. Emerging economies are projected to increase their demand for U.S. firms providing services abroad. Currently, the U.S. accounts for only 20% of global construction market expenditures. This represents opportunity for Norwel online via e-commerce as its used equipment and parts business has no geographic boundary, the international market has historically been a large buyer of used equipment from the United States, and New Orleans is a major shipping port.

Over the last few years the industry has seen the emergence of large rental equipment companies who provide a broad array of products to end customers, from manual shovels and small generators to traffic safety equipment and 20-ton excavators. With more than $21 billion in sales, the equipment rental market has grown nearly 22% annually since 1982. It now accounts for approximately 14%-19% of all equipment used in the industry. Driving this growth has been the viewpoint of many contractors that renting for some of their equipment needs is more cost effective than buying. The industry has seen rapid and significant consolidation among equipment rental companies (the top 10 rental companies now account for approximately 16% of the rental industry's volume) but is still highly fragmented with more than 14,000 locations, most of which are mom and pop operations. Norwel sells equipment and parts to rental companies, as well as services their equipment, but also competes directly with them for rental business, as do all other John Deere dealers. This industry segment is expected to continue consolidating and growing over time (Roher 2000, pp. 10-12).

Within the company's AOR, the company competes against distributors of equipment produced by manufacturers other than John Deere including Caterpillar, Komatsu, and Case. The company also faces competition from distributors of manufacturers of specific types of industrial equipment, including Kobelco, Bobcat, Timberjack, JCB, Volvo Construction, Hitachi and Toyota Industrial. Outside of its AOR the Company competes for used parts and components sales against AIS Construction, a Komatsu dealer in Grand Rapids, Michigan and Empire Equipment, Norwel's sister company in Sacramento, California. For used equipment business, the Company competes against numerous companies around the world (p. 37).

The six primary markets that Norwel serves are the construction, rental, governmental, forestry, industrial, and agricultural markets. Approximately 88% of total revenues are derived from in-state sales with the balance of 12% coming from buyers outside of Louisiana. Out-of-state sales consist mostly of used equipment and used parts and components. Until very recently, Norwel concentrated all of its efforts on selling almost exclusively to customers in Louisiana. With the development of its Used Parts business in the late 1980s and formation of a Used Equipment department in early 1991 Norwel has begun to aggressively market used parts and equipment throughout North America. This has opened up a sizeable market that Norwel was previously uninvolved in. Management believes that it provides for very significant future growth (p. 38).

SETTING THE STAGE

In support of these business areas, Norwel uses a diverse set of information systems and services, many of which were initiated by the Parts Operation Manager, Eddie King, about three years ago. In May, 1996, he set up the company's wide area network (WAN) giving terminal access outside the Baton Rouge HQ office. In December, 1996, Patty DuFour was hired as the first full-time electronic data processing (EDP) manager. DuFour migrated most of the office staff to Microsoft (MS) Windows personal computers (PCs) from either dumb terminals or no computer access.

The major systems used by Norwel are described in the next section, Major Information Systems. The following section, Computer Networks, describes the WAN and local area networks (LANs) that provide connectivity between the desktops and most of the major systems. In addition to the major systems, a number of people have e-mail and Internet connectivity through dial-up modems. The company pays for several dial-up accounts and most employees now have access to the Microsoft (MS) Office program suite.

Major Information Systems

Several information systems are used by Norwel and are supported by DuFour. The primary business operations and accounting system is Dealer Management System, a comprehensive business system. Automatic Data Processing (ADP) provides payroll and human resource management (HRM) systems. Specialized systems are used within different departments including JD Vision and sales support database. Global Sourcing Network, Inc. (GSNet) provides Web page hosting. These systems are summarized in Table 1.

The most important system used by the company is DMS developed by Ontario based PFW Systems Corp. (PFW) and running on an AS/400 is available company wide through a WAN. The system, first set up by King in 1996, supports accounting and the major business functions. PFW's Web page describes their product as offering "a totally integrated management system that allows industrial, construction and agricultural dealerships to record,

Table 1. Norwel's major information systems

System Name	System Function	Vendor	Computer Platform*
Dealer Management System (DMS)	Comprehensive system including accounting, inventory, and sales	PFW	AS/400
Automatic Data Processing (ADP)	Human resource management (HRM)	ADP	Windows 95/ Intel
ADP	Payroll	ADP	service
JDVision	Visual parts catalog	John Deer Information Systems (JDIS)	Windows NT/ Intel
Sales Support Database	Used equipment inventory, sales packet generation	Developed in-house	Windows 98/ MS Access/ Intel
Web Page	Advertising	Global Sourcing Network, Inc. (GSNet)	service

* The term "service" in the platform column refers to systems that are maintained offsite by the vendor.

analyze and access real-time information about their dealership — when it's needed, where it's needed" (PFW, 2000). The system is quite comprehensive in that it stores all customer information, complete accounting information and all transactions. The system also supports management of the Norwel's major business lines including new and used equipment, new and used parts, and service. For equipment and parts it tracks inventory as well as transactions. The system used character based interface and was accessible through terminals and PCs with terminal emulation software (screenshots can be seen at the case supplement Web site).

DMS was selected following Eddie King's recommendation because Norwel needed software that could communicate with John Deere. Only DMS and Deere's own system would work. King's contacts at other dealers recommended DMS. Today, King says "the PFW software is good." He continues, "They keep taking our suggestions and making changes and we download the changes." King attends PFW workshops to learn about the latest features and make suggestions to them for improvement.

The DMS software runs on an IBM AS/400 9402 which has been in use for three years. The AS/400 is the primary system for which PFW develops and the only system on which DMS runs. Fran DeDecker of PFW says, "The 400 is strategic to us." Norwel's AS/400 is connected to the headquarters LAN.

ADP provides payroll services and personnel software. The payroll services are inexpensive and would be hard to duplicate as cheaply in house. The human resource package holds all of the company's personnel data. The company had discussions with ADP about allowing greater access to human resource package. They would like at least one person at each office to have read access to the personnel database. ADP responded by saying that they would need duplicate licenses of the software, making it cost prohibitive, so the system must be managed on a single machine at headquarters.

A new-parts catalog application called JD Vision, developed by John Deere Information Systems (JDIS), is used to help customers choose parts. The system can show the customers images of how equipment is put together so that they can select the right parts. The JD Vision system is sold to dealers as a complete hardware and software solution running on a dedicated Windows NT Server with a multi-CD disk tower. Every part of every John Deere piece of equipment, even including the decals, is stored on a set of CDs loaded into the tower. Norwel has one of these systems at each of its dealer locations. JD Vision client software communicates with the server over the LAN.

Norwel does not generally develop software in-house, but Used Equipment Division Manager Jim Erbesti developed a sales support database in Microsoft Access. The application helps the sales staff put together a sales kit for a perspective customer. A sales kit includes several pieces of information, often including a digital picture, about a piece of used equipment in Norwel's inventory. Previously this information was difficult for the sales staff to assemble. Much of the information is in the AS/400 but it is difficult to access and assemble with other information. Erbesti said "The AS/400 is accounting driven not sales driven. If you wanted to do a sales kit on the AS/400, you could have somebody come in and program it. The customer contact information, for example, only has the accounts payable contact. We need the sales contact like a contact management system." Erbesti developed the application on his laptop and connects it to a modem in the office so the salespeople can dial-in and run and download reports he has designed. Erbesti, has been giving the sales force help in transferring digital photos. The sale force has access to a digital camera and can use the application and upload and download images by dialing into a laptop on Erbesti's desk.

Norwel's Web presence is outsourced to GSNet, "the premiere search engine for heavy equipment and trucks" (Global, 2000). They host Norwel's domain name and Web pages. GSNet provides a Web presence at a cost of $150/month and automatically collects used equipment inventory data from John Deere's dealer support system (JDSS), a system implemented by Deere to facilitate exchange of dealer inventory information. Pictures are uploaded separately to GSNet and are matched to the inventory based on directory and file name. The data flow of equipment information is shown in Figure 3. Along with providing registration with search engines, GSNet is becoming a central Web site for heavy equipment.

Computer Networks

Providing connectivity to these systems is a company WAN, LANs at each site, and a satellite link to John Deere.

King was involved with the first efforts to connect the Norwel offices. When asked why Frame Relay was chosen he said, "We started with modems and analog." MCI was giving me several ways to go. They had point to point. We were thinking of putting voice and data on the lines. We started looking at the hardware and found it was cheaper to leave the Frame and voice separate. They had a good price. It has been good and MCI has been good. They have a proactive system. MCI will call and tell us Monroe is down [before we know it]." The eight out-of-town sites are tied into the Frame Relay cloud. The two local sites are tied into headquarters through local dedicated analog connections. The Frame Relay is configured with a 112 Kbps connection to headquarters and 56 Kbps connections to the other site. Norwel's WAN is shown in Figure 4. King also selected and configured CISCO 2501 routers to tie the network together. The Frame Relay links and the analog lines tie into headquarters in the telecommunications closet. The EDP Manager now manages the WAN, along with other information systems.

Data communication with John Deere is done through a satellite link at headquarters. The satellite dish is tied into the AS/400 directly.

The EDP manager, Patty DuFour, has been managing the AS/400 and installing network PCs. When DuFour first came to Norwel, users were on dumb terminals. She has now installed

Figure 3. Used equipment data flow

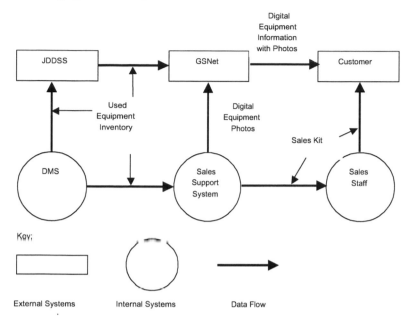

PCs on most desktops and networked through an Ethernet network. The PCs use a number of version of Microsoft Windows including Windows 3.1, 95 and NT. The PCs communicate with AS/400 across the network and access AS/400 software using terminal emulation software.

The headquarters LAN is 10BaseT with hubs located in the computer room near the AS/400. The AS/400 also ties into the hubs. The LAN connects to the WAN through CISCO 2501 routers located in the telecommunications closet as shown in Figure 5.

The networks are primarily used for AS/400 connectivity, although some people have begun using print sharing to print documents at other locations.

CASE DESCRIPTION

In August 1999, executives at Norwel met to discuss computer and network issues. Chris Baker, Controller, and Patty DuFour, Systems Operations Manager, had been hearing a number of complaints and requests from the systems' users and suggested to Richard Hevey that many of the requests were valid and could be fulfilled at minimal cost, but they were not sure how. DuFour told the group of a recent memo from John Deere requiring dealers to specify sales contact e-mail address by October 1, 1999. Deere would begin using e-mail to send updates on equipment and availability. Hevey himself said that PFW was developing a sales system for Lotus Notes that looked promising. He could envision equipping most of the sales force with laptops eventually.

PFW was contacted and they described the system they were developing called Mobile Salesman, an IBM/Lotus Notes application that provides inventory data to the sales staff. The

Figure 4. Louisana frame relay WAN

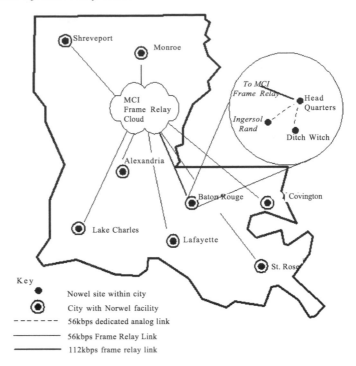

system can run on a laptop computer for the sales force to take the information in the field. The AS/400 can run the Notes server and sales staff can access it over the Internet to replicate the database on their laptops. The AS/400 can also be used as a Web server. IBM claims that reliability of the AS/400 and its operating system make an ideal platform for Web systems, substantially surpassing a Wintel strategy in performance and robustness. DeDecker says, "We ran Notes on NT and it was down twice a week." PFW was also releasing a new version of DMS, but it would require an upgrade to a new AS/400. Baker and Dufour concurred that Mobile Salesman and an upgraded DMS would be useful. Hevey and Baker thought they needed to postpone an investment in a new AS/400.

Deere was also contacted. John Howell of Deere says, "In the future we'll stop publishing on paper." The parts catalog will no longer be published on microfiche. Dealers should already rely on the NT server for a parts catalog. Deere is officially recommending that all dealers be connected to the Internet and requiring that they have e-mail capability. Deere will begin sending updates to dealers via e-mail and each dealer will need to designate three e-mail addresses per location to receive the updates. All Deere systems will eventually move to the Web to make better use of public networks. Until conversions are made, satellite and cross corporate frame relay will continue to be used; however, dealers should be aware that these networks will be phased out. In early 1999, Deere had suggested that dealers could gain access to the Internet through their cross-corporate connection to Deere. Since that time Deere has changed its policy to recommending dealers use a public Internet provider. In limited cases

Figure 5. Headquarters LAN

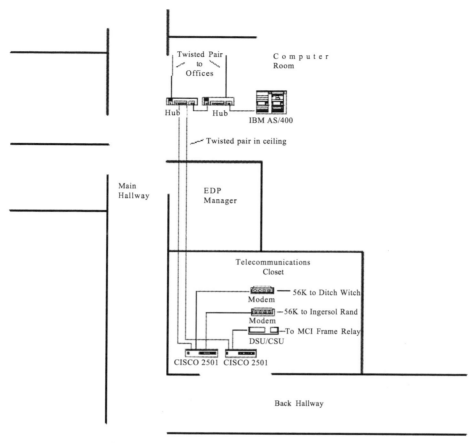

Deere may still support an Internet connection for a dealer. The satellite system will be dropped soon and the Frame Relay should be dropped in two years when all systems are made Web compatible.

Deere is also looking for other ways to help dealers help themselves. For instance, Deere does not want to maintain used equipment data. Deere is looking at an open file format where the dealers can specify the fields and maintain the data themselves. XML may be a technology that will support this.

Several at Norwel were consulted on computer issues. Eddy King said, when asked about computer problems, "The biggest problem in my job has been the speed of the AS/400, which causes headaches and lost time. I run reports from my house to get them while nobody is on the computer. A stock report takes all day. At home it takes 2 hours." He also runs queries with similar speed problems and says the parts clerks are slowed down during peak time. He suggests putting the used and new parts lists on the Web, particularly unreturnables. These parts are essentially worthless as they cannot be returned and are sitting in the warehouse, not needed locally. But nationally, they could be sold.

King continues, "If you can imagine everybody waiting it costs a lot of money. My stock ordering used to take 20 minutes and it takes as much 2 hours now." "It's the biggest gripe. In the stock order program, the computer goes out and looks at every branch with a bunch of criteria. And says you need to order this much for each location. And then it sets up a transfer from branches that are overstocked. I edit the transfer and release it and it creates POs and invoices on the system and they pull and ship the parts. I then submit it to Deere at night."

King also noted that "We have a customer up in Canada with the COPS.[1] They dial in see their prices. There are other companies inquiring. We have another company that wants to do it but it's a long distance call so they don't stay connected. We have a sister company in California but we can't see their system. Empire, based in California, is also owned by the same general partner. In the future we'd like to get into their AS/400 directly. We need to talk to PFW about how to get parts masters from Empire. It doesn't have to be real time. They deal with our used parts department by phone."

Erbesti says, "I have two issues, one internally with our AS/400 system and one with the Internet. We have very limited capabilities to put together sales information. We have it in an Access database on my PC. My PC is not on the network. We need people to get into Access and print reports and get the digital pictures. Salesmen need to dial into the system and print the reports. There are a few formats set up. They need to take it out of Baton Rouge and get it into the field where they need it. We have a lot of information on both new and used machines, but they are dependant on us to compile a report and mail it them."

Erbesti continues, "Now JD is going to let us have four pictures. If I could upload my Access database to them then we could have it automatically sent to the Web site and other places. We edit the Web site with same information. I send it to John Deere then to GSNet and then it's on the ERS [JDDSS] site. The JD ERS [JDDSS] Web site is one site where we enter the information. We need to be able to upload from a file. Our Web site has been the same for 18 months."

SUMMARY

Management has decided that Internet connectivity is needed and should be pursued immediately, particularly in light of Deere's requirement to communicate updates via e-mail. Management has also decided the purchase of a new AS/400 should be delayed for at least six months. A consultant was hired to make an initial assessment for Norwel. The following is the consultant's brief that was used as a handout for a management review meeting.

Excerpts from Consultant's Brief

Preliminary Assessment Revised and Submitted: 08/06/1999
Prepared by: Network Consultants
For: Norwel Equipment Co. L.P.

The following report was developed as a discussion draft after completing an initial analysis for the purpose of management review.

New Capabilities Desired

The following new capabilities to the Norwel Computing environment were identified.

- Share digital pictures of equipment (file sharing across WAN)
- Improve computing capability of field sales staff
- Provide e-mail capabilities
- Provide Internet access over the WAN (Deere is moving to Internet-based applications)
- Provide local access to personnel files
- Improve Web presence
- Sell used equipment on the Web
- Sell parts on the Web
- Maintain faster response from AS/400

Implications

The new capabilities desired by Norwel appear feasible and cost effective. However, the increased use of PCs, file sharing, and other network software greatly increases Norwel's dependence on these systems and the network support staff. Corresponding upgrades in network security, network management software, and network support skills need to be considered along with the expanding capabilities. Therefore, the cost of each new feature will underestimate the cost to the organization, as the support staff will have to learn the new products and take corresponding security, backup and other management actions to make the new feature reliable.

With the exception of a few immediate needs, real business benefits will not be achieved without business process reengineering. Business process reengineering documents the current business process and develops improvement ideas based on new technological capabilities. With just a few basic steps, as documented in the Phase 1 recommendation below, Norwel will be capable of full data communications capability within the company and across the Internet. This allows relatively simple file sharing, document sharing, picture sharing, e-mail, and intranet (using Web pages internally). This creates the opportunity to reduce administrative paperwork, and in some cases increase sales, but it will not happen without changing business practice and training users on the new methods.

Recommendations

Some strategic decisions need to be made relevant to outsourcing and the current selection of vendors. It is our recommendation that the current set of information systems partners is adequate and Norwel should actively maintain good communication with those providers. However, as Norwel adds new computing features, both their current computing vendor and at least two others should be consulted to ensure that the current vendors remain competitive.

If it is Norwel's strategy to be leading edge, then they will require more in-house staff and possibly new vendors, but this is probably not current business strategy. However, there is an opportunity in this strategy to gain first-mover advantage in online businesses in parts or used equipment. It can be hard to catch up if others do it well first. Watch out for Deere's strategy that may include an increase in direct sales to leverage their own Web infrastructure.

A second important decision is the use of the Notes Sales system from PFW. This has implications for both servers and e-mail systems because Notes needs a larger server than your other system needs and can include e-mail. If you go with Notes, you should probably take their e-mail. If not, it is not worth buying their e-mail. It would be our guess that you should move to Notes eventually, but it needs a good evaluation by the people that will use it (used equipment sales).

John Deere is currently recommending that dealers set up a WAN, get on the Internet, get a direct Deere connection, and get Internet e-mail. The current WAN meets this need and is capable of using the Frame Relay for a direct Deere connection. Deere wants to phase out the frame relay connections in two years and just use the Internet directly. However, some Deere systems only run on the frame relay. [As a technical note, Deere has dropped its recommendation that IP addresses follow the 10 dot standard they laid out last year. You should now use IPs coordinated with your ISP. JDIS has made a strategic shift away from being the network provider of the industry to outsourcing and utilizing public networks whenever possible.]

We have divided our recommendation into five phases ordered based on priority; however, some of the steps could be conducted in parallel. Phase one essentially establishes a corporate network infrastructure, which appears should be looked at as a high priority. Phase II essentially addresses issues of making the basic infrastructure to all employees that can use it. Phase III considers the AS/400 upgrade and makes no real change to current thinking. Phases IV and V briefly address issues in bringing Norwel into Web-based business. These issues are only briefly addressed as a strategic decision needs to be made as to whether or not the opportunity is worth a major commitment.

Phase I: Internet and Network Infrastructure

With Deere going towards the Internet, Phase 1, with the exception of steps 5 and 6, appears necessary. Steps 6 and 7 are highly recommended and should proceed with relatively high priority.

1. Support staff training in Windows NT and TCP/IP networking.
2. Revise current wiring plans.
3. Choose an Internet service provider (ISP).
4. Set up a firewall if you set up access through your network.
5. Set up a short-term e-mail scheme with GSNET.
6. Set a machine up as an NT file server and put the pictures on it.
7. Thoroughly evaluate the PFW Notes Sale software.
8. Eliminate Y2K issues
9. Improve Web presence.

Phase II: Notes and Company-Wide E-Mail

1. If the Notes evaluation is positive, pick a server and install.
2. Set up company e-mail.
3. Train users on new features.
4. Implement remote workstation management software.

Phase III: AS/400
The AS/400 plan described by Norwel appears on target.

Phase IV: Improved Web Presence
Improve information aspects of Web presence.

Phase V: E-Commerce
It does appear that there is an opportunity for doing business on the Web.

Management met to discuss the consultants brief and recommended that a part-time technical support person be hired to help DuFour implement Phase I of the brief, after which other phases could be considered. The skills of the technical support person should include Windows NT configuration, TCP/IP, and server applications including e-mail, FTP, and Web server.

REFERENCES

Global Sourcing Network Inc. (2000). Home Page. Retrieved February 23, 2000, from http://www.gsnet.com/

Norwel Equipment Co. LP. (2000). *Business plan.* Baton Rouge, LA.

PFW Systems Corporation. (2000) Home Page. Retrieved February 23, 2000, from http://www.pfw.com/

Roher Capital Group, LLC. (2000, January). Norwel Equipment Co, LP. The Product Support Company.

FURTHER READING

Cisco Systems, Inc. (1997). *Cisco 2500 series overview.* Available at http://www.cisco.com/univercd/cc/td/doc/product/access/acs_fix/cis2500/2505/2500him/77414.htm. accessed 2/23/2000.

Cisco Systems, Inc. (1997). *Cisco 2500 series router installation and configuration guide.* Available at http://www.cisco.com/univercd/cc/td/doc/product/access/acs_fix/cis2500/2501/2500ug/conf.htm. accessed 2/23/2000

Cisco Systems, Inc. (1998). *Router products release notes for Cisco IOS release 11.0.* Retrieved February 23, 2000, from http://www.cisco.com/univercd/cc/td/doc/product/software/ios11/rnrt110/rnrt110.htm

Comer, D. E. (1995). *Internetworking with TCP/IP: Volume I, Principles, protocols, and architecture.* Upper Saddle River, NJ: Prentice Hall.

Helmstetter, G. (1997). *Increasing hits and selling more on your Web site.* New York: John Wiley and Sons.

Machine Mart Home Page. Available at http://MachineMart.com/

Machine Trader Home Page. Available at http://MachineryTrader.com/

Motorolla, Inc. (1991). *Technical summary: Second-generation 32-bit enhanced embedded controller.* Retrieved February 23, 2000, from http://www.mot.com/SPS/HPESD/aesop/680X0/030/EC030_ts.pdf

Norwel Case Supplemental Web site. Available at http://isds.bus.lsu.edu/cvoc/projects/
    norwel/
Shapiro, C., & Varian, H. R. (1999). *Information rules: A strategic guide to the network
    economy.* Boston: Harvard Business School Press.

ENDNOTES

[1]   COPS is a software package that allows authorized customers to browse the DMS
       inventory by dialing in to the AS/400.
[2]   Note that there are not actually 256 valid numbers per octet. But for simplicity we
       assume in this example that all numbers 0 to 255 would result in a valid, routable
       IP address.

APPENDIX: NETWORK ROUTING AND THE CISCO 2501

by Van Goodwin

Routers, also known as "gateways," are small, dedicated computers that provide an efficient means to connect multiple networks to each other. By maintaining a table of information about the networks they connect to, routers allow for transmitting data across large networks, including the largest network of all, the Internet.

A router's basic job seems simple. It looks at the destination *network* of a packet of data, attempts to determine the quickest route to that network via other routers, and sends the packet on its way. Packets are transferred from router to router until they arrive at a router that can deliver the packets directly to the intended host. One must keep in mind that since the router catalogs only the locations of networks, it does not need to keep track of every host that can be contacted. Furthermore, any one router does not need to know the full path between the source and the destination. It only needs to know the address of the destination host and the next router in line to get to that address. For example, in Figure A1, if a server on router A sends a packet to a PC on router C, router A needs only to know that router B is the next hop to get to router C. This accounts for how routers can perform their function with such limited storage.

On many larger networks, a concept known as subnetting is required. This allows the wider Internet to view a collection of physical networks as a single physical network with only the local router knowing the true local network topology. For example, suppose a server needs to send packets to 130.39.100.122. This IP address is a member of a physical network that is a member of several other physical networks that are connected to a router. But to the wider Internet, this whole collection of networks is a single physical network. To determine what physical network to send packets to, the local router can divide the IP address similar to the illustration in Figure A2.

This would tell the router that the destination is host 122 of subnet 100. This configuration allows for up to 256^2 networks, each with 256 hosts in this configuration. As far as the wider

Figure A1. Routing a packet from Server to PC

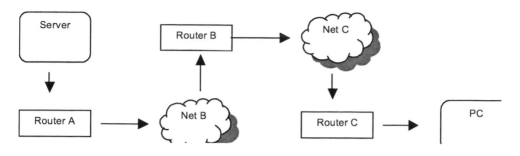

Figure A2. A subnetted IP address

Wider Internet Prefix	Local Network	Host
130.39.	100.	122

Internet is concerned, all possible 65,636 hosts reside on the same physical network. To accommodate different network topologies, routers can divide the IP address in various other ways. So, the company that requires 10 physical networks with 300 hosts each or 300 physical networks with 10 hosts each, can still interpret an IP address differently to provide for its needs. It might be useful to think of IP addresses like U.S. phone numbers. Take the telephone number "225.334.3724". In this case, the "225" tells the telephone network an area code, much like the "130.39" in the above IP address. The "334" tells the network which exchange it needs to access, as the "100" in the IP address identifies what network to go to. And from there, the "3724" tells the exchange specifically what telephone should be reached like the "122" tells the router which host to access.

To accomplish all of these tasks, routers must have a set method for deciding what the "next hop" router should be to eventually arrive at the destination host. Some routers rely on a static method with a predefined table that tells them what the appropriate next hop is for a given IP address. Others use a more complex approach by looking at various options from their table and compensating for such factors as current line conditions and type of packet being transmitted before making a decision. Since the tables are seldom changed manually, routers periodically exchange information about routes to accommodate changes in network topography. In this manner, changes in routing requirements are propagated throughout the entire collection of networks. Routers vary widely in the amount of sophistication with which they deal with packets. Thus, many classes of routers exist to accommodate different needs of various companies.

Cisco™ 2500 series routers, for example, allow TCP/IP connections over a variety of network interfaces including Ethernet, Token Ring, Synchronous Serial, Integrated Services

Figure A3. Rear view of a cisco 2501 router

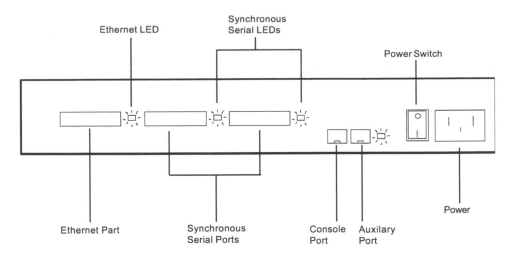

Digital Network (ISDN) and Basic Rate Interface (BRI). Some models can be connected to each other via 68-pin SCSI interfaces. Figure A3 is a rear-plate diagram of a Cisco 2501 router that shows its various network interfaces.

At the heart of 2500 series routers is a 32-bit 20-MHz Motorola processor. The level of router complexity is determined primarily by the router's Internetwork Operating System (IOS) version. The IOS is to the Cisco router as a suite of software is to a PC; it determines how the router behaves in its function. The types of protocols handled, packet routing methods, and most other rules for handling information is determined by the IOS.

Kenneth R. Walsh is an assistant professor of information systems in the Information Systems and Decision Sciences (ISDS) Department at Louisiana State University. He received a PhD in information systems from the University of Arizona in 1996. His research interests include structured modeling group support systems (SM-GSS), Internet support for education, and motivations and methods in software development, and he has published several articles. He was also a founding researcher of LSU's Center for Virtual Organization and Commerce and has received more than $500,000 in funding. He teaches network information systems and Internet systems development as well as a number of IS courses.

Van Goodwin is an information systems student at Louisiana State University. He serves as system administrator for the ISDS Department, an independent consultant to several firms and a technical advisor to various research projects.

This case was previously published in the *Annals of Cases on Information Technology Applications and Management in Organizations*, Volume 3/2001, pp. 179-195, © 2001.

Chapter XIX

Integration of Third-Party Applications and Web Clients by Means of an Enterprise Layer

Wilfried Lemahieu, Katholieke Universiteit Leuven, Belgium

Monique Snoeck, Katholieke Universiteit Leuven, Belgium

Cindy Michiels, Katholieke Universiteit Leuven, Belgium

EXECUTIVE SUMMARY

This case study presents an experience report on an enterprise modelling and application integration project for a young company, starting in the telecommunications business area. The company positions itself as a broadband application provider for the SME market. Whereas its original information infrastructure consisted of a number of stand-alone business and operational support system (BSS/OSS) applications, the project's aim was to define and implement an enterprise layer, serving as an integration layer on top of which these existing BSS/OSSs would function independently and in parallel. This integration approach was to be nonintrusive and was to use the business applications as-is. The scope of the case entails the conception of a unifying enterprise model and the formulation of an implementation architecture for the enterprise layer, based on the enterprise JavaBeans framework.

BACKGROUND

This case study deals with a company acting as supplier of fixed telephony and of broadband data communication and Internet services. A particular feature of the company is that all the telecom services it offers are facilitated via "Unbundling of the Local Loop" (ULL).

ULL is the process where the incumbent operator makes its local network (the copper cables that run from customers' premises to the telephone exchange) available to other operators. These operators are then able to use individual lines to offer services such as high-speed Internet access directly to the customer. The European Union regulation on ULL requires incumbents to offer shared access (or line sharing). Line sharing enables operators and the incumbent to share the same line. Consumers can acquire data services from an operator while retaining the voice services of the incumbent. Some operators may choose to offer data services only, so with line sharing consumers can retain their national PTT service for voice calls while getting higher bandwidth services from another operator without needing to install a second line.

The regulation on ULL can have a significant impact on the competing forces in the telecom industry: it offers a great opportunity for new companies to enter the telecom market and compete with the incumbent operator. Indeed, by means of ULL the sunk cost of installing a countrywide network infrastructure is not an obstacle anymore for new entrance in the telecom market.

A large telecommunication company immediately understood the business opportunities behind this new regulation and decided to exploit the ULL benefits in all European countries. In a first step, it has created a starter company that is the subject of this case study. As a means to differentiate from the services offered by the incumbent operator, the new company focuses on telecom services for the business market, the small- and medium-sized sector in particular. The main headquarters are located in the first European country where ULL is possible. The starter company was set up in September 1999. In March 2000, it succeeded in acquiring two major investors from the U.S., which are both specialists in new media. The company has evolved rapidly, and in August 2001 it already surpassed 2000 customers and employed about 150 people. Presently, the company offers its services in two countries. Gradually, the company will extend its coverage of the European Union by opening new premises in the countries that enable ULL.

SETTING THE STAGE

Business Units with Stand-Alone Software Packages

The company is organised around four key business units: Sales & Marketing, Service Provisioning, Finance, and Customer Services. The business unit Sales & Marketing is responsible for identifying emerging trends in the telecom industry and offering new telecom services in response. They are in charge of PR activities, contact potential customers and complete sales transactions. The business unit Service Provisioning is responsible for the delivery of the sales order and organises the provisioning of all telecommunication services the customer ordered. They have to coordinate the installation of network components at the customer's site and the configuration of these components according to the type of service requested. The business unit Finance takes care of the financial counterpart of sales transactions and keeps track of the payments for the requested services. The business unit Customer Services is responsible for the service after sales. They have access to the entire network infrastructure, and on request

Figure 1. Main business process

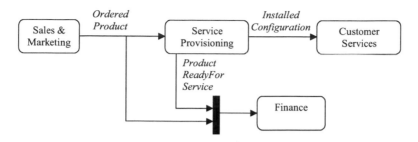

they can inform a customer about the progress of a service provisioning activity, about the network status or possible network breakdowns. The main business process is shown in Figure 1.

It is a company policy that the amount of in-house developed software must be limited. The company has therefore acquired a number of off-the-shelf software packages. Apart from the Sales & Marketing business unit that relies only on elementary office software, each of these business units relies on different business and operational support systems (BSS/OSS) that are tailored to the problems at hand. Until recently, all BSS/OSS were functioning independently from each other. The company quickly understood that the integration of the BSS/OSS could improve its competitive position in three ways. On the short run, the company could benefit both from a *better performance* in transaction processing and *more transparency* in its business process. On the long run, the activities of the company should become *better scalable* to an international setup with multiple business units. Now is presented in some more detail how the integration of the BSS/OSS can realise these benefits.

Better Performance

To differentiate from the services offered by the incumbent operator, the company has targeted her services towards the SME market segment. As opposed to the segment of multinational companies, the SME segment typically involves large volumes of relatively small sales orders. But the performance of the present information infrastructure with stand-alone BSS/OSS deteriorates significantly when a large number of transactions have to be processed simultaneously. A main factor explaining this performance decrease is that the BSS/OSS applications are essentially used as stand-alone tools: each application supports only part of the value chain and is treated in isolation. Although each package is very well suited for supporting a specific business unit, the lack of integration between the different applications causes problems: each application looks only after its own data storage and business rules, resulting in a state of huge data duplication and severe risks for data inconsistency on a company-wide level. As a result a substantial amount of manual work is required to coordinate the data stored in the different packages. To avoid these problems in the future, an integration approach should be adopted guaranteeing a centralised data storage and organising the automatic coordination between the different BSS/OSS.

More Transparent

By integrating the BSS/OSS applications, company-wide business rules can be maintained that give a more formal structure to the company's business organisation. As a result, the business process will become more transparent and better controllable. Previously, the co-operation between different business units was rather ad hoc and not yet formalised on a company-wide level. As a result, a number of business opportunities/ risks could often not be detected by the company. For example, the company had problems with a customer company not paying her bills. Later on the commercial contact of this company reappeared as financial contact person in another customer company, and again bills didn't get paid. This repetition of the same problem could have been avoided if information on people had been centralised. In the current system, data on people is stored in all four business units. Sales & Marketing stores data on in-house sales people, out-house distributors and commercial contacts. Service Provisioning and Customer Services both maintain data on technical contacts. The Finance application keeps track of financial contacts. Since the company mainly deals with SME, an individual often takes several of these roles simultaneously, so that with the current information infrastructure data about one person will be distributed and replicated across several business units. Nevertheless, the company could benefit from an approach ensuring that data on a single person is stored and maintained in one place. Indeed, if a person is not loyal in his or her role of commercial contact, the company should take extra care in the future when assigning this person the role of financial contact. Such a policy is only maintainable if all information on individuals is centralised. (In the integrated information infrastructure such a person will be altered to the state "blacklisted," and as long as he resides in this state he cannot be assigned to one of the above roles.)

Better Scalable

As an early adopter of the ULL opportunity, the company is very likely to become one of the major players targeted to the telecom SME market in the European Union. Flexibility in both the adoption of new products, to keep pace with the evolving technology, and in the adaptation of existing products, to conform to different company standards, will be one of the cornerstones to realise this objective. However, the current information infrastructure cannot guarantee this level of flexibility.

The *Sales & Marketing* business unit is responsible for conducting market research and offering appropriate telecom solutions to keep pace with evolving business opportunities. What they sell as one single product can be further decomposed into a number of parts to install and parameters to configure by the *Service Provisioning* business unit. An example of this is depicted in Table 1.

In fact, the business units need a different view on products. Sales & Marketing and Finance need a high-level product view and are not interested in low-level issues related

Table 1. Commercial and technical view on products

Product	Parts and Parameters
bidirectional link of 256 kbps	installation of an unbundled line (2x)
	installation and configuration of a router (2x)
	configuration of a virtual circuit with bandwidth of 256 kbps

to the installation and configuration activities. Service Provisioning needs to know which parts where to install and to configure and does not bother about the high-level sales and marketing issues. In an attempt to keep a unified view on products while accommodating for their different needs, people of the business units tend to twist the product definitions in their respective software packages. By abusing attributes and fields for cross-referencing purposes, they try to maintain a more or less integrated approach.

However, as the set of products in the product catalogue will increase, a unified view is no longer sustainable. Also in an international setup with multiple business units a unified view can be held no longer: what if a single product from a sales point of view requires different technical configuration activities depending on the business unit. An example of the latter is Internet access: whereas it can be implemented by means of ULL in the Netherlands and the UK, it must be provided with a leased line in Belgium where ULL is not (yet) possible. The company would certainly benefit from a scalable design reconciling the commercial view and the technical view that can be taken on a single product, the former important for the Sales & Marketing and Finance business units and the latter important for the Service Provisioning business unit.

CASE DESCRIPTION

Goals of the Project

The goal of the project was to develop a solution for the integration of the BSS/OSS applications, so as to achieve fully automated processes that are able to handle scalable business processes, systems and transaction volumes. The same software will be used in all European divisions of the company. The company has contacted a team of researchers at a university to consult them for the development of a solution. A consultancy company, unknown at the start of the project, will do the implementation. The time frame of the project is roughly five months for requirements engineering, about two months for the technical design of the solution's architecture and another four months for implementation. During the requirements engineering period, the company will look for an adequate implementation environment and contract a consultancy company for the implementation part. The remainder of this section outlines the proposal as the university team developed it.

Choosing the Right Integration Approach

A Bridging Approach

A possible approach to the integration problem would be to build "bridges" between the different software packages. Such a bridging approach is usually based on the "flows" of information through the different business units. From the business process shown in Figure 1, we can derive such an architecture for the case at hand (see Figure 2).

The major advantage of this approach is its simplicity. If one chooses to automate the existing manual procedures, the amount of required requirements analysis is limited.

Figure 2. "Stove-pipe" architecture derived from the information flow defined in the business process

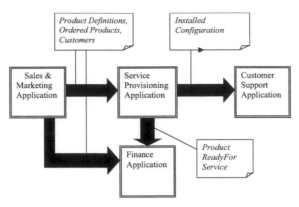

In addition, one can rely on the expertise of the providers of the BSS/OSS software to build the bridges. Such architecture does however not resolve data mapping and data duplication problems: information about customers, products, and other common entities is still found in different places. Although it is unlikely that data replication can be completely avoided, another major problem with this kind of architecture is that the business process is hard-coded into the information management architecture. Reengineering of the business processes inevitably leads to a reorganisation of the information management architecture. Such a reorganisation of IT systems is a time-consuming task and impedes the swift adaptation of a company to the ever-changing environment it operates in.

An Integration Approach Based on an Enterprise Layer

An approach that does not hard code the underlying business processes is to define a common layer serving as a foundation layer on top of which the stand-alone BSS/OSS can function independently and in parallel (see figure 3). This so-called enterprise layer has to coordinate the data storage by defining a unified view on key business entities such as CUSTOMER and PRODUCT. The common information is stored into a shared object database that can be accessed through an event-handling layer. This event-handling layer shapes the manipulation of enterprise layer objects through the definition of business events, their effect on enterprise objects and the related business rules. From an integration point of view this is certainly a more flexible approach: it does not hard code the current business processes and more easily allows for business process redesign. In addition, the replacement of a particular BSS/OSS service will not affect the other packages: all interaction is accomplished through the enterprise layer: they never interact directly with each other. In this way the company is more independent of the vendors of the packages. On the other hand, this solution has a number of factors that will make it more costly than the bridging approach:

•   It requires a thorough domain analysis in order to integrate the concepts of the four functional domains;

Figure 3. "Enterprise layer" architecture derived from the information flow defined in the business process

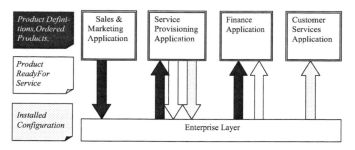

- Relying on the expertise of the software vendors will be more difficult;
- There is no experience in the company with such an approach.

Phases of the Automation Strategy and Global Architecture

After considering the advantages and disadvantages of both the bridging and the enterprise layer approach, the company has opted for the latter. An additional motivation is that apart from the integration aspects, the enterprise layer will allow the future development of e-business functionality directly on top of the enterprise layer, rather than via one of the BSS/OSS applications. The automation strategy of the company is hence summarised by the following four steps:

Step 1: Roll out of industry-proven BSS/OSS applications with out-of-the-box functionality. Each application is treated in isolation (total lack of integration).

Step 2: Specification and development of an enterprise layer that will support all applications and user interfaces. The enterprise layer is by definition passive and not aware of its users.

Step 3: Integration of the existing BSS/OSS applications by "plugging" them in on the enterprise layer. The interface between these applications and the enterprise layer will be realised by the definition of agents, responsible for coordinating the information exchange.

Step 4: Development of user interfaces on top of the BSS/OSS applications or directly on top of the enterprise layer.

At the start of the project, step 1 had been realised: three out of the four main business domains (as represented in Figure 1) were supported by a standalone software package. There existed no automated support for the sales domain: sales business processes were mainly paper-based or used standard office software such as word processors and spreadsheets. The university team was responsible for requirements engineering and technical design for steps 2 and 3.

Modelling the Enterprise Layer

Development of an Enterprise Model: Choice of the Methodology

In order to specify an enterprise layer that acts as an integration layer of the existing information infrastructure, it is key that all the company's business rules are formalised and modelled as a subset of the enterprise model. This requires the choice of a systems analysis method, preferably one that is suited for domain modelling. Two options were considered:

Enterprise Modelling with UML

UML has become a de facto standard for object-oriented analysis. It has a very broad scope of application and the major techniques (class diagram, state machines, use cases) can be used for enterprise modelling purposes. Many CASE tools are available that facilitate the transition to design and implementation. The language is known to practically all IT people, which will facilitate the communication with the implementation team.

On the other hand, UML is not an approach but a *language*. It is not specifically tailored to enterprise modelling either. As a result, UML is not a great help for the modelling task. Another disadvantage of UML is its informal semantics. Because of the informal semantics, UML offers no support for the consistency checking and quality control of specifications. In addition, the informal semantics also increase the risk of misinterpretation by the implementation team.

Enterprise Modelling with MERODE

A particular feature of the object-oriented enterprise modelling method MERODE (Snoeck et al., 1999; Snoeck et al., 2000) is that it is specifically tailored towards the development of enterprise models. MERODE advocates a clear separation of concerns, in particular a separation between the information systems services and the enterprise model. The information systems services are defined as a layer on top of the enterprise model, what perfectly fits with the set-up of the project. The enterprise layer can be considered as a foundation layer on which the acquired off-the-shelf software can act as information services.

A second interesting feature of MERODE is that, although it follows an object-oriented approach, it does not rely on message passing to model interaction between domain object classes as in classical approaches to object-oriented analysis (Booch et al., 1999; D'Souza & Wills, 1999; Jacobson et al., 1997). Instead, *business events* are identified as independent concepts. An object-event table (OET) allows the definition of which types of objects are affected by which types of events. When an object type is involved in an event, a method is required to implement the effect of the event on instances of this class. Whenever an event actually occurs, it is broadcast to all involved domain object classes. This broadcasting paradigm requires the implementation of an event-handling layer between the information system services and the enterprise layer. An interesting consequence is that this allows the implementation of the enterprise layer both with object-oriented and non-object-oriented technologies. Since some of the acquired software is built with non-object-oriented technology, this is a substantial advantage.

A third feature of MERODE is that the semantics of the techniques have been formalised by means of process algebra. This allows the checking of the semantic integrity of the data model with the behavioural models.

On the negative side, MERODE has a narrow scope of application: it is only intended for enterprise modelling purposes. Although some hints are given on how to implement an enterprise layer, the MERODE approach gives but very little support for the specification and implementation of services on top of the enterprise layer. A second major disadvantage of the method is that it is unknown to most people. Because of this and of the peculiar way of treating events and interobject communication, difficulties are to be expected for communicating the requirements to the implementation team. It is likely that the implementation team will not be familiar with the approach and that additional MERODE training will be required.

Based on the above considerations, the project manager decided to go on with MERODE as the modelling approach. The major reasons for this choice are the consideration that quality of the enterprise model is a key factor for the success of the enterprise layer approach and the possibilities for implementation with a mixed OO and non-OO approach.

Key Concepts of the Enterprise Model

To better illustrate the responsibilities of the different layers, objects in the domain layer and the event handling layer are exemplified by considering an example of an order handling system. Let us assume that the domain model contains the four object types CUSTOMER, ORDER, ORDER LINE and PRODUCT. The corresponding UML (Booch et al., 1999) Class diagram is given in Figure 4.

Business event types are *create_customer*, *modify_customer*, *end_customer*, *create_ order*, *modify_order*, *end_order*, *create_orderline*, *modify_orderline*, *end_orderline*, *cr_product*, *modify_product*, *end_product*. The object-event table (see Table 2) shows which object types are affected by which types of events and also indicates the type of involvement: C for creation, M for modification and E for terminating an object's life. For example, the *create_orderline* event creates a new occurrence of the class ORDERLINE, modifies an occurrence of the class PRODUCT because it requires adjustment of the stock-level of the ordered product, modifies the state of the order to which it belongs and modifies the state of the customer of the order. Notice that Table 2 shows a maximal number of object-event involvements. If one does not want to record a state change in the customer object when an order line is added to one of his/her orders, it suffices to simply remove the corresponding object-event participation in the object-event table. Full details of how to construct such an object-event table and validate it against the data model and the behavioural model is beyond the scope of this case study but can be found in Snocck and Dedene (1998) and Snoeck et al. (1999).

Figure 4. Class-diagram for the order handling system

Table 2. Object-event table for the order handling system

	CUSTOMER	ORDER	ORDER LINE	PRODUCT
create_customer	C			
modify_customer	M			
end_customer	E			
create_order	M	C		
modify_order	M	M		
end_order	M	E		
create_orderline	M	M	C	M
modify_orderline	M	M	M	M
end_orderline	M	M	E	M
create_product				C
modify_product				M
end_product				E

The enterprise layer developed in the project covers the company's four main business domains: People, Products, Orders and Configuration.

The *People* domain concerns both the customers and the salespersons. Information about people is found in all four business processes. The Sales & Marketing process stores data on salespersons (both in-house and distributors) and on commercial contacts for customers. The Service Provisioning application and the Customer Services application both maintain data on technical contacts. Finally, the Financial application keeps track of data about financial contacts. Since the company mainly deals with SME, a single person often takes several roles simultaneously, so that information about the same person can be distributed and replicated across several business processes. The enterprise layer ensures that information about an individual person is stored and maintained in a single place.

The *Products* domain describes the products sold by the company. The four business processes have their own particular view on the definition of the concept "product," and each BSS/OSS maintains its own product catalogue. Again, the enterprise layer will be responsible for tying together the description of products. Initially, it was assumed that a single company-wide definition per product would be possible, but soon it appeared that each functional domain had to work with a particular view on this definition. Therefore, the final enterprise layer describes products both from a marketing/sales perspective (what products can be sold and for which price) and from a technical perspective (what are the technical elements needed to deliver an item). The sales perspective is variable over time depending on sales marketing campaigns, both from a price and descriptive standpoint. The technical description is constant over time, but new technology may cause a redefinition without changing sales description.

The *Orders* domain encompasses all business objects related to sales orders. The sellable products are defined and registered in the *Products* domain; the actual ordered products are registered in the *Orders* domain. Finance will use this domain.

The *Configuration* domain keeps track of all business objects related to the technical configuration that is build at a customer's site during provisioning activities. The parts to install and the parameters to configure are defined and registered in the *Products* domain; the actual installed parts and configured parameters are registered in the *Configuration* domain. The information of this domain is provided by the Service Provisioning application and is consulted by the Customer Services application.

Business Rules and Constraints

Whereas the above mainly dealt with information *structure*, the MERODE enterprise model also incorporates *behavioural* specifications. Indeed, as can be seen from the description of the business process, activities of the different departments must be coordinated in a proper way. Notice how in this type of approach the business process is not hard-coded in the architecture. All information flows through the enterprise layer. In this way, the integration approach is deliberately kept away from a document-based, flow-oriented, "stovepipe"-like system. Interaction between respective applications is not based on feeding the output document of one application as input to the next in line, but on the concurrent updating of data in the shared, underlying enterprise layer. This results in a maximum of flexibility in the interaction between users and system. However, wherever certain workflow-related aspects in the business model necessitate a strict flow of information, the correct consecution of business events can be monitored by the *sequence constraints* enforced in the enterprise layer.

Indeed, sequences between activities are enforced in the enterprise layer by allowing domain objects to put event sequence constraints on their corresponding business events. For example, when a customer orders a product, a new order line is created. In terms of enterprise modelling, this requires the following business events: *create_salesorder, modify_salesorder, create_orderline, modify_orderline, end_orderline*. The *customer_sign* event models the fact that a final agreement with the customer has been reached (signature of sales order form by the customer). At the same time this event signals that installation activities can be started. These kinds of sequence constraints can be modelled as part of the life cycle of business objects. In the given example this would be in the life cycle of the sales order domain object. As long as it is not signed, a sales order stays in the state "existing." The *customer_sign* event moves the sales order into the state "registered." From then on the sales order has the status of a contract with the customer, and it cannot be modified anymore. This means that the events *create_orderline, modify_orderline* and *end_orderline* are no longer possible for this sales order. The resulting finite state machine is shown in Figure 5.

A Layered Infrastructure: Coordination Agents and User Interfaces

Once the enterprise layer is developed, the integration of the business unit applications is realised by plugging them in on the enterprise layer. The previously stand-alone BSS/OSS applications now constitute the middle layer. An important feature of the enterprise layer is that it is a *passive* layer that is not aware of its possible users. Therefore, an *active* interface between the enterprise layer on the one hand and the business unit applications on the other hand had to be defined. This interface can be realised by developing agents responsible for coordinating both sides. Such *coordination agents* will listen to the applications and generate the appropriate events in the enterprise layer so that state changes in business classes are always reflected in the enterprise model.

For example, by listening to the *customer_sign* event, the coordination agent for the Service Provisioning application knows when a sales order is ready for processing

Figure 5. State machine for sales order

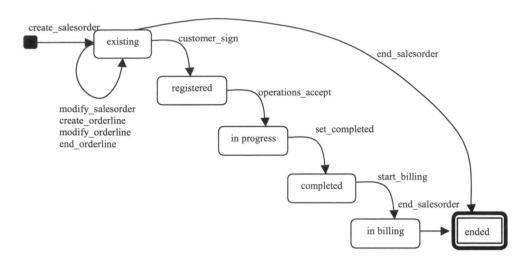

for Service Provisioning. In a similar way, the enterprise layer allows to monitor signals from the Service Provisioning area (such as the completion of installations) and to use these signals to steer the completion of sales orders. At its turn, the completion of a sales order (event *set_completed*) can be used by the billing agent to start the billing process. In this way, the enterprise layer allows for an automated coordination of behavioural aspects of different business areas. A redesign of the business process requires the redesign of the event sequencing in the enterprise layer but leaves the coordination mechanisms unchanged.

The top layer is established by the development of user interfaces that offer a Web interface to the BSS/OSS applications or directly to the enterprise layer. For example, functions related to the sales process (entering of Sales Order Forms) and to customer self-care functions (access and maintenance of personalised content) are input and output functions performed directly on the enterprise layer. The resulting layered infrastructure is depicted in Figure 6.

Implementation of the Enterprise Layer

Choosing the Implementation Architecture

The enterprise layer consists of the MERODE enterprise objects, as defined above. These objects embody a combination of data and behaviour. As to the pure data, a relational database (Oracle) seemed to be the obvious implementation choice. However, to implement the entirety of data and behaviour, three alternatives were considered:

A Combination of Stored Procedures and Relational Tables

A first option consisted of a combination of *stored procedures* (for the functionality) and *relational tables* (for the data). Indeed, although MERODE is an object-

Figure 6. Enterprise layer integration approach

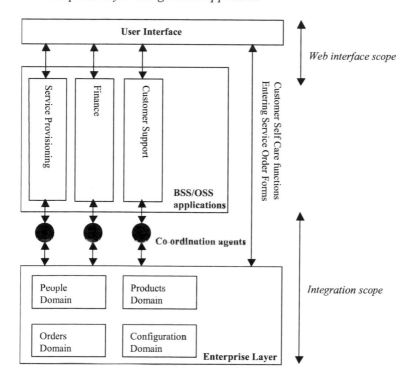

oriented analysis and design methodology, the implementation is not restricted to an object-oriented environment. The object definitions can be mapped to relational structures, whereas the object-event table can be translated to stored procedures. Each horizontal line in the object-event table will define a single stored procedure, resulting in one procedure per event. Each such procedure enforces the necessary constraints upon all objects participating in the event and manipulates the relational database tables to which these objects are mapped.

An Object-Relational Approach

A second possibility was to build upon the *object-relational capabilities of Oracle*. Here, the MERODE enterprise objects would be implemented as user-defined types in Oracle, which combine data and behaviour into "object types," stored in the Oracle database.

A Distributed Object Architecture

A third option was to turn to a *component-based distributed object architecture* such as EJB, CORBA or DCOM. In this approach, the business objects would be represented as distributed objects, which are mapped transparently into relational tables. The external view, however, remains fully object-oriented, such that interaction can be based on method invocation or event handling.

The eventual preference went out to an approach based on a *component-based distributed object architecture* for several reasons. First, it allows for the enterprise layer to be perceived as real objects. In this way, all constraints defined in MERODE can be mapped directly to their implementation. The constraints are not only attributed to *events*, but also to *objects* and can be enforced and maintained at the object level (i.e., the vertical columns of the MERODE object-event table: each column represents an object that reacts to different events). This is not possible with a purely relational design or with an object-relational implementation, where object-oriented concepts are still tightly interwoven with relational table structures. As a consequence, a distributed object approach offers the best guarantee with regard to maintainability and scalability. Moreover, a number of "low-level" services such as concurrency control, load balancing, security, session management and transaction management are made virtually transparent to the developer, as they can be offered at the level of the application server.

As to the choice between CORBA, EJB and DCOM, the Java-based EJB (Enterprise JavaBeans) was chosen, offering a simpler architecture and being easier to implement than CORBA. Moreover, unlike DCOM, it is open to non-Microsoft platforms. The EJB architecture guarantees *flexibility*: all enterprise Beans are components and can be maintained with little or no consequences for other Beans. Also, such environment is easily *scalable*, as enterprise Beans can be migrated transparently to another server, e.g., for load-balancing purposes. Finally, in this way, the enterprise layer would easily exploit Java's opportunities for Web-based access such as JSP (Java Server Pages) and servlets. The proposed architecture conformed the *n-tier* paradigm, utilising the database server only for passive data storage and moving all actual functionality to a separate application server, where the enterprise Beans represent the "active" aspects of the enterprise layer (see Figure 7). The MERODE enterprise objects are implemented as so-called *entity Beans*. These are a specific type of enterprise JavaBean that essentially denote object-oriented representations of relational data, which are mapped transparently into relational tables. Updates to an entity Bean's attributes are propagated automatically to the corresponding table(s). Hence, although a relational database is used for object persistence, the external view is fully object-oriented, such that interaction can be based on (remote) method invocation or event handling.

All functionality is available in the enterprise Beans on the application server. External applications are to make calls upon the enterprise Beans by means of Java's RMI (Remote Method Invocation). A business event is dispatched to each entity Bean that represents a business object that participates in the event. An enterprise object handles such event by executing a *method*, in which constraints pertaining to that particular (object type, event type) combination also are enforced. Hence, rather than enforcing integrity constraints at the level of the relational database, they are enforced in the entity Beans' *methods*. Clients only interact with the enterprise Beans; the relational database is never accessed directly. In this way, the MERODE specification can be mapped practically one-to-one to the implementation architecture.

Interaction Between Applications and Enterprise Layer

Although the data in the enterprise layer can be queried directly by means of purpose-built user-interface components, its primary focus is to offer a unified view on the data objects observed by the respective BSS/OSS applications, to which the enterprise layer serves as a foundation layer.

Figure 7. Proposed architecture

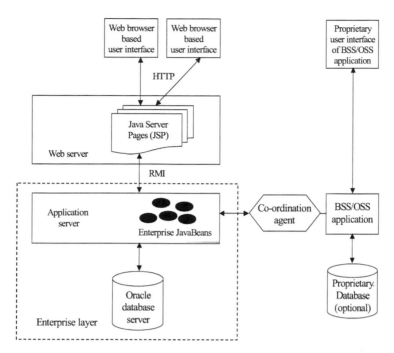

Each BSS/OSS application deals with two potential types of data: its *proprietary data* that are only relevant to that particular application and the *shared data* that are relevant to multiple applications and that are also present as attribute values to the objects in the enterprise layer. Whereas the proprietary data are handled by means of the application's own mechanisms (such data are not relevant outside the application anyway), it is the task of a coordination agent to provide the application with the relevant shared data and to ensure consistency between the application's view on these data and the enterprise layer's. External applications can interact with the enterprise layer in two ways: by *inspecting attribute values* of enterprise objects and by generating *business events that affect the state of enterprise objects*. These two mechanisms correspond roughly to "*reading from*" and "*writing to*" the enterprise layer (see Figure 8).

"Reading" from the enterprise layer, i.e., information is passed from the enterprise layer to an application, is rather straightforward: the coordination agent inspects the relevant attributes of one or more enterprise objects and passes these values to the application. As already mentioned, the enterprise objects are deployed as *entity Beans*. These can be seen as persistent objects, i.e., they are an object-oriented representation of data in an underlying relational database. Entity Beans have predefined *setAttribute()* and *getAttribute()* methods, which can be published in their remote interface. In this way, these methods can be called remotely through RMI for, respectively, reading and updating the value of a particular attribute. Hence, attribute inspections come down to calling a *getAttribute()* method on the corresponding entity Bean. Applications (or their coordination agents) can call directly upon published *getAttribute()* methods to retrieve data from the enterprise layer.

The situation where information is passed from the application to the enterprise layer (the application "writes" to the enterprise layer) is a bit more complex: because the updates that result from a given business event are to be coordinated throughout the entirety of the enterprise layer (they can be considered as a single transaction), coordination agents should never just *update* individual attributes of enterprise objects. Hence, in EJB terminology, *setAttribute()* methods should never be published in an entity Bean's public interface. Changes to the enterprise layer are only to be induced by

Figure 8. Interaction between applications and enterprise layer

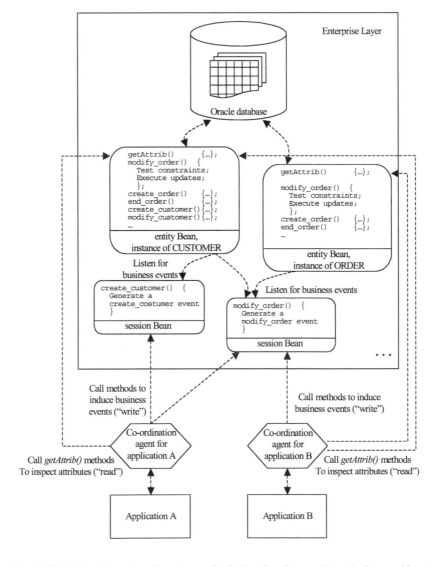

generating *business events* that affect the state of the enterprise objects, as stated in the MERODE specification.

A business event corresponds to a row in the object-event table and affects all enterprise objects whose column is marked for this row. They are generated by means of another kind of enterprise JavaBeans: *session Beans*. These represent nonpersistent objects that only last as long as a particular client session. Each business event type is represented by a session Bean. The latter publishes a method for triggering an event of the corresponding type. A coordination agent acknowledges relevant events that occur in its associated application and that have to be propagated to the enterprise layer. The agent "writes" to the enterprise layer through a single *(remote) method invocation* on a session Bean. The session Bean's method generates a "real" Java event, which represents the business event, to which the entity Beans that take part in the event (as denoted in the object-event table) can subscribe.

The entity Beans have a method for each event type in which they participate. If a relevant event occurs, they execute the corresponding method. This method checks constraints pertinent to the (object instance, event type) combination and executes the necessary updates to attributes of that particular object if all constraints are satisfied (the effect may also be the creation or deletion of objects). If not all constraints are satisfied, an exception is raised.

For example, when a *create_orderline* event is triggered, four domain objects are involved that each might impose some preconditions on the event:

- The order line checks that the line number is unique;
- The product it refers to checks its availability;
- The order it is part of checks whether it is still modifiable; and
- The customer object validates the event against a limit for total cost of outstanding orders.

The global result of the business event corresponds to the combined method executions in the individual objects: the update is only committed if none of the objects that take part in the event have raised an exception. Otherwise, a rollback is induced.

Implementation of the Coordination Agents

The BSS/OSS applications that are currently used have very diverse APIs and ways of interacting with external data. The latter will affect the way in which an agent actually mediates between the application and the enterprise layer. Some applications only feature an in-memory representation of their data, whereas others have their own local database with proprietary data. Also, not every application's API will allow for the shared data to be accessed directly from the enterprise layer in real time. This section discusses three concrete categories of implementations for a coordination agent and, consequently, interaction mechanisms between an application and the enterprise layer. Virtually all kinds of applications will fit in at least one of these (as, obviously, did the applications that were used in the project). The categories can be characterised by means of the *moment in time* on which the reading from and writing to the enterprise layer takes place. One can discern between synchronous interaction without replication of shared data, synchronous interaction with replication of shared data and asynchronous interaction.

Synchronous Interaction, Without Replication in a Local Database

If the application's API allows for data beyond its local database to be accessed *directly* through a database gateway, all shared data can be accessed straight from the enterprise layer, without the need for replication in the application's local database. The coordination agent interacts directly with the application's in-memory data structures for storage to and retrieval from the enterprise layer. This is the preferable approach, as the shared data are not replicated, hence cannot give rise to any inconsistencies between application and enterprise layer. The interaction is *synchronous*, as all updates are visible in both the enterprise layer and the application without any delay.

Information is passed from the enterprise layer to the application on the actual moment when the application wants to access the shared data: the application issues a request, which is translated by the coordination agent to attribute inspections on the entity Beans representing the enterprise objects. The result is returned to the application.

An information flow from the application to the enterprise layer exists on the moment when the application updates (part of) the shared data: the application issues a request, which is translated by the coordination agent to a method call on a session Bean that corresponds to the appropriate business event. The session Bean generates an event in the enterprise layer, to which each entity Bean that participates in the event responds by executing a corresponding method, which executes the necessary updates to that particular entity Bean. These updates are only committed if none of the entity Beans raises an exception because of a constraint violation.

Synchronous Interaction, with Replication in a Local Database

If the application's API does not allow for data beyond its local, proprietary database to be accessed directly through a database gateway, all shared data have to be replicated in the local database for access by the application. The enterprise layer then contains the "primary" copy of the shared data, and the coordination agent is responsible for "pumping" the relevant shared data from the enterprise layer into the local database and vice versa. A crucial task of the coordination agent will be to guarantee a satisfactory degree of consistency between enterprise layer and replicated data, especially given the situation of concurring business events.

Even in this case, the interaction can still be synchronous if the local database can be kept in consistency with the enterprise layer at any time, i.e., all updates are visible in both the enterprise layer and the application without any delay. However, the moment of reading from the enterprise layer differs from the situation without replication: when the application accesses shared information, all data (both shared and proprietary) are retrieved from the local database. An information flow from the enterprise layer to the application now takes place *when shared attributes in the enterprise layer are updated by another application*. For that purpose, the application's coordination agent *listens* to events (generated by another application) in the enterprise layer that cause updates to data relevant to its corresponding application. The mechanism can be implemented by means of the observer pattern (Gamma et al., 1999). When a create, modify or end event on relevant data in the enterprise layer is detected by the agent, the information is propagated by issuing inserts, updates or deletes in the application's local database. Note that the enterprise layer itself can never take the initiative to propagate updates to an application: it is not aware of its existence.

Similarly to the case without replication, information is passed from the application to the enterprise layer when the application updates shared data. These updates are initially issued on the replicated data in the local database. The coordination agent listens for updates in the local database. If such update concerns data that is also present in the enterprise layer, it triggers a business event (again by means of a session Bean) in the enterprise layer, which provokes the necessary updates.

Asynchronous Interaction

When the interaction between application and enterprise layer is *asynchronous*, a continuous consistency between updates in the application and updates in the enterprise layer cannot be guaranteed. Again, shared data will have to be replicated in a local database, but this time updates on one side will only be propagated *periodically* instead of in real time. In terms of the risk of inconsistencies, this is of course the least desirable approach. However, it has the advantage of allowing for a very loose coupling between application and enterprise layer, whenever a synchronous approach with a much tighter coupling is not feasible. One can distinguish two situations that call for asynchronous interaction; a first cause, when "writing" to the enterprise layer, could be the fact that the application's API doesn't allow the coordination agent to listen for the application's internal events. Consequently, the coordination agent is not immediately aware of the application updating shared data replicated in the local database. These updates can only be detected by means of periodical polling. Hence, an information flow from the application to the enterprise layer exists when, during periodical polling, the coordination agent detects an update in the shared data and propagates this update to the enterprise layer.

With regard to "reading" from the enterprise layer, a coordination agent can always listen in the enterprise layer for relevant events (triggered by another application) causing updates to shared data. This is the same situation as for synchronous interaction with replicated data, where the updates cannot be propagated right away to the application's local database. Indeed, one could imagine situations where the corresponding application is not always able to immediately process these updates, e.g., because it is too slow, because the connection is unreliable, etc. This is a second cause for asynchronous interaction. In that case, the propagation of updates will be packaged by the coordination agent as *(XML-) messages*, which are *queued* until the application is able to process the updates (in particular, this situation would occur if in the future part of the functionality is delegated to an application service provider, who publishes its software as a Web Service). Information is passed from the enterprise layer to the application when the queued updates are processed by the application, resulting in updates to the local database.

Table 3 summarises the different interaction modalities between application and enterprise layer. "A → E" represents an information flow from application to enterprise layer. "A ← E" represents an information flow from enterprise layer to application.

Transaction Management

Transaction management can easily be directed from the session Bean where a business event was triggered. The processing of a single MERODE event is to be considered as one atomic transaction, which is to be executed in its entirety or not at all. The event is generated by a method call on a session Bean and results in methods being

Table 3. Summary of the possible interaction mechanisms between application and enterprise layer

	Data access by application	Update by application	Update in Enterprise Layer, caused by other application	Polling on application's local database	Application processes message in queue
Synchronous interaction, no replication	A ← E	A → E			
Synchronous interaction, with replication		A → E *	A ← E		
Asynchronous interaction				A → E *	A ← E *

executed in all objects that participate in the event. If none of these raises an exception, i.e., no constraints are violated in any object, the transaction can be committed. If at least one such method *does* generate an exception, the entire transaction, and not only the updates to the object where the exception was raised, is to be rolled back. In this way, the enterprise layer itself always remains in a consistent state.

However, if the business event is the consequence of an application first updating shared data replicated in its local database, the coordination agent is to inform the application about the success or failure of the propagation of the update to the enterprise layer by means of a return value. In case of the event that corresponds to the update being refused in the enterprise layer, the application should roll back its own local updates, too, in order to preserve consistency with the enterprise layer, which contains the primary copy of the data. Cells marked with a "*" in Table 3 denote situations where there is a risk of inconsistency between application and enterprise layer. As can be seen, the issue is particularly critical in the case of asynchronous interaction.

Applications and User Interfaces

The existing BSS/OSS applications interact with the enterprise layer through the coordination agents. They are virtually unaffected by being plugged into the enterprise layer. Coordination agents are to be rewritten when an application is replaced, such that the applications themselves and the enterprise layer can remain untouched.

The existing user interfaces of the third-party BSS/OSS applications are proprietary to these applications and were of no concern during the integration project. Apart from these, new Web browser-based user interfaces have to be developed in-house for direct interaction with the enterprise layer. This can be accomplished fairly easily by means of Java Server Page technology. This issue is, however, beyond the scope of this case study.

CURRENT CHALLENGES/PROBLEMS FACING THE ORGANISATION

Once the solution was specified, the main challenge resided in the successful implementation of the proposed solution. The choice of the implementation environment was a key influencing factor for success. It is the company's policy to outsource IT development efforts as much as possible. The requirements engineering task is done in-house, but system realisation should be done by buying off-the-shelf software or by outsourcing. In addition, coding efforts should be limited by utilising code generators where possible.

Considerable complications resulted from the implementation environment that was chosen by the company: an application development tool that generates Enterprise Bean code, but that is nonetheless not fully object-oriented. In fact, by no means can it conceal its origin in the *relational* paradigm. Its so-called "objects" actually behave like single rows in a relational table: they are interrelated by means of foreign keys (instead of object identifiers) and lack the concepts of subtyping, inheritance and real behaviour. Moreover, whereas MERODE defines *shared participation in business events* as the enterprise objects' means of interacting, the tool only supports event notification at the level of *inserts*, *updates* and *deletes*. These are not very suitable for modelling business events: they pertain to the actual changes to rows in the database but do not carry information about *which business event* induced the change, ignoring the possibility of different underlying semantics, preconditions and constraints. Therefore, whereas an almost perfect mapping existed between concepts from the MERODE-based enterprise model and the proposed EJB architecture, several of the MERODE model's subtleties were lost when being translated into the design environment.

Furthermore, MERODE advises a "transformation"-based approach to implementation: the transformation is defined by developing code templates that define how specifications have to be transformed into code. The actual coding of an enterprise model is then mainly a "fill in the blanks" effort. MERODE provides a number of examples for transformations to purely object-oriented program code. Also a lot of experience exists with transformations to a COBOL+DB2 environment. Transformation with a code-generator had never been done before.

Two additional elements were important in further evolution of the project. The company had explicitly opted for the development method MERODE by asking the MERODE team for assistance in the project. Although the method is in use by several companies and has in this way proven its value, it is not widespread enough to assume that it is known by the programmers of the implementation team. In fact, no members of the implementation team were acknowledged with the chosen modelling approach MERODE. The same was true for the members of the consultancy team of the company providing the implementation environment.

Finally, on the longer term there might be a problem of domain model expertise. Initially, it was agreed that the company would appoint one full-time analyst to actually do the requirements engineering. The university team would advise this person. However, due to shortage on the labour market, the company did not succeed to find a suitable candidate. As a result, the requirements engineering has completely been done by the university team. Consequently, valuable domain model expertise resides outside the company.

REFERENCES

Booch, G., Rumbaugh, J., & Jacobson, I. (1999). *The unified modeling language user guide.* Reading: Addison-Wesley.

D'Souza, D. F., & Wills, A. C. (1999). *Objects, components and frameworks with UML: The catalysis approach*. Reading: Addison-Wesley.

Fowler, M., & Kendall, S. (1998). *UML distilled: Applying the standard object modeling language*. Reading: Addison-Wesley.

Gamma, E., Helm, R., & Johnson, R. (1999). *Design patterns: Elements of reusable object-oriented software*. Reading: Addison-Wesley.

Jacobson, I., Christerson, M., & Johnson, P. (1997). *Object-oriented software engineering: A use case driven approach*. Reading: Addison-Wesley.

Snoeck, M., & Dedene, G. (1998). Existence dependency: The key to semantic integrity between structural and behavioral aspects of object types. *IEEE Transactions on Software Engineering, 24* (24), 233-251.

Snoeck, M., Dedene, G., Verhelst, M., & Depuydt, A. M. (1999). *Object-oriented enterprise modeling with MERODE*. Leuven: Leuven University Press.

Snoeck, M., Poelmans, S., & Dedene, G. (2000). A layered software specification architecture. In A. H. F. Laendler, S. W. Liddle, & V. C. Storey (Eds.), *Conceptual Modeling-ER2000, 19th International Conference on Conceptual Modeling. Lecture Notes in Computer Science 1920* Salt Lake City, Utah (pp. 454-469). Springer.

Wilfried Lemahieu is a post-doctoral researcher at the Katholieke Universiteit Leuven, Belgium. He holds a PhD in applied economic sciences from K.U. Leuven (1999). His teaching includes database management and data storage architectures. His research interests comprise hypermedia systems, object-relational and object-oriented database systems, distributed object architectures and Web services.

Monique Snoeck obtained her PhD in May 1995 from the Computer Science Department of the Katholieke Universiteit Leuven, Belgium, with a thesis that lays the formal foundations of MERODE. Since then she has done further research in the area of formal methods for object-oriented conceptual modelling. She now is an associate professor with the Management Information Systems Group of the Department of Applied Economic Sciences at the Katholieke Universiteit Leuven in Belgium. In addition, she is an invited lecturer at the Université Catholique de Louvain-la-Neuve since 1997. She has also been involved in several industrial conceptual modeling projects. Her research interest are object oriented modelling, software architecture and gender aspects of ICT.

Cindy Michiels is a PhD student at the Katholieke Universiteit Leuven (KUL), Belgium. She holds a bachelor's degree in applied economics and a master's degree in management informatics (2000). Since 2001 she has worked for the Management Information Systems Group of the Department of Applied Economics of the KUL as a doctoral researcher. Her research interests are related to domain modelling, software analysis and design, and automatic code generation.

This case was previously published in the *Annals of Cases on Information Technology*, Volume 5/2003, pp. 213-233, © 2003.

Chapter XX

Shared Workspace for Collaborative Engineering

Dirk Trossen, Nokia Research Center, USA

André Schüppen, Aachen University of Technology, Germany

Michael Wallbaum, Aachen University of Technology, Germany

EXECUTIVE SUMMARY

In the broad context of the collaborative research center project IMPROVE (Information Technology Support for Collaborative and Distributed Design Processes in Chemical Engineering), the presented case study has been concentrating on the provision of appropriate communication technology, specifically shared workspace means, to enable collaborative working between distributed engineering teams. Issues like distributed developer meetings, sharing common data, or even sharing entire workspaces including off-the-shelf tools being used for the development process are the driving forces for the studies on how to provide appropriate technology means in collaborative engineering. The considered case in the field of chemical engineering and development represents a difficult candidate for collaborative engineering due to the variety of proprietary data and tools to be integrated in a shared workspace. Furthermore, common aspects of cooperative working among development teams have to be considered as well. The resulting architecture-based on the findings of the current stage of the case is presented, trying to use as many existing software as possible. Drawbacks and challenges being encountered during the case study due to the a-posteriori approach are outlined, leading to a revised architecture proposal to be used in the future as a common platform for the information technology support within the context of the research project. Expected benefits and problems of the introduction of the new architecture are drawn.

BACKGROUND

The collaborative research center project 476 IMPROVE (Information Technology Support For Collaborative and Distributed Design Processes in Chemical Engineering), sponsored by the German Research Foundation, is a cooperation between chemical engineers and computer scientists to improve development processes in chemical and plastic engineering and to study information technology aspects to support development processes by means of evolving multimedia, planning, and modeling tools.

A development process in chemical and plastic engineering is concentrating on designing a new or revising an existing production process, including its implementation (Nagl & Westfechtl, 1999). A development process in process engineering, similar to other engineering sciences, is divided into different phases. In a first phase, the requests for the material products are determined on the basis of a market analysis. Afterwards, the concept design and the basic engineering take place, finally leading to a definition of the basic data of the engineering process. It follows the detail engineering, deepening the data determined in the basic engineering. Then it is transferred into apparatuses, machines, piping including measuring technique, control engineering and automatic control. Finally, the development process is finished, and the procurement of all units, the assembly of the system, their line-up and the system operation take place.

As outlined in Nagl and Marquadt (1997), the first phases of a development process, i.e., basic and detail engineering, are of utmost importance due to several reasons. Firstly, around 80% of the production costs are defined in these phases. Secondly, the complexity of these phases is fairly high due to the different aspects of the production process to be considered. Hence, a proper support of these phases by appropriate methodologies as well as information technology is crucial to guarantee a successful completion of the development process.

SETTING THE STAGE

The focus of the research project IMPROVE is on some subtasks of the concept design, which seem particularly important, to improve and accelerate future development processes. Typical characteristics, among others, of these subtasks are the following (Nagl & Westfechtl, 1999):

- *Number and background of developers:* Different tasks of the concept design are usually solved by different developers and development teams. The number and size of these teams as well as the personal background of the individual team members usually complicates information exchange and understanding of solutions, additionally caused by different terminology and lack of command tool support.
- *Geographical and institutional distribution:* Due to the globalization of institutions, the aspect of geographically distributed teams becomes more and more important. Intra- and interteam meetings become a challenging task for the supporting information technology since arranging physical meetings at a single place adds significant overhead to the development process in terms of additional

journeys. Hence, appropriate synchronous as well as asynchronous communication means are desired to support the widely dispersed developer teams.

- *Team coordination and planning of development process:* Planning and management tools are desired for team coordination and planning purposes inherently supporting the dynamic nature of a development process.
- *Cooperation and information exchange:* Information of all sorts, such as documents and planning information, has to be exchanged among all developers while ensuring the information's consistency. This task places a burden on the supporting information technology, specifically on the version and database management of the project.
- *Reusability:* Due to the desire to reduce development costs, reusing well-known techniques becomes more and more important. Specific techniques as well as generalized patterns for solutions are typical candidates for reusability. Appropriate modeling and documentation means are required for this issue.

Within IMPROVE, a chemical development process is used as a specific case of the different tasks to be covered by the project. The project is divided into four main parts which are further divided into subtasks. The first part is dealing with development processes for chemical engineering, the second one covers methods and tools to support development processes, while the third one is investigating the mapping onto new or existing information technology platforms. Finally, the fourth part is responsible for integrating the project to disseminate the results towards industrial partners.

Figure 1 shows an overview of the different parts and their subtasks within IMPROVE (Nagl & Marquadt, 1997).

The considered case study of this chapter is covering the information technology support for the multimedia communication within IMPROVE. This subproject (dashed box in Figure 1) is aiming to improve synchronous multimedia-based interaction among distributed developer teams as one of the core elements of development processes nowadays.

Figure 1. Structure of IMPROVE

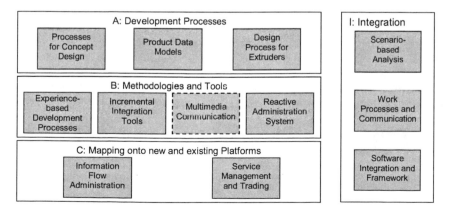

CASE DESCRIPTION

As mentioned in the previous section, multimedia communication and interaction among distributed development teams is crucial to improve the overall design process. Hence, the provision of an appropriate architecture to enable collaborative synchronous working among the teams will be the considered case in the following. For that, the development scenario within IMPROVE is outlined, leading to a requirements definition for the communication architecture. Finally, the current architecture within IMPROVE is presented to fulfill the previously-defined requirements.

Collaborative Development Scenario

The typical engineer in a chemical development process is a specialist, being only responsible for a small part, e.g., special chemical reactors, of the whole project. Usually, specialized applications are used within his knowledge domain. Input of a different sort, such as specifications and parameters, has to be worked in, but also results from his specific part have to be distributed to other experts to be used in their specific tasks. At certain points in the development process, meetings are used to discuss the results and to synchronize the work. In addition to, mostly pre-announced, meetings, ad hoc coordination might be necessary due to several ambiguities or errors during the engineering process, which makes it necessary to contact another involved specialist. For instance, the specialist for the degassing has problems reaching the specifications caused by a special input parameter. He can ask the expert of the chemical reactors to modify the reactor's specification by changing this parameter. Besides audiovisual exchange of information, the data-sharing aspect is of utmost importance, specifically concerning off-the-shelf applications used by the individual engineers.

Requirements for the Information Technology Support

From a user's perspective, the major requirement of the information technology support within IMPROVE is to provide multimedia conferencing among distributed groups of developers by means of audiovisual real-time communication and shared workspace facilities, allowing different groups to use their existing local tools. In the following, a non-formal description of the required functionality from a user's perspective will be presented to define requirements for the selection process.

As mentioned above, the considered conferencing scenario is typically a project meeting among developers which takes place mostly in small groups. For that, facilities like special conferencing rooms with special audio/video equipment (Trivedi & Rao, 1999) might be used as well as standard PC equipment. The potential group of users is usually well-known, i.e., the conference might be announced beforehand or ad hoc meetings might be established among this group. A conference database might reflect the participating members together with additional information, e.g., site address or phone number. Functionality like authentication and authorization is needed to provide secure and restricted access conferences due to the closed group character of the case.

As a minimal functionality for a virtual meeting, users should be able to exchange audiovisual information, either mixed from the streams of all participants or selected according to a given access scheme. Furthermore, one participant is often defined to conduct the meeting and enforce the adherence to certain social rules (Walker, 1982) for the interaction among the participants, e.g., indicating a question before sending the own audiovisual data. Thus, a rule-based communication among the conference members takes place which has to be supported by the underlying system.

Besides the aspect of sharing audiovisual information among the users, distributing data of legacy applications such as specific development tools is crucial to share specific knowledge among the group. For that, legacy applications have to be shared among different sites while ensuring consistency of the different copies of the application. Each local change of this application is displayed on each user's terminal and is appropriately synchronized. For interactive communication, the system should support, giving control of the shared application to certain users.

In the following, this non-formal description of required functionality is broken down to technical requirements of the desired architecture to highlight the issues to be addressed during the technology selection process.

Conference Management

The issue of managing conferences is divided into several aspects to be covered, namely management of the environment, provision of a conference database, handling security aspects and managing network resources. These aspects are described in detail in the following:

- *Management of the conference environment:* Functionality to announce and to initiate a conference is needed. For that, advertisement functionality is required. Furthermore, a conference setup functionality is crucial, including features to invite other users to a (running) conference or to merge conferences by appending or inviting entire conferences. Additionally, splitting capabilities can be used to form new sub-conferences out of existing ones. In general, sophisticated reconfiguration features have to be provided by the management services. Furthermore, the conference management should enable conducted conferences in the sense that a dedicated user is defined having special management rights, like expelling other users from a running conference or denying the access to a conference.
- *Provision of a common conference database:* A common conference database is crucial for the provision of membership information containing commonly used information (like naming and location information) as well as extended, scenario-specific data. Note that this database is not handling any legacy-specific data which is addressed by a separate database management system.
- *Handle security aspects:* For the provision of secure conferences, authorization and authentication have to be provided to support secure transfer of user data.
- *Manage network resources:* The conferencing architecture should provide facilities to control the network resources, e.g., by providing a defined and guaranteed quality of service.

Multipoint Transfer

The aspect of secure, location-transparent exchange of user data among several developers independent of specific networks and end systems has to be supported by appropriate multipoint transfer means. It should provide communication patterns ranging from one to-one to many-to-many communication. Furthermore, secure transfer of the data has to be implemented. For the location transparency of the transfer service, a transport layer abstraction has to be provided to address a certain user group in the conference. This abstraction should also enable the use of different kinds of transport facilities depending on the transmitted data type, e.g., streaming protocols for real-time audiovisual data. Moreover, the transfer of user data among heterogeneous end systems has to be provided, i.e., appropriate marshaling capabilities (Trossen, 2000a) should be provided by a common transfer syntax of the exchanged user data.

Floor Control

This topic covers functions for the operation of distributed applications in general. This means in particular the provision of means to synchronize the application state and to enable a concurrent access to shared resources such as storage devices or printers but also to objects or streams. For that, the access to resources is abstracted by the notion of a floor which was introduced by turn-taking psychological studies (Walker, 1982). Dommel et al. proposed a floor control protocol with basic operations like allocate, ask for, and pass a floor. These floors might be defined exclusively or non-exclusively. Additionally, with a member query operation, it was shown that these operations are sufficient to abstract access control and conventional turn-taking in conferencing scenarios. Thus, this functionality should be provided for the access control in distributed applications.

Legacy Support

Since development processes mostly use specific tools during the lifetime of the process cycle, the support of legacy software and its distribution among the different sites is a crucial aspect for the provision of information technology in development processes. Hence, the desired architecture should specifically enable sharing of any legacy software among the sites to preserve the given pool of software in use within the different processes. Specifically for the considered case, the exchange of chemical process data and its rendering information is desired.

Process Integration

Besides the support of legacy software, an overall integration of the shared workspace functionality in the engineering process is of utmost importance. This includes organizational challenges, such as planning, as well as integration with other information technology means, such as calendar and database systems. From these points, the calendar and scheduling aspect directly influences the architecture for the considered case, since the initiation of collaborative meetings has to be coordinated with appropriate scheduling systems.

Organizational Issues

Although some of the technical requirements above contain organizational aspects like the integration of legacy data and applications, the overall organizational paradigm for the selection process as such is following an *a-posteriori* approach, i.e., as many existing software as possible are to be integrated in the solution. However, specialized hardware to provide some functionality is not desired due to the increasing cost effect of this solution and thus should be avoided, if possible.

Chosen Architecture

Based on the findings concerning information technology requirements, the current architecture is outlined in the following section. For that, the main components of the architecture are sketched with respect to their provided functionality before giving a more technical insight in the chosen technology in terms of implemented protocol stacks.

Component Overview

The current workspace within IMPROVE provides transfer of audiovisual information. Currently available audiovisual conferencing products, e.g., H.323-based systems (ITU, 1996), requires highly expensive Multipoint Control Units for performing mixing capabilities in multi-user conferences, which is contradictory to the goal to keep costs low. Furthermore, these centralized solutions increase network load due to traffic duplication which restricts the applicability due to the high bandwidth usage. As a consequence, self-developed audio/video tools are used to deliver the audiovisual information to all participants using the Internet's multicast capability and performing the mixing at the end systems. Floor controlled conferences are currently not supported. For the presentation of non-real-time data such as slides or whiteboard contents, but also for displaying local application's output, the shared application functionality of a commercial tool is used based on the current standard of the International Telecommunications Union (ITU, 1998) for this functionality. This solution facilitates sharing any local application with a set of users.

Figure 2 shows the different components of the chosen architecture. It can be seen that the workspace comprises three components providing audio, video, and data transfer functionality.

All components are realized either as separate tools, i.e., for audio and video, or by using commercial products, i.e., for data transfer, enabling a simple replacement of the different components. Additionally, a user interface was realized for launching and terminating the individual components, abstracted in Figure 2 by the gray block spanning around the different tools. Hence, the user is confronted with an integrated user interface which transparently executes the desired tool.

Implementation Overview

After outlining the components in Figure 2, this section gives a brief technical insight into the components. For that, the realized protocol stack is depicted in Figure 3 outlining the functionality of the different layers to implement the workspace architec-

Figure 2. Current workspace architecture

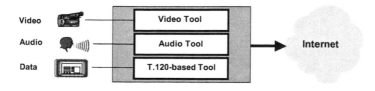

Figure 3. Implemented protocol stacks

Starter Tool		
Audio tool	Video tool	T.120-based tool
RTP/RTCP		T.120 Protocols
UDP		TCP
IP/IP Multicast		

ture of Figure 2. Since this section only gives a rough overview of each layer, please see the specific references for a more technical description.

The gray blocks outline the different tools residing on top of transport-level protocols. The starter tool is executing the different underlying tools. The audio/video tools have been developed within internal research projects about audio/video transfer over the Internet using adaptive transmission technologies (Wallbaum & Meggers, 1999). With this technique, the transmission rate of the audiovisual information is adapted to the changing bandwidth of an IP-based network using rate adaptation techniques such as adaptive nodes (Meggers, 1999). Both tools use the Real-time Transfer Protocol (RTP) (Schulzrinne et al., 1996) for exchanging the audiovisual information either using a point-to-point connection to a peer station or based on the IP multicast capability (Eriksson, 1994).

For realizing the desired shared application functionality, an ITU-T.120 (ITU, 1998) standard-conform application sharing tool is used. For that, a parallel data conference is established using the management functionality provided by this ITU protocol suite. The shared application as well as the conference management information for the data conference part is transmitted using the multipoint protocol defined in the protocol suite (T.122, see ITU, 1998). It is worth mentioning that this protocol maps the multipoint communication of the shared application protocol onto a point-to-point transfer independent of the underlying network, even if a multicast-capable network is available.

Hence, it can be seen that the chosen implementation realizes a mixture of multicast transfer using IP multicast tools and point-to-point transfer by integrating ITU-based commercial products.

Current Problems

The main motivation for the current realization of the IMPROVE workspace is the integration of existing software following the *a-posteriori* approach. This minimizes the in-house development of the workspace to the integration of these tools in the common interface. However, this approach has several drawbacks and problems with respect to the specific requirements for the considered application scenario, i.e., enabling shared distributed engineering. These problems are presented in the following section before outlining a proposal for a revised architecture to be realized in the future.

Drawbacks and Problems of the Current Workspace

Regarding the derived requirements for the considered case, several drawbacks and problems have been encountered when using the current workspace.

Lack of Common Conference Management

The current workspace does not provide sophisticated conference management facilities. Basic conference management features are only supported for the data conferencing part. A common conference database and appropriate invitation and initiation mechanisms are not provided for the entire workspace. This directly affects the integration in calendar and scheduling tools, and therefore the integration in other sub-processes within the engineering process. Moreover, reconfiguration functionality, such as merging or splitting conferences, is not supported. Furthermore, management for closed user groups, such as authorization and user management mechanisms, is not covered by the chosen architecture. Hence, the current architecture does not provide a rich functionality to establish ad hoc or even scheduled closed group meetings.

Lack of Floor Control

Floor control mechanisms (Dommel, 1995) are currently not supported, i.e., role-based communication, such as conducted conferencing, is not feasible. Furthermore, access control to distributed resources, such as files, is not supported. For the data conferencing part, the access control for the shared application is realized by a simple centralized token mechanism which has proved to scale poorly (Trossen, 2000b) with respect to the response time for a token request.

Performance of Shared Application

The performance of the shared application functionality being provided by the ITU-conform commercial product is very poor. This is mainly caused by two reasons. Firstly, the mapping of multipoint communication onto point-to-point transfer leads to a bottleneck in transfer performance, even in multicast-capable environments. This is caused by the multipoint transfer part of the T.120 protocol (T.122), even if there is a multicast-capable network available. Currently, there is no implementation available based on pure multicast transfer capabilities.

Secondly, the chosen implementation paradigm of the used application sharing protocol is following the shared GUI approach (Trossen, 2001). This means that a local application is shared by transferring every local output change to the set of users and sending input events from the users back to the application hosting end system for control purposes. This approach causes significant overhead at the hosting end system due to the interception delay (Trossen, 2001) being necessary for capturing the desired GUI output data. Furthermore, transferring this graphical output leads to a steadily high network load.

This performance drawback becomes even worse in our case since the chemical developer teams are often working with graphic-intensive visualization tools, which cause a high workload on the local machine due to the rendering operations. Sharing this large amount of graphical output makes the situation even worse due to the additional workload on the hosting machine and due to the large amount of data to be transferred over the network. Hence, these tools cannot be shared effectively due to the large amount of data to be distributed among the set of users.

As a summary of the detailed outline of the encountered problems, it can be stated that the lack of conference management and mediation support, i.e., floor control functionality, indicates shortcomings of the underlying conferencing system in general, while the lack of performance specifically affects the considered case due to the necessity of using off-the-shelf software for the engineering process. From a user's perspective, the outlined problems lead to a certain reluctance of using the provided systems, since the user's perception of the workspace functionality suffers from the described problems.

In the following, a revised architecture is presented addressing these problems by introducing new or revised components to the current solution.

Revised Architecture

The encountered drawbacks and problems described in the previous section indicate a lack of core functionality to be required in the considered case. This could be solved partially by additional hardware, e.g., for mixing purpose, which is not desired due to the enormous costs for these components. However, some of the missing functionality is not even provided in the fully equipped environment due to shortcomings of the standard as such (Trossen, 2000b).

Hence, a revised architecture is proposed, introducing a common middleware layer for the provision of a common conference control infrastructure while remaining the component architecture of Figure 2 intact. With this middleware layer, the missing conference management and floor control is introduced to the shared workspace.

For that, the Scalable Conference Control Service, as proposed in Trossen (2000a) (see also Trossen, 2000b), is used to realize generic conference control functionality. The generic functionality of the service allows for future developments in the area of conference course control (Borman et al., 2001) as well as the usage of commercial tools, if this is desired. The following functionality is provided by SCCS:

- *Conference management features:* Membership control, reconfiguration of conferences during runtime, and invitation of users is provided by SCCS. Moreover, a conference database containing information like name, location, or phone numbers is supported which can be accessed by each participant. Conducted conferences can be implemented using privileged functions for expelling users or denying join requests. Moreover, conferences can be set up based on external events, which allows for the integration of calendar and scheduling means. Therefore, the integration in the development process is improved.
- *Multipoint transfer abstraction:* A channel addressing scheme is used to abstract from the specific transport layer interface. The protocol to be used is specified during creation of the channel. Since the abstraction is independent of the specific protocol, multicast-capable transport protocols can easily be supported.
- *Floor control:* This service is to be used for the implementation of application state synchronization or access control in distributed environments. Thus, interaction rules for specific conference scenarios can be realized. The current implementation of SCCS uses a self-developed routing scheme to scale the service to a

Figure 4. Future workspace protocol stack

Management Tool		
Audio tool	Video tool	Application Sharing tool
SCCS		
RTP/RTCP	TCP	MTP-SO
UDP		
IP/IP Multicast		

larger number of users without degrading the response time of the floor control request (Trossen, 2000a).

On protocol level, SCCS establishes a conference control tree topology interconnecting the different conference participants for control purposes. However, a multicast transport protocol is also used to improve the performance and scalability of the conference database management. It has to be emphasized that any user data like audiovisual information is not transferred using the control topology. This is realized by the specific transport protocols which are only abstracted by SCCS.

In Figure 4, the resulting protocol stack of the revised architecture is depicted. SCCS resides on top of TCP and MTP-SO (Multicast Transport Protocol – Self Organizing, see Borman et al., 1997), which are used for transferring the protocol's control data. The former is used for transfer within the control tree topology while the latter is used for distributing the conference database content.

The functionality of the simple starter tool in the current architecture is extended by integrating conference management functionality like conference database access, inviting users or merging conferences, leading to a common management tool.

The audio/video tools are built on top of SCCS using its services for management, transfer abstraction and floor control. Different schemes are to be implemented controlling the access onto the audiovisual stream by using the floor control facilities of SCCS. For that, the core of the tools are subject to minor changes since only the addressing has to be revised. The additional functionality for access control, i.e., the floor control support, has to be added to the tools from scratch, of course.

The second change in the current IMPROVE architecture is addressing the performance drawback of the shared application part. For that, a new application sharing module is proposed based on the event sharing paradigm (Trossen, 2001) instead of using the T.120-based solution. Using this paradigm, graphic-intensive applications are to be shared across host boundaries by running the applications locally and sharing the generated input events only. Despite the restrictions of this paradigm, which are shown in Trossen (2001), this approach is very promising for our case since the engineering teams are mostly working on a common data basis so that the data inconsistency problem can easily be avoided.

FURTHER READING

A detailed overview of the goals, approaches, and structure of the entire IMPROVE research project and its subprojects can be found in Nagl and Marquadt (1997) and Nagl and Westfechtl (1999).

In Meggers (1999) and Meggers and Wallbaum (1999), an overview and more technical description of adaptive networking techniques is given including a presentation of the audio and video tools that are used in the current architecture of IMPROVE. In Ingvaldsen (1998), studies concerning the importance of delay factors in collaborative computing can be found.

An overview of the generic conferencing service SCCS is presented in Trossen (2000a), while a lengthy technical description of the service and its implementation can be found in Trossen (2000b). The chosen implementation architecture as well as activity diagrams, protocol sequence charts, and modeling means for evaluation are depicted.

Other examples for collaborative environments are presented in Altenhofen et al. (1993), Chen et al. (1992), and Gutekunst and Plattner (1993), while Schooler (1996) and Lubich (1995) give a more general introduction into the area of collaborative computing. The notion of a floor to realize access control in distributed environments is presented in Walker (1982) and Dommel (1995).

CONCLUSION

This case study presented the architectural findings of the IMPROVE project based on requirements for collaborative engineering in chemical development processes. For that, the current as well as a proposed revised architecture were presented, trying to meet the defined requirements.

Four issues can be pinpointed to summarize the major challenges encountered in the current phase of the project. Firstly, currently available solutions for conferencing usually require additional costly hardware like centralized mixing units. Secondly, the integration of legacy data and applications in a common workspace is still a research topic and is only poorly supported in systems nowadays. The considered application case aggravates the situation due to its variety of different off-the-shelf software that has to be shared among the development groups. Thirdly, integration of access control and social rules in the conferencing scenario, e.g., to provide more sophisticated conducted meetings, is yet poorly supported by available commercial systems. Fourthly, current systems lack efficiency concerning their bandwidth usage due to inefficient usage of multicast capabilities of the underlying network. This is especially important in the considered case, which usually connects a small number of local groups usually located in a multicast-capable local area network.

While the first issue was avoided by our requirements, the proposed revised architecture, to be realized in our future work within IMPROVE, specifically addressed the three latter issues. It introduced a middleware layer approach which allows for future development in this area to be replaced by more sophisticated mechanisms. Additionally, a paradigm shift was proposed for the problem of sharing off-the-shelf software by using the sharing event paradigm, which is specifically suited for the usually graphics-intensive chemical development process.

As a concluding remark, it can be stated that cooperative engineering is still a challenging task once it comes to its realization in real life. Lack of efficiency and integration of legacy software as well as missing provision of sophisticated floor control, reduces users' perception and makes the introduction of information technology for collaborative engineering difficult.

REFERENCES

Altenhofen, M. et al. (1993). The BERKOM multimedia collaboration service. *Proceedings of the 1st ACM Conference on Multimedia* (pp. 457-463).

Borman, C., Kutscher, D., Ott, J., & Trossen, D. (2001). *Simple conferencing control protocol — Service specification* (Internet Draft) [Work in progress]. Retrieved from http://globecom.net/ietf/draft/draft-bormann-mtp-so-01.html

Borman, C., Ott, J., & Seifert, N. (1997). *MTP/SO: Self-organizing multicast* (Internet Draft) [Work in progress]. Retrieved from ftp://ftp.ietf.org/internet-drafts/draft-ietf-mtp-so-01.txt

Chen, M., Barzilai, T., & Vin, H. M. (1992). Software architecture of DiCE: A distributed collaboration environment. *Proceedings of the 4th IEEE ComSoc International Workshop on Multimedia Communication* (pp. 172-185).

Dommel, H.-P., & Garcia-Luna-Aceves, J. J. (1995). Floor control for activity coordination in networked multimedia applications. *Proceedings of Asian-Pacific Conference on Communications.*

Eriksson, H. (1994). MBONE: The multicast backbone. *Communications of the ACM, 37*(8), 54-60.

Gutekunst, T., & Plattner, B. (1993). Sharing multimedia applications among heterogeneous workstations. *Proceedings of 2nd European Conference on Broadband Islands* (pp. 103-114).

Ingvaldsen, T., Klovning, E., & Wilkins, M. (1998). A study of delay factors in CSCW applications and their importance. *Proceedings of 5th International Workshop on Interactive Multimedia Systems and Telecommunication Services* (pp. 159-170).

ITU-T. (1996). *Visual telephone systems and equipment for local area networks which provide a non-guaranteed quality of service* (ITU-T Recommendation H.323).

ITU-T. (1998). *Data protocols for multimedia conferencing* (ITU-T Recommendation T.120).

Meggers, J. (1999). Adaptive admission control and scheduling for wireless packet communication. *Proceedings of IEEE International Conference on Networks.*

Lohmann, B., & Marquardt, W. (1996). On the systematization of the process of model development. *Computers Chem. Eng., 20,* 213-218.

Lubich, H. P. (1995). *Towards a CSCW framework for scientific cooperation in Europe. Lecture notes in computer science 889.* Berlin: Springer.

Nagl, M., & Marquadt, W. (1997). SFB 476 IMPROVE: Informatische Unterstützung übergreifender Entwicklungsprozesse in der Verfahrenstechnik. In M. Jarke, K. Pasedach, & K. Pohl (Eds.), *Informatik 97: Informatik als Innovationsmotor* (pp. 143-154), Berlin: Springer Verlag.

Nagl, M., & Westfechtel, B. (Eds.). (1999). *Integration von Entwicklungssystemen in Ingenieuranwendungen.* Berlin: Springer-Verlag.

Schooler, E. M. (1996). Conferencing and collaborative computing. *ACM Multimedia Systems, 4*(5), 210-225.

Schulzrinne, H., Casner, S., Frederick, R., & Jacobson, V. (1996). *RTP: A transport protocol for real-time applications* (IETF Request for Comment 1889).

Trivedi, M. M., & Rao, B. D. (1999). Camera networks and microphone arrays for video conferencing applications. *Proceedings of SPIE International Symposium Voice, Video & Data Communications* (pp. 384-390).

Trossen, D. (2000a). Scalable conferencing support for tightly-coupled environments: Services, mechanisms, and implementation design. *Proceedings of IEEE International Conference on Communications* (pp. 889-893).

Trossen, D. (2000b). *Scalable group communication in tightly coupled environments.* Dissertation thesis, University of Technology Aachen, Germany

Trossen, D. (2001). Application sharing technology: Sharing the application or its GUI? *Proceedings of International Resource Management Association Conference* (pp. 657-661).

Wallbaum, M., & Meggers, J. (1999). Voice/data integration in wireless communication networks. *Proceedings of the 50ᵗʰ Vehicular Technology Conference.*

Walker, M. B. (1982). Smooth transitions in conversational turn-taking: Implications for theory. *Journal of Psychology, 110*(1), 31-37.

Dirk Trossen has been a researcher with Nokia Research since July 2000. In 1996, he graduated as a MSc in mathematics from University of Technology in Aachen, Germany. Until June 2000, he was with the same university as a researcher, obtaining his PhD in computer science in the area of scalable group communication implementation and modeling. His research interests include evolving IP-based services and protocols for the future Internet, group communication architectures, as well as simulation and modeling of distributed systems.

André Schüppen received his diploma degree in computer science from the Aachen University of Technology (RWTH) in 1996. In the same year, he joined ELSA AG as an engineer for analog modems. Since late 1999, he has been a researcher at the Computer Science Department of the RWTH. His research interests include mobile communication and real-time multimedia technology. He is currently involved in the collaborative research center project 476 IMPROVE, sponsored by the German Research Foundation.

Michael Wallbaum received his diploma in computer science from the Aachen University of Technology in 1998. He wrote his thesis at the University of Cape Town, South Africa, on the topic of security in communication systems. Since 1998, he is a researcher and PhD candidate at the Department of Computer Science in Aachen. In the past, he has participated in several European research projects and is currently involved in the European IST-project ParcelCall. His research interests include mobile communications, multimedia, Internet telephony and active network technology.

This case was previously published in the *Annals of Cases on Information Technology*, Volume 4/2002, pp. 119-130, © 2002.

Chapter XXI

GlobeRanger Corporation

Hanns-Christian L. Hanebeck, GlobeRanger Corporation, USA

Stan Kroder, University of Dallas, USA

John Nugent, University of Dallas, USA

Mahesh S. Raisinghani, Texas Woman's University, USA

EXECUTIVE SUMMARY

This case traces the dynamic evolution/revolution of an e-commerce entity from concept through first-round venture financing. It details the critical thought processes and decisions that made this enterprise a key player in the explosive field of supply chain logistics. It also provides a highly valuable view of lessons learned and closes with key discussion points that may be used in the classroom in order to provoke thoughtful and meaningful discussion of important business issues.

INTRODUCTION

GlobeRanger's business concept originated from the idea that wireless technology today allows for easy tracking of people, but that there are no means to comprehensively track assets and goods in real time as they move through the supply chain. Initial discussions about the value proposition focused on hardware-based mobile tracking systems as they are predominantly found in the market today. A solution necessarily would have consisted of a Global Positioning System (GPS) receiver chip linked to the asset being tracked, which in turn is connected to a wireless device that sends information to a central tracking center. Such a system would provide great benefits to those involved with the asset's transportation or dependent on its arrival.

After an extensive analysis, GlobeRanger made an early decision that hardware-based systems will substantially extend the time-to-market, tie-up resources, and limit the company's ability to adapt to its fast changing environment and unique customer requirements. Prior to funding, the executives decided that the greatest value was in location information and its derivatives that could add value in supply chain management. As a result, the decision, to drop hardware and concentrate on information, at a very early stage led to a novel and innovative business model that is, to this day, unparalleled in the market place. This strategic shift from tools to solutions is the major focus of this case.

GlobeRanger's management team's insights regarding risks and pitfalls of pursuing a hardware strategy and, in turn, its early shift in direction within the same field is the heart of this study. Tracking is time and information sensitive. An organization's supply chain management function is increasingly important in today's highly competitive environment with compressed business cycles. Many firms are reengineering their logistics operations and moving to just-in-time manufacturing, distribution and delivery. In effect, GlobeRanger shifted from the data gathering function to that of providing real-time data mining and analysis with the attendant opportunity of offering proprietary solutions to clients by using advanced IT techniques and technologies.

Today, GlobeRanger is a facilitator of B2B e-commerce solutions that add value to supply chain management initiatives by creating visibility in an area that has been largely overlooked. GlobeRanger's proprietary solutions offer its clients improvements in efficiency, effectiveness, customer satisfaction, and competitive advantage while creating barriers to entry.

One of the fundamental observations that led to GlobeRanger's current business model was that location information alone is of very limited value. Rather than settling for information about the mere location of an asset or goods, GlobeRanger has deliberately taken a much more comprehensive approach of integrating various other sources of information that provide context to location. For example, delivery trucks could be rerouted dynamically based on ever-changing traffic and road conditions. GlobeRanger serves as a facilitator in improving complex logistics systems. Traffic is getting more congested. Payroll and fuel costs are increasing, thus providing the needed business case for the GlobeRanger system named eLocate™.

GENESIS OF IDEAS

Several things came together rather serendipitously to bring GlobeRanger to life. As is often the case, the company originated as much from coincidence, right timing and luck, as it did from its founders' experiences and track records. Without several key individuals meeting at the right time, each one with ideas that ultimately melded into an innovative business concept, the company would have never been founded. A senior executive from Nortel Networks, a successful entrepreneur, and a seasoned IBM researcher put all of their experience and knowledge together to create the foundation for GlobeRanger.

It all started when George Brody left his post as vice president at Nortel Networks to become an entrepreneur and founder of a company that was to operate in the converged areas of wireless and Internet technologies. He was contemplating several possible venture ideas. George met with Roy Varghese, a Dallas entrepreneur who also had ideas of starting a high-tech venture. The relationship between the two men had initially been formed when Roy invited George to meet an Indian government official for lunch. About a year later they met again to set off into what was to change the course of their lives.

Roy, at that time, had already worked closely with a third individual — Dr. Shrikant Parikh, a seasoned researcher at IBM — for several years. In early 1999, Shrikant started to think about novel applications of GPS technology that he felt would lastingly transform the way people live, work, and commute. At the time, he was studying novel telecommunications applications that would come to market several years later. Roy immediately grasped the vision and potential of Shikant's ideas. So he pitched them to George for his insight and thoughts. George, in turn, discussed his own ideas and where he wanted to go.

Once all three of them met, their thoughts quickly crystallized into the beginnings of a plan. It was George who first grasped the potential that lay in combining of his own ideas with those of Roy and Shrikant. During his years as a senior executive at Nortel Networks, he founded the wireless business of the company and, beginning in 1979, personally oversaw the development of wireless technologies that would forever change the way we communicate. In his role at Nortel, George had been actively involved in the development of the very first wireless standards and knew one thing that the others had overlooked. He knew that the standard ignored one critical piece of information — the precise location of a caller.

From the outset, George's vision for wireless communications had focused on an intelligent network. One that would locate callers no matter where they were, and would allow a person to carry a single phone number independent of where in the world he or she traveled. In addition, George envisioned a network smart enough to intelligently route a call to its designated recipient. This could be the case in a household with multiple potential recipients. The network would be smart enough to route a given call to its correct destination without involving all other potential recipients and members of that household. To a large extent, we now see George's vision materialize after more than 20 years. As George knew that the precise location of a caller had been lost or crudely approximated while developing the first wireless networks, he was aware that things like 911 emergency calls would not work since the network did not know to which police or fire station to transfer the call. Hence, when George talked to Roy and Shrikant, he immediately grasped the importance of locating not just a person, but also an asset such as a truck or a railcar.

George, Roy and Shrikant, the initial group of founders, decided to launch a company that subsequently leveraged wireless communications, the Internet and GPS technology to deliver value solutions to end users and businesses. While it was immediately obvious that the combination of wireless and GPS technologies would lead to tracking something or someone, there still was a fairly large field of possible applications. Supplying the business-to-consumer (B2C) market with tracking or location devices similar to wristwatches, for example, at first seemed to have the advantage of sheer market size. At that time, B2C Internet businesses appeared to be taking off. Both factors argued for jumping in. It was George who quickly saw that the possible target market was narrower than they had originally thought, due to the fact that there only were few applications, such as tracking people within amusements parks. In addition, this business case would have needed substantial funding to cover the exorbitantly high marketing expenses. This conclusion drove the discussion towards a business-to-business (B2B) model that could eventually provide a much larger market size and thus, leave enough room to scale the company later on. In addition, it would allow the founders to initially concentrate on one segment, which is an essential criterion for any small company entering an existing market or creating a new one.

George's point of providing tracking capabilities to B2B customers had the added benefit of leaving few alternatives. While this may sound odd, it immediately focused the group on the area of tracking assets and goods as they move through a given supply chain. This foundation of the business model is shown in Figure 1. The new company would equip trucks, trailers, containers or railcars with hardware devices that included a GPS receiver chip and a wireless modem. As these assets move through the supply chain, their location can be reported back to a central database.

The founders then set out to discuss whether developing and manufacturing a device that housed a GPS chip and a wireless modem would make sense. Again, it was George with his background in wireless telecommunications, who created an ingenious turn of events when he had the idea of being device independent. He argued that producing hardware would tie up considerable amounts of capital and require a lengthy development cycle. Rather, the new company should utilize any device in the market and readily deploy it based upon unique customer requirements. George's idea marked the most important strategic decision in the company's short life. It allowed GlobeRanger to enter the tracking market faster and with substantially less capital than any other player in this space. It provided a thread that would tie all ends together and create a business model that to this day is unique in this market space.

George's initial argument centered around the benefits to the new company, which mainly lay in a much shorter time-to-market and tailored customer solutions — versus the "one size fits all" model. Moreover, he saw great advantages for customers in that they could choose between different technologies and communication standards as their needs expanded while still being served by GlobeRanger's offerings. Unlike most of the world, where GSM is the prevailing narrowband wireless standard, the U.S. to this day is characterized by a multiplicity of competing wireless communication standards leading

Figure 1. The business model foundation (© GlobeRanger, 2000)

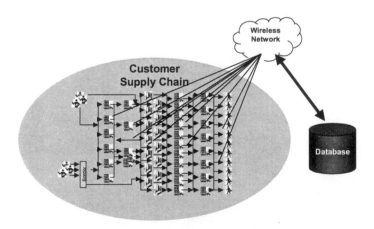

to fractured coverage for many users. His point was that a customer could buy a satellite solution with full coverage within the continental U.S. while that customer effectively needed coverage only within one larger metropolitan area. Conversely, the opposite case could appear that a customer would buy a cellular solution and eventually miss out on location information whenever he was out of range. In addition, satellite communication is much more expensive than cellular or paging. In summary, there is no single, "one size fits all" solution for disparate classes of customer applications. As a result, GlobeRanger is now device and hardware independent. The company is able to sell different solutions to different customers based on their needs, while employing largely common software modules to all customers. In this way, there is an economy of scale that allows for "mass customization." Yet, George's idea had another equally important dimension. His approach allows "future-proofing" of existing customers. That is, GlobeRanger can provide its functionality by simply changing its interfaces to new hardware and communications technology. Customers can immediately replace older technologies with new ones as they become available. Last, his idea also proved to be a very strong competitive weapon since GlobeRanger was able to tap into the existing market of assets that already had devices deployed. George argued that they could literally take on a competitor's customer and use the devices installed by the competitor while providing all GlobeRanger services to that customer and thus, reaping the benefits of the true value creation.

A first market analysis led to the realization that there was a fairly large field of players in the tracking market, each one supplying its own hardware solution based on one communication technology or another. This effectively validated George's idea of independence from a given hardware platform. Now the founding group was convinced that its focus on supplying location information was a very attractive target of opportunity. In addition, the analysis turned up that almost all of the existing players sold proprietary client software that customers had to use in order to access information about the location of their assets and goods as they moved through the supply chain. This, too, was an initial concern to the group, and it readily decided on a fully Internet-based solution and an Application Service Provider (ASP) model that would alleviate the need to buy software while making the information available and accessible anywhere and at any time. The approach also had a second advantage that potentially any authorized party could access information about the location of an asset or a particular shipment from any Internet connection — provided, of course, they had the appropriate security codes. This would enable transportation carriers to provide real-time monitoring of the goods being shipped to the originator and/or recipient and thus, drastically increase their level of customer service in a real-time environment. In addition, there was a seemingly simple but very powerful advantage in terms of time-to-market. While client software-based players had to ship literally thousands of CD-ROMs for every change in their functionality, GlobeRanger could change features and functionality virtually overnight and make it available to a single customer, groups of customers or to every user of eLocate™.

It was Shrikant, with the mind of a scientist, who saw a vitally important relationship resulting in an important differentiator between GlobeRanger and its main competitors. In his view, it was not enough to show the position of an asset or particular shipment on a map over the Internet. Rather, he argued, it would make sense to integrate location information into enterprise- level software such as Enterprise Resource Planning (ERP) systems. According to his vision, there had to be an advantage to the reduction of work steps, for a company to garner added value. Hence, if a given customer's employee did not have to look at an ERP screen for shipment details and then switch over to a second system and screen to access the current position of that exact shipment, this would provide a distinct competitive advantage and customer value add. Both George and Roy immediately bought off on the idea and established it along with hardware independence and full Internet access as the main pillars of service positioning for the new company.

The extensions to the business model foundation are shown in Figure 2. George's idea of being hardware and device independent is reflected in that the company was able to tailor the wireless network to specific customer requirements. Shrikant's idea of ERP integration is shown as a two-way link between the knowledge base and a cloud representing ERP software. In addition, the founders argued that the ubiquitous nature of the Internet would permit specific reporting and analysis capabilities that could directly feed into customers' systems. This would greatly enhance product offerings. These reports could be used by transportation carriers to enhance efficiency of daily operations and by customers such as originators or recipients of goods to provide better services.

Shortly thereafter, George and Shrikant had the idea of calling the company GlobeRanger to bring out the dimensions of being global and being a "trusted agent" for location-based services. The founding team then went out and convinced well-connected individuals to take positions on the board or as senior advisors to the company. For example, Ambassador James Jones, former Chief of Staff to President Lyndon B. Johnson, Chairman of the House Budget Committee, U.S. Ambassador to Mexico and President of the American Stock Exchange, was asked to serve as Chairman of the Board of Directors. The team then created a first business plan and raised an angel round of over a million dollars to get the company off the ground. In addition, the founders brought in the first employees with software, marketing and supply chain management expertise to round out the senior team.

Figure 2. Extensions to the business model foundation (© GlobeRanger, 2000)

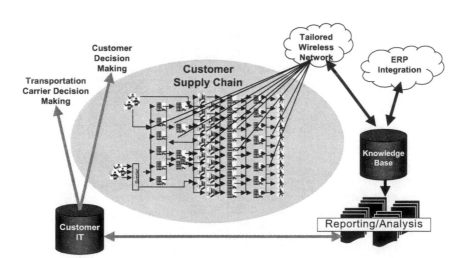

In addition to George, Roy and Shrikant, five employees worked at GlobeRanger to prepare for the launch. Rather than hiring all vice presidents right from the outset, with one exception George focused on hiring employees who had 5-10 years of operational and management experience. This decision substantially helped him to save on salaries and also kept the team highly agile in terms of unanimously executing decisions once they were made. In early 2000, the core team started out with three key priorities in mind: (1) to secure seed and eventually first-round VC funding, (2) to define the value proposition and segment the market space, and (3) to evolve the founders' ideas into a first product release. It marked the beginning of a long and eventful journey, which led to highly competitive services and solutions for customers in diverse industries and with a variety of individual requirements.

EVOLUTION OF GLOBERANGER CORPORATION

Once the founders had assembled a core team, they started to focus on their three key priorities. All of the earlier work now merged seamlessly into the efforts that started in January of 2000. George knew that he needed what he called "smart money" — a funding source that would be able to provide much more than just capital. In addition to getting funding, he was looking for investors who could not only add value to corporate strategy through their deep-rooted expertise and business savvy, but who could also contribute substantial contacts to potential customers and to highly attractive alliance partners. In order to get funded, George turned to premier venture capitalists (VCs). He needed to provide a bulletproof business plan that consisted of a detailed market analysis and demand curve, a novel and hard to imitate business model, and rock solid financials to show a strong ramp through the first five years. In addition, George had to have the first customers lined up who were willing to take a bet on GlobeRanger's new technology and ideas.

For the market analysis, John Sweitzer was brought in as the VP of marketing. John had more than 30 years of strong telecommunication and marketing expertise when he joined the team. Like George, John had spent a good part of his career as a senior executive at Nortel and had actually assumed some of George's responsibilities when George left. The two had been working closely together for a number of years, and so it was natural for both to immediately snap into a highly productive working relationship. John's first step was to identify the benefits of the product for prospective customers and to talk to as many of them as possible. Within weeks, he had contacted virtually everyone he knew that was affiliated with transportation or asset management. In these early days a passerby visiting John's office would see huge piles of market segmentation reports and analysis books on his desk. It made it almost impossible to spot John amidst of all of his reading. More often than not, his studies lasted well into the night and turned up an incredible wealth of information on how to "attack" the market. The core team, assisting John in his marketing activities, took his findings to create collateral, presentation slides and amend their ever-growing business plan document.

In regard to the business model, it was the team of Rakesh Garg and Chris Hanebeck, both in their early thirties and eager to add a lasting success to their professional experience, who worked closely with George to devise the next iterations to the business model. The two complemented each other in an almost perfect way. Rakesh came straight from Netscape with a strong background in Internet technologies and software architecture, while Chris covered strategy, ERPs and supply chain management after eight years as a senior management consultant on three continents. In addition, both had startup experience and knew all too well how much blood, sweat and tears it would take for the team to build a business like GlobeRanger from ground up. Especially in the early days, the two frequently seemed to have their heads glued together and cherished their daily exercise of bouncing ideas back and forth until they became so intertwined that no one was able to make out a single originator. One of their main tasks lay in defining key features and functionality around the extended business model as it is shown in Figure 2. Rakesh concentrated on the technical aspects such as the general architecture, partner software integration and finding a highly skilled team of developers to implement the first release. Chris, on the other hand, built a first business case based on the architecture, and constantly worked on presentations that were to be given to investors.

All along, the founding team was closely involved in everything that went on. George, as the most experienced senior executive, naturally had taken on the role of President & CEO. Roy, equipped with the most positive outlook on life that one could possibly imagine, was ideally suited to work on strategic alliances and key partnerships. Shrikant, too soon, had to retreat from his active duty as chief technologist for personal reasons, but has stayed close to the company as a senior advisor.

Once the first iteration of business plan and presentations was completed, George started to look for an incubator that could provide seed funding and help him establish relationships to top VC firms. During his tenure at Nortel, George had been instrumental in establishing STARTech, a Richardson, Texas-based incubator specializing in early stage seed funding and startup support. Thus, his first priority was to visit Matt Blanton, the CEO of STARTech. After hearing about GlobeRanger's extended business model and after providing several key inputs that were accepted, Matt agreed to take GlobeRanger on and run due diligence for seed funding. Together with one of his key principals, Paul Nichols, who, at the time, was responsible for operations, Matt proved instrumental in guiding and coaching the GlobeRanger team forward. They immediately liked the extended business model (as shown in Figure 2) for its applicability to logistics and, specifically, supply chain management.

Chris, who had been brought in for his expertise in supply chain management, had devised several application areas for GlobeRanger's technology. He quickly realized that the simple tracking solutions that were on the market would not suffice to win large customers. Mainly, Shrikant's idea of integration between GlobeRanger and enterprise-level applications such as ERP systems resonated well with his knowledge of the market. This integration could also be extended into more production and material management related systems such as Supply Chain Planning (SCP), Supply Chain Execution (SCE) and Event Management (EM) systems. The central question revolved around whether GlobeRanger could close the information gap that existed for material flows between companies. Throughout the nineties, companies had consolidated information within their own operations and across departments with ERP software. In the late nineties and fostered by the proliferation of the Internet, supply chain applications between companies became more available. Yet, they only covered sending documents such as orders and confirmations back and forth. No one had yet tried to create a system that was based on real-time information about materials and goods as they moved from one place to another. The obvious problem many companies experienced was that an information "black hole" existed in the supply chain with respect to physical movements of goods.

Just-in-time manufacturers, for example, will always schedule their production to maximize efficiency and then set up delivery times for all input materials as they become needed during production runs. Rather than solving the true problem of knowing where these necessary materials are, they often set up contracts with stiff penalties for missing the promised delivery date. While this practice has proven to enforce discipline, it has not addressed the underlying problem of uncertainty inherent in the physical deliveries. More precisely, if the manufacturer knew where all of his expected shipments were and when they would arrive, he could schedule his production runs more efficiently. This, in turn, would create flexibility and provide the ability to react faster to events such as truck breakdowns, cargo theft or simply traffic jams.

Furthermore, there was hardly any historical information about the quality and sophistication of transportation service providers that could be used to evaluate and rank them. A second opportunity in regard to historical information was that it could be used for business process improvements. Yet, historical information has not been available to manufacturers willing to invest in programs such as process improvements. George had already decided to offer advanced data warehousing and mining capabilities in GlobeRanger's product and had brought in a small team of outside consultants specialized in this area. This decision, once again, turned out to be a very visionary one as prospective customers looking for supply chain management services now were able to base process improvements and transportation carrier evaluations on historical data rather than intuition.

Once these ideas crystallized, it became apparent that previous efforts to track assets and goods throughout the supply chain had neglected a major component of the value proposition. They all focused on the asset owner rather than the originator or recipient of the goods being shipped. The result — they only provided very limited information such as pointing out an asset on a map. The existing companies in this field did not even provide an estimated time of arrival, which is crucial to a recipient of goods who is not interested in knowing about the specific location of an asset. To GlobeRanger's founding team and first employees, backed by the senior team at STARTech, this realization provided enough confidence to go into the first meetings with VCs. Again, it was Matt and Paul from STARTech who proved to be invaluable sources of knowledge about the ins and outs within the VC community. They tirelessly worked with GlobeRanger to provide feedback for presentations and carve out how to communicate a clear and unique value proposition. As it turned out, GlobeRanger, in addressing not only asset owners and operators with their tracking solution, but also the originators and recipients of goods, had found a market that was previously untapped. The value proposition resulting out of these meetings is described in the following section.

CREATION OF THE VALUE PROPOSITION

As the evolution of GlobeRanger illustrates, the founding team continually refined and enhanced the business model. This proved to be a crucial success factor as the company matured and sought initial customers. Yet, before the team was ready to deploy an initial release of its asset tracking and supply chain management services, by now named eLocate™, a precise value proposition needed to be shaped so that it could be communicated to VCs and prospective customers. This section provides a detailed overview of the resulting value proposition.

Definition of Location

Whenever we talk about a location that we have visited and appreciated, we have explicit and rather precise mental images. The location of that little restaurant in Rome or of the Eiffel tower in Paris are burned into our memories forever and are intertwined into a variety of other emotions that we have experienced while being in those exact places. Yet, much less romantic, vector coordinates of latitude and longitude can also define these very same locations. In the technical environment of logistics this would be the preferred way to describe and communicate information about a location. Independent of how we see it, a main question is whether the information about location by itself contains value?

A simple and straightforward answer is "No." Information about location by itself does not and cannot contain value. Only when location is put into a relevant context, does it become valuable and can it be utilized for a variety of purposes. Knowing precisely where we are on a highway is of little value unless particular knowledge about a traffic jam two miles down the road is overlaid. Likewise, it is of little or no value to know someone else's location in relation to our own unless a specific circumstance arises. For example, the location of a delivery truck is only relevant to the production manager who is holding up the manufacturing line and making a decision on next production runs while waiting for a critical part. To summarize, information about the location of a person or an asset is of little value unless circumstances or other information enhance and enrich it. With the above said, location is defined as a unique geographical position. This naturally translates into a unique and identifiable location for every spot on our planet.

In regard to the above, a classification for location information is presented in Figure 3. It is based on a differentiation between data, information and knowledge as three levels of increasing value.

Figure 3. Levels of location value (© GlobeRanger, 2000)

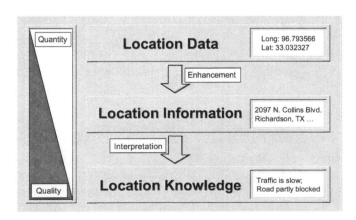

Figure 4. Example of location knowledge (© GlobeRanger, 2000)

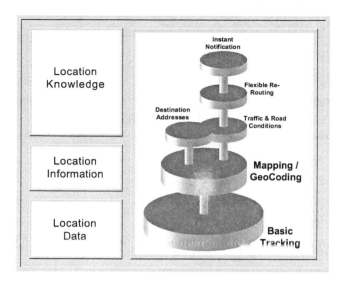

As Figure 3 shows, there are three levels in the creation of value in conjunction with location. First, location data in and of itself does not contain value other than serving as the foundation upon which the next two levels are built. Second, location information results from adding facts to location data such as a street address that will increases its value. Third, location knowledge results from overlaying other relevant information such as traffic and road conditions that can be utilized to evade an unpleasant circumstance such as a traffic jam.

To broaden the concept of location knowledge, we can now begin to define all the areas in which other, relevant information can be overlaid to increase the value of location information. As previously stated, any information or knowledge that brings location data into a meaningful context for its user can be integrated. Figure 4 illustrates this point by a simple example of utilizing the precise location of a mobile asset such as a truck to flexibly adjust its path from origin to destination by adjusting to varying conditions along the way.

In Figure 4, basic tracking refers to the ability to identify and communicate the precise location of the mobile asset via GPS or triangulation in periodic time intervals. Seen in the light of our classification, basic tracking provides us with location data. On top of this, there are a multitude of combinations to generate and utilize other relevant information that lead to location information. In the above example, these are asset demand, asset status, geographical mapping, and automated regulatory and tax report generation. Geographical mapping relates to the ability to pinpoint and communicate the location of a mobile asset in natural language, such as a street address and an exact point on a geographical map. Based on available location data and information, location knowledge can now be created. The example in Figure 4 shows four possible combinations of destination address, traffic and road conditions, flexible rerouting and instant notification. In this example, destination addresses can be used to determine the distance between the truck and each of its drop-off locations. Based on these distances and their relationship to each other, software can then calculate the optimal route. Once we overlay actual traffic and road conditions, factors previously unavailable to us are taken into the equation to determine an optimum route. For example, there could be a slowdown or road closing along the way that would make it more efficient to change the order of destination addresses as the truck moves. The newly derived routing information is made available to the dispatcher who can then send an instant notification via text messaging to the truck operator. Of course, this process would repeat itself as often as economically feasible and necessary. Moreover, because location information is known, such information can feed reporting systems such as those required for regulatory reasons.

Leveraging Location for Intelligent Management

The above examples are very simple ones. There are, however, highly interesting combinations of possibilities within the interface of location data and Supply Chain Management (SCM). This holds especially true for the latest developments of electronic exchanges enabled by the Internet. Most prominent SCM software providers today are establishing electronic market places that will enable the smooth and fluent integration of inter-company business processes across the whole supply chain from producer of raw materials all the way down to the retailer.

In terms of integration into logistics SCM software systems and electronic exchanges, location information and location knowledge can be applied and also be generated. As previously stated, location data needs overlay information to be valuable. SCM software is one valuable source and target for such overlay information. As a result, the relation between SCM software and location information is two-fold: location information creates value for SCM software and applications and at the same time, SCM software provides overlay information that is extremely useful in the creation of location knowledge. Figure 5 shows this relationship.

GlobeRanger's eLocate™ software interfaces with a variety of Internet-based sources that can add value to location data. In addition, eLocate™ interoperates with SCM software to receive overlay information and transform it into location information and knowledge, which is fed back to the SCM software. The very same relationship for eLocate™ and electronic exchanges holds true. In the case below, overlay information could be detailed scheduling or routing information, planned shipping dates and arrival times, specific commitments to customers or simply generated bills of lading. Internet-based real-time information services are flexibly deployable, as they are needed. GlobeRanger's eLocate™ allows for maximum flexibility in the sense that any service can be interfaced as long as it allows for access to its data. If it is not directly available in real time, a cached source can be created in the GlobeRanger database.

What becomes apparent in Figure 5 is that neither the originator nor recipient of overlay information by itself can create the complex relationships that arise from their integration. eLocate™ functions as a clearinghouse between all parties involved so that efficient and effective aggregation and retrieval of overlay information can be achieved.

The following brief example, illustrated in Figures 6 and 7, will explain the interplay between overlay information and location knowledge in the case of a typical manufacturing process. Once a product such as a golf shirt is designed, it is offered to professional buyers who, in turn, take it to wholesalers or retailers for sale to consumers. These buyers

Figure 5. Relation between SCM software and location information (© GlobeRanger, 2000)

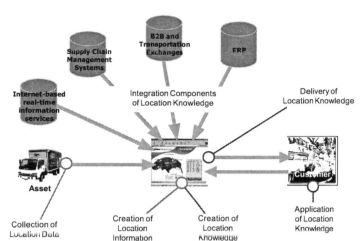

will meet with their customers and receive orders that they pass on to the manufacturer. The manufacturer will already have established a Bill of Materials (BOM) based on the design of the golf shirt. The BOM specifies the quantities and qualities of fabric, thread and buttons that are needed for any given size of the shirt. Thus, as the manufacturer looks at the orders from buyers, it will be able to determine the materials needed based on the BOM to manufacture the golf shirts. Next, the manufacturer will start two parallel processes of Production Planning and Control (PPC) and Procurement. As Figure 6 shows, both parallel processes have to be completed before the actual production can begin. It is intuitively apparent that the speed of procurement will determine the overall process from design to delivery. A closer look into procurement reveals that it consists of several steps, from a Request for Procurement (RFP) to a Purchase Requisition (PR) and finally a Purchase Order (PO) that is sent to the suppliers of materials needed to produce the golf shirt.

Location knowledge, as illustrated in the first section of this chapter, certainly has a strong impact on the efficiency of asset utilization by being able to transport goods along more effective routes, through less traffic and without delays typically caused by factors such as wait times or scheduling discrepancies. In addition, it is highly important to know the Estimated Time of Arrival (ETA) for all materials as the production process can only start when every needed part of the golf shirt is available on site. Knowing the ETA and integrating it into SCM software the shirt manufacturer uses will allow the SCM software to schedule and run production according to real-time information flows of how materials stream into the plant facility. This leads to lower inventory levels, an overall lower supply chain cost and less stockouts or prolonged manufacturing cycle times. There are also secondary saving effects such as cost savings from reduced warehouses, insurance, and personnel, as well as less rescheduling time due to accurate ETA information, and reduced capital expenditures.

Figure 6. Location knowledge effects on inbound supply chain processes (© GlobeRanger, 2000)

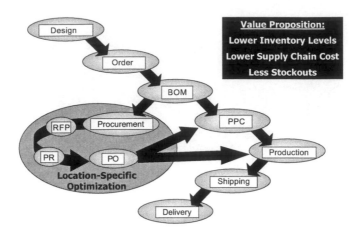

Figure 7. Location knowledge effects on outbound supply chain processes (© GlobeRanger, 2000)

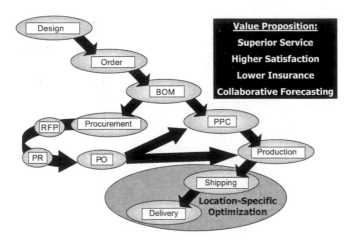

In addition to cost and time savings on inbound material flows, location knowledge also leads to greatly enhanced efficiencies in outbound logistics as shown in Figure 7. The very same advantages that the golf shirt manufacturer captured during inbound logistics now become available to the manufacturer's customers during the outbound movement of goods. For the manufacturer, this leads to providing superior service to its customers, such as offering guaranteed deliveries, immediately notifying the customer when parameters change, or the ability to flexibly reschedule deliveries. It also leads to lower insurance premiums as the real-time tracking of assets and goods in transit will drastically reduce theft, especially when sensors for door opening, ignition or deviations from the scheduled route are deployed. Most important of all, the sharing of location knowledge between our manufacturer and its customers enables collaborative forecasting that, in turn, creates efficiencies on all ends.

Visibility into the movement and status of assets and goods throughout the supply chain brings savings of cost and time for all parties involved and, at the same time, raises the quality of the transportation process while enabling greater flexibility. The next section details these effects further by illustrating package tracking based on GPS and scanning technologies.

Visibility into the Supply Chain

As we had previously stated, knowledge of the whereabouts of a mobile asset such as a truck contains value in and of itself. It enables a scheduling and routing flexibility previously not possible. In addition, location information helps us overcome capacity bottlenecks, roadside emergencies, and even theft. Yet, there are limitations to tracking a mobile asset only. These limitations become apparent when we want to take into consideration that a truck, trailer or container is nothing more than a vessel that contains objects of interest such as materials or finished goods. A deeper look reveals that location information and knowledge are only relevant in the context of what is actually being shipped at any given point in time. Thus, the contents of a truck or trailer become overlay information just like traffic or road conditions.

Figure 8 illustrates a typical route from a west coast manufacturing facility to east coast retail locations. As packages are loaded into a truck at the manufacturers outbound warehouse in San Diego, they are scanned based on a bar code or using Radio Frequency Identification (RFID) tags. Should the manufacturer use RFID tags, a RFID reader would be installed within each of his trucks or in the staging area, eliminating the need to manually scan each package. The information about a trucks' content is then "married" to the truck identification number, which allows GlobeRanger's eLocate™ to instantly display both the asset location and content location. As the truck progresses towards the cross-dock in Dallas, users such as the manufacturer, freight agent and customer at

Figure 8. Example route for a package tracking application (© GlobeRanger, 2000)

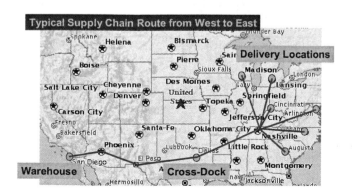

the delivery location can immediately see where the goods are. Regular scheduled or unscheduled ad hoc GPS readings can be taken by any authorized user of eLocate™ at any time. In addition, eLocate™ will provide the estimated arrival time (ETA) of the goods at the cross-dock and their delivery locations taking traffic and road conditions, as well as routing and scheduling information, into consideration.

Once the truck reaches the cross-dock, its contents are unloaded and scanned as they enter the cross-dock. This scan is necessary to ensure the complete arrival of all goods that were loaded onto the truck at its location of origin. It also provides GlobeRanger with the necessary information to "divorce" truck ID and contents in its eLocate™ database. The location of individual goods is now shown at the cross-dock in eLocate™ and may be entered into a Warehouse Management System (WMS). Usually, while at the cross-dock, whole truckloads and palettes of goods are disassembled according to their final destination and then put onto new palettes for their recipients. When they move out of the cross-dock, another scan is performed at the warehouse: the contents of a truck are "re-married" to another truck ID that can immediately be seen in GlobeRanger's eLocate™.

This second truck will now cover the remaining route between the cross-dock and the delivery locations. Just as on the first leg of the route, any authorized user can see the precise location of goods by simply requesting the location of the truck. As our truck delivers goods to the recipients, the truck operator will use a handheld scanner to report those goods that are "divorced" from the truck as they are unloaded. This information can be kept on the truck to be downloaded for later use or can be wirelessly transmitted into GlobeRanger's eLocate™ database in real-time. With RFID tags and a reader inside the truck, manual scans can be eliminated thus reducing the time and effort spent to deliver goods. Figure 9 shows the process and how GPS readings of the truck relate to content information about the goods onboard the truck.

The example process described earlier illustrates how a bill of lading can be applied as overlay information to enhance and interpret location data into relevant location

Figure 9. Example process for GlobeRanger's package tracking solution (© GlobeRanger, 2000

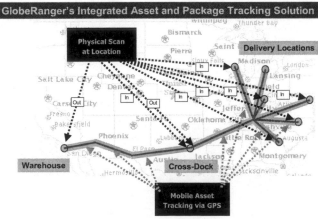

information and knowledge. The way in which the necessary information is integrated into GlobeRanger's eLocate™ software is through application of devices that may use the wireless application protocol (WAP), Windows CE, JAVA, Palm OS or other platforms. These devices can be used for two-way text messaging communication with the driver as well as to send data such as package information to GlobeRanger's eLocate™.

Figure 10 shows how GlobeRanger's information technology components work together to enable seamless asset and package level tracking with integration of scanning or RFID technologies. A wireless device such as an enhanced Palm Pilot is used to scan package information. Once the device transmits data, it is received by a protocol server such as a WAP server for WAP devices, a Palm Server for Palm OS devices and so on. GlobeRanger's eLocate™ is fully device independent and, as such, can accommodate the device most suited for a specific requirement rather than a narrow choice of devices. The protocol server can be understood as a means of translating the data from a proprietary protocol language into one that Internet-based systems can understand. In this function, the protocol server sends the data to a Web Server, which in turn populates GlobeRanger's eLocate™ database. Once stored in eLocate™, package and mobile asset tracking knowledge is displayed through a standard Web browser so that users such as a dispatcher or materials manager and customers such as the recipient at the delivery location or the owner of the goods in transit can access it at any time. In addition, it might become necessary to intersect a separate package-tracking database. This could be the case when the warehouse and cross-dock scans are taken at each location and not via a handheld device. Should this be the case, a direct connection between these location scanners and a central package-tracking database is created. The database then feeds eLocate™ so that asset IDs and package tracking information can be flexibly "married" and "divorced" throughout the supply chain. GlobeRanger's eLocate™ database in Figure 10 also contains a link to decision support systems (DSS) that can be applied for later data mining and warehousing, as well as into

Figure 10. Package tracking integration in eLocate™ (© GlobeRanger, 2000)

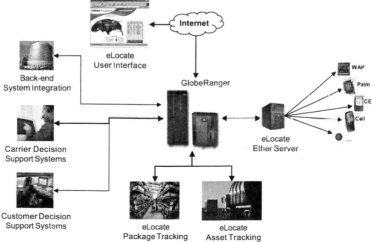

back-end systems such as SCM software or Enterprise Resource Planning (ERP) systems.

Benefits Capture

The benefits of tracking packages are easily identifiable and range from enhanced customer satisfaction to security considerations and the ability to flexibly reschedule deliveries "on the fly." One of the biggest advantages stems from the difference between used and available capacity of mobile assets. Both freight operators and manufacturers can easily auction or sell available capacity of mobile assets if they have access to real-time availability data and an electronic exchange.

Location information and knowledge generated through a clearinghouse such as GlobeRanger will become a valuable addition to exchanges' offerings for their customers. One of the great potentials of asset and package tracking is the ability to learn from previous business processes by deriving improvements that can be implemented through a continuous process improvement cycle. It will, in effect, allow for optimization of a large part of supply chain management features in real-time and with much more information than is available today.

MATURING OF GLOBERANGER CORPORATION

Once the detailed value proposition had been created, the GlobeRanger team was well prepared to present its ideas to the VC community. Paul Nichols and his team at STARTech set up a kick-off meeting to introduce GlobeRanger to several top notch VCs in their network — most notably to Centerpoint, Sevin Rosen Funds and HO2. During the presentation, George and his team showed the extended business model, market segments and supply chain applications. They also had a first demo of the product that the group of data warehousing experts had put together. Overall, the presentation was well received, yet only marked the beginning of the funding round. The VCs present at the kick-off meeting were very interested and grasped the novelty of GlobeRanger's value proposition immediately. Especially two of them, Terry Rock, the leading partner at Centerpoint Ventures, and Victor Liu from Sevin Rosen Funds, took to GlobeRanger and agreed to schedule follow-up meetings. During the weeks and months that followed, George and his team worked relentlessly on preparing presentations, reworking assumptions and detailing the business plan so that it would meet the growth and market requirements set by the VCs.

For the complex task of establishing a detailed financial analysis and models, John Nugent, a professor at the University of Dallas and former President of several AT&T subsidiaries, was brought into the team as a consultant. John took on the role of acting CFO and chief financial strategist naturally and proved instrumental in steering through difficult financial terrain in all meetings extremely well. His lasting contribution to the company still remains, as he established the financial planning foundation throughout many meetings with George, John and Chris. Primarily, a five-year financial plan documented anticipated revenues over various customer segments and cost of goods sold, all the way down to depreciation and amortization. The company valuation was then tied into the financial plan and reached over the same period in time. From the outset, George and John Nugent strongly focused on a financial model that included a fast track to profitability and sound, yet defensive assumptions about the adoption of eLocate™ in the market. Both men knew all too well that the times when startup companies had to have large losses in their business plan were over. Once the senior team at GlobeRanger had bought off on the financial plan, each projected year-end result became an objective of financial performance for the years to follow.

It was not until three months later that George and his team finalized presentations to three leading VC firms in the Dallas area, as well as to Marsh McLennan Capital, the VC arm of Marsh Inc. in New York. During these last days of first-round funding, everyone at GlobeRanger had one single goal in mind–to conclude presentations and successfully close the first round of funding. Then, finally, came the day the whole team gathered in George's living room where George opened a bottle of champagne to share the news he had just received: four VCs, led by Centerpoint Ventures and Sevin Rosen Funds, had agreed to move ahead and several other parties willing to invest had to be deferred to a following round. Including the other participating investors — HO2 and MMC Capital — GlobeRanger raised $10.0 million in its first round. This certainly was much better news than many on George's team could have hoped for, given that many technology startups were not being funded due to the drastic downturn of the NASDAQ and Internet-based companies during the first half of 2000. It marked the beginning of GlobeRanger's first strong phase of growth and provided enough financial stability to kick off the intense product and market launch in the months to follow.

One of the first initiatives after funding was to validate the extended business model as shown in Figure 2. With the help of Victor Liu, the well-known consulting firm McKinsey & Co. was hired and promptly set up a team of senior consultants at GlobeRanger's offices. Together with John and Chris, the consultants researched and tested all underlying assumptions of GlobeRanger's business model. Overall, they

Figure 11. Strategic sales filters (© GlobeRanger, 2000)

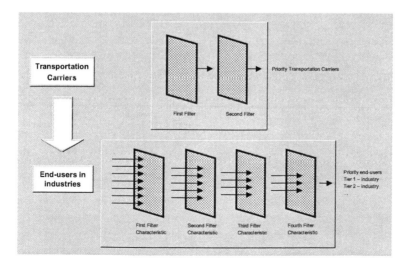

proved instrumental in detailing and communicating the market approach. As the discussion of GlobeRanger's value proposition had shown, the company differentiated itself early on by looking at supply chain solutions rather than simple track and trace applications. The strategy consultants set out to broaden the sales model into the two-staged approach of how GlobeRanger could identify possible sales targets, as shown in Figure 11.

As Figure 11 illustrates, GlobeRanger would go into a sales meeting with a two-staged approach. In the first stage, possible sales targets within the transportation industry are identified and can be chosen according to different criteria shown as two separate filters above. The main reasoning behind this first stage was that GlobeRanger obviously needed to deploy a hardware device on an asset or good. Thus, it made perfect sense to start out with the owners and/or operators of assets. During the second stage, the sales team would look beyond the transportation carrier and towards that carrier's customers. Here, a grid of four filter characteristics helps sales personnel to establish which carrier customers and industries to go after. The second stage model is actually based on characteristics of the industry and of the goods being shipped.

While the strategic assessment went on, George was its most keen observer and prolific contributor. Yet, he also shifted his main focus to product development and to rounding out his executive team. It had been clear to him from the beginning that the development team needed much of his time, insight and experience. Together with Rakesh, he supervised many of the technology decisions and talked to most developers on a daily basis. During the months after first-round funding, the product grew from a software demo to a full-fledged first release for several different modes of transportation. One of the early adaptors had been a large U.S. railroad company that used the eLocate™ service to monitor the utilization of its locomotives in real-time. In addition, several transportation companies became customers over time. With these customers, it became apparent that the strategic decision to custom-tailor solutions for each of them had been a very wise one. GlobeRanger was even able to replace an existing asset tracking system with the eLocate™ service, mainly due to its ability to listen and understand customers as opposed to merely arguing why a certain technology would be most beneficial for everybody in the market.

The hiring of executives took much more time since George wanted to ensure that each of them was not only highly qualified but also a good fit in terms of organizational culture. He knew well that his startup posed very different challenges from those his executives would face in any large corporation. In addition to high tolerance for ambiguity, all of his team members had to be very adaptive to the tight work environment where every member knew everyone else. In addition, executives had to do a lot of tasks for which they would normally have a support staff. To name just one trivial example, first class flights were and still are completely out of the question for everyone, including George himself. George felt it imperative that the leader "walk the talk," and serve as an example. George also fostered diversity right from the start. The first team at GlobeRanger was already comprised of members born on three different continents. He encouraged cultural ambiguity wherever he could and saw to it that people reached consent rather than obeying top-down decisions. Most of all, George knew that having fun can be a tremendous motivator and, equipped with a very intelligent sense of humor, certainly contributed more than his own fair share.

Figure 12. George's time management chart based on the Covey Model

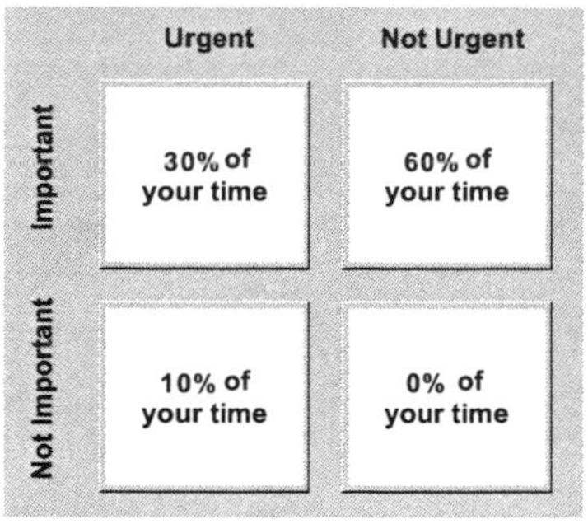

	Urgent	Not Urgent
Important	30% of your time	60% of your time
Not Important	10% of your time	0% of your time

Another truly important issue for George was the clear definition of processes. Right from the outset, he had been very diligent about setting up structured lightweight business processes for GlobeRanger and has continued to foster this issue tirelessly. He had brought in Ramses Girgis, a widely known and acknowledged expert in performance and process management, as an outside trainer for the starting team, and continued to invite Ramses back at every major stepping stone. Every member of the senior team set out to define his or her own business processes, performance indicators and deliverables towards every other member of the team. This effectively proved to be of tremendous value since it structured activities and guided senior team members in their own actions, but in a manner such that all activities were integrated. At the same time, it depleted redundancy in processes and ensured that every activity was performed for a reason. To individual team members, having business processes meant that they were able to prioritize their tasks and tune their own work towards the overall goal and vision of the company.

As part of the process and performance training, George took a standard quadrant chart and organized it according to urgency and importance of a task, as shown in Figure 12 . He then explained that most people spend almost all of their time in the two Urgent quadrants, but hardly any time in the Not Urgent-Important quadrant. This very much reflects reactive behavior to things as they occur. Especially in a startup company, people tend to do 50 things in parallel, and they hardly ever have the time to sit down and reflect on what would make their work easier in the long run. Thus, everyone reacts to reality rather than actively shaping it. George went on to argue that truly successful managers would do the exact opposite. Most of their time is spent on preempting future work and requirements. Thus, everyone should be proactive and save time by anticipating future events, needs and requirements. This approach to management certainly was reflected in George's insistence on structured processes, which he knew would help the company later on once the business grew more complex and less transparent than it had been when the team consisted of merely seven core members.

As a founder, Roy had naturally attended all funding meetings and was well informed of everything that went on within the company. Yet, he also aggressively drove his own responsibility of forming strategic alliances with a number of different companies forward. As the business model implied, there were a variety of different companies that would either supply goods and services to GlobeRanger or would foster the distribution of its eLocate™ service. Very early on, Roy drafted a priority list of possible alliance partners and put them into a priority order. He knew very well that some partners would automatically exclude others as they were competing in the same market space. Yet, his timing was crucial to keep new entrants out of GlobeRanger's market and to decrease the time to market for its service. Based on George's time management chart in Figure 12, Roy differentiated possible strategic alliance partners into three groups and then spent his time accordingly. The A-group of *not urgent*, yet *important* strategic alliance partners consisted of cellular network partners, device manufacturers and enterprise-level software vendors. Due to Roy's very early efforts and hard work, GlobeRanger today has a variety of marquee strategic alliances in place that cover some of the most prominent names in their respective industries.

Over time, a fourth type of partner entered Roy's A-group: insurance companies. Together with John Coghlin, a lawyer by training and Director at MMC Capital, Roy strove to utilize the ties that Marsh Inc. had in the insurance industry and to define the service offering that not only consisted of the value proposition described in this chapter, but also catered to insurance companies and underwriters. His reasoning was that the eLocate™ service would sell substantially better to GlobeRanger customers if he could offer clients a reduction in insurance premium. Roy's reasoning, of course, immediately made sense to the team at GlobeRanger. Yet, to insurers and underwriters there had to be tangible proof that GlobeRanger's eLocate™ would, in fact, reduce the number of claims due to theft and shrinkage. Together with Chris, he worked diligently on refining their pitch to the insurance industry, had numerous meetings with all sides involved in typical insurance transactions, and never rested to prove that his vision would eventually materialize. While his insurance-related effort certainly was one of the hardest tasks, it also seemed to motivate him the most in his daily work. Not a day went by in which Roy did not solve one issue or another to fully integrate insurance relevant features into eLocate™. In small and large steps at times, he continually strove to achieve his grand vision. GlobeRanger proved to be the first company in the asset and supply chain management market to truly integrate insurance companies into the definition of its location-based service.

As Figure 13 illustrates, GlobeRanger saw the development of its headcount increase more than six-fold during the second half of 2000. The executive team had been rounded out with seasoned executives from technology and transportation companies. In addition, several key directors had been hired for product development and management responsibilities. Both, the development and solution consulting teams had been fully staffed and everyone started to work tirelessly to achieve the vision of the three founders: to create a location-based asset management and supply chain visibility company.

Everyone had that sense of urgency that many large corporations lack — to build what had not been built in the decade since asset tracking first emerged. GlobeRanger's goal is now a reality, a fully integrated suite of services that allows customers to choose how they want their assets to be managed and that enable visibility throughout the supply chain.

Figure 13. Headcount development throughout 2000 (© GlobeRanger, 2000)

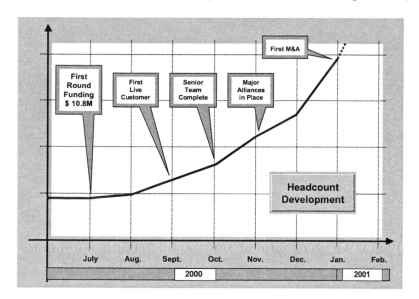

SUMMARY

The company GlobeRanger came to life when three individuals with different backgrounds and various ideas met. Each of the three founders had a unique perspective on business and technology. All three managed to bridge their diversity and strike a balance between vision and feasibility. They set out to create their own market and to change the rules of business in one area of supply chain management. Backed by the trust of their employees and investors, they are well on their way to succeed.

GlobeRanger's case is hardly the only one that has proven successful over a long run. It has not been the first such company and will certainly not be the last. It can, however, serve as an example in that some of the lessons learned are highly valuable. During the first year, the company saw its product architecture materialize, became familiar with its first set of supply chain applications, experienced its first round of VC funding and met its first real-life customers in a highly competitive marketplace. During that time, the core team learned several valuable lessons about how to build and grow a startup business. In summary, these nine lessons stand out as being extremely important:

- Establish the business model very early on, and keep improving on it.
- Assemble a team of very experienced and flexible people.
- Create a culture of continuous change, ambiguity and adoption.
- Provide strategic direction to the team continuously.
- Focus on "smart money" provided by VCs.
- Build lightweight and structured business processes early on.

- Complement technologists with strong management talent.
- Avoid politics, intrigues and infighting at all cost.
- Create and communicate a clear value proposition to customers.

In regard to the individual decision to join and augment the team of a startup company, it is true that some personalities get long better in situations of uncertainly and risk. Yet, there is hardly a typical startup personality. We have found that these companies are always comprised of people from many diverse backgrounds and with different ideas, dream, goals, and skillsets. One thing that all of them share is a unique enthusiasm for creating a new and better way of doing things. They are not afraid to venture far out of the ordinary, and they do not fear to believe in their dreams.

In summary, to be successful as a startup company, as Tom Hedrick, Senior Partner of the Dallas' McKinsey practice quoted one of his customers, all you have to do is:

Think big, start smart and scale fast.

BUSINESS CASE DISCUSSION TOPICS

1. How important do you think a good management team and Board of Directors is in raising venture capital? Do you think GlobeRanger did a good job here?
2. How important do you think the founding members are in shaping the initial directions of the business? Did they have the right skill sets? Might this determine success or failure?
3. How important is it to recruit and retain a very experienced senior staff in a start-up? Do you think this was of particular importance to the VCs in making an investment decision in GlobeRanger?
4. Do you think it is very important to attract leading VC firms as investors, or is money just money? Do such leading firms contribute any other value? Expertise? Relationships? Access to customers?
5. Do you believe GlobeRanger made an appropriate decision to offer customized solutions, value propositions, to its clients by remaining technology agnostic? That is, by tailoring solutions to unique customer requirements by providing a service versus selling a hardware product.
6. Do you think GlobeRanger's vision regarding "technology refreshment" is important in today's marketplace?
7. With the information you have, how effectively do you believe GlobeRanger is in addressing the 4 Ps of Marketing (Product, Price, Promotion, Placement)?
8. Do you believe GlobeRanger is correct in offering its clients open interfaces to its solutions, accessible by the Internet, anywhere, anytime, versus closed, proprietary approaches taken by its competitors? How important do you weigh this aspect relative to market demands and the enterprise's success?
9. Do you believe it is important, as GlobeRanger did, to reduce cycle times and time to market, as well as capital requirements and other costs by providing an integrated solution based on commercially available products? Is this similar to the U.S. government's move to Commercial Off the Shelf (COTS) acquisition practices? What impact should such actions have on a start-up enterprise?

10. Did GlobeRanger appear to address Porter's Five Forces Model in its business case based on the information you have at hand?

11. How would you segment GlobeRanger's markets? Value of the goods? Perishable nature of the goods? Other segmentation or metrics?

12. Do you think GlobeRanger made the right decisions regarding its business case and direction concerning the imperfect nature of information available in a new market? What information would you like to have available in order to prepare a good business plan for a company in this market segment?

13. What skill sets do you think a founding team would be required to have in order to launch a successful company in GlobeRanger's market?

14. Do you think GlobeRanger is correct in its perception of the value of location information?

15. Do you agree with GlobeRanger that location "content" information is valued information? More so than geographic coordinates alone?

16. Would you integrate such technology solutions into your ERP systems if you had such a need? What do you foresee as potential downfalls of such a system?

Do you see any additional benefits deriving from the deployment of such a GlobeRanger solution?

Hanns-Christian L. Hanebeck, Director of Enterprise Applications at GlobeRanger Corporation, is an expert in supply chain management, marketing and business strategy. Prior to joining the company as one of the first employees Mr. Hanebeck worked extensively as a consulting director in the U.S., Europe, and Asia. His professional experiences include founding and management of joint ventures, establishing strategic alliances and key account management as well as project leads in business process engineering, enterprise resource planning and software implementations. His areas of interest are focused on electronic commerce, intercompany collaboration, corporate performance management, supply chain planning and optimization, and knowledge management.

Stanley L. Kroder is an associate professor, Graduate School of Management, University of Dallas and founding director of both the MBA program in telecommunications management and the Center of Distance Learning. He joined the University in 1989 after a 29-year career with IBM. In 1993, he received the Douglass Award for teaching excellence and entrepreneurial spirit for the Graduate School of Management. In 1999, he was chosen as the Michael A. Haggar Fellow, one of the two top honors for faculty of the entire university. Dr. Kroder received a Bachelor of Science in industrial management from MIT's Sloan School of Management and a Master of Science in operations research from Case - Western Reserve University. He holds a PhD in organization theory, business strategy and international management from the School of Management, University of Texas at Dallas.

John H. Nugent serves as an assistant professor in the Graduate School of Management at the University of Dallas, Irving, TX, where he teaches in the telecommunications and entrepreneurship concentrations. Dr. Nugent concurrently serves as CEO of the Hilliard Consulting Group, Inc., a leading strategy consulting firm in the telecommunications and IT industry segments. Previously, Dr. Nugent served as president of a number of AT&T subsidiaries where he won the Defense Electronics "10 Rising Stars" award.

At the time that this chapter was originally published, Mahesh S. Raisinghani served as an assistant professor in the Graduate School of Management at the University of Dallas, teaching MBA courses in information technology and e-commerce. He also served as the director of research for the Center for Applied Information Technology and is the founder and CEO of Raisinghani and Associates International, Inc., a diversified global firm with interests in software & strategy consulting and technology options trading. Dr. Raisinghani was the recipient of the 2001 Faculty Achievement Award for excellence in teaching, research and service. He has published in numerous leading scholarly and practitioner journals and presented at leading world-level scholarly conferences. Dr. Raisinghani was also selected by the National Science Foundation after a nationwide search to serve as a panelist on the Information Technology Research Panel and the E-Commerce Research Panel. He is included in the millennium edition of Who's Who in the World, Who's Who Among America's Teachers *and* Who's Who in Information Technology.

This case was previously published in M. Raisinghani (Ed.), *Cases on Worldwide E-Commerce: Theory in Action*, pp. 1-31, © 2002.

Chapter XXII

Added Value Benefits of Application of Internet Technologies to Subject Delivery

Stephen Burgess, Victoria University, Australia

Paul Darbyshire, Victoria University, Australia

EXECUTIVE SUMMARY

The application of Internet technologies towards distance education is widely discussed in the literature. This case applies Porter's "added value" theory relating to the use of IT to the application of Internet technologies used as a supplement to traditional classroom subject delivery. Most of the reported advantages of this type from online course and subject delivery relate to cost savings in terms of efficiency, flexibility and/ or convenience for the students. The case study examines a range of subjects taught in the School of Information Systems at Victoria University, Australia. Each subject uses Internet technologies for different "added value" benefits. Subject coordinators comment upon the use of the Internet technologies for both academic and administrative aspects. Students are surveyed to determine the value of Internet technologies from their perspective. Student responses indicated the applications were perceived to be at least "useful," and findings supported Porter's theory. The challenge for the faculty is to demonstrate the "business" benefits to faculty staff of adopting Internet technology for teaching. The case studies have shown that the use of Internet technologies by students seems to be higher where the coordinator actively encourages it.

SETTING THE STAGE

The application of Internet technologies towards distance education is widely discussed in the literature, however the overwhelming majority of educators use the Internet to supplement existing modes of delivery. Importantly, the Internet is providing a number of 'added value' supplemental benefits for subjects and courses delivered using this new, hybrid teaching mode.

This case study examines a range of subjects taught in the School of Information Systems at Victoria University, Melbourne, Australia. The case study involv^es the examination of four separate subjects (two undergraduate and two postgraduate) offered by the school. Each subject uses Internet technologies (as a supplement to traditional teaching methods) in a different way, for different "added value" benefits. Subject coordinators comment upon the "added value" provided to them and to the School by the use of the Internet technologies in both the academic and administrative aspects of subject delivery. Students of the subjects are surveyed to determine the value of the application of the Internet technologies from their viewpoint.

Information Technology: Efficiency and Added Value

There are a number of reasons for using IT in organisations today (O'Brien, 1999):

- *For the support of business operations*. This is usually to make the business operation more efficient (by making it faster, cheaper and more accurate). Typical uses of IT in this way are to record customer purchases, track inventories, pay employees and so forth. Most uses of IT in this area are internal to the organisation.
- *For the support of managerial decision making*. To assist with decisions such as whether to add or delete lines of merchandise, expand the business or employ more staff. This is done by the simplification allowing more sophisticated cost-benefit analyses, providing decision support tools and so forth.
- *For the support of strategic advantage*. This final reason for using IT is not as well known as the other two areas (especially in small businesses). It refers to the use of Porter's three generic strategies (low-cost producer, differentiation and niche market provider) as a means of using information technology to improve competitiveness by adding value to products and services. By their very nature such systems need to refer to forces external to the organisations (customers and sometime competitors).

It has been recognised for a number of decades that the use of computers can provide cost savings and improvements in efficiencies in many organisations. Michael Porter (refer to publications such as Porter,1980; Porter & Millar, 1985) has generally been credited with recognising that the capabilities of information technology can extend further to providing organisations with the opportunity to add value to their goods. Value is measured by the amount that buyers are willing to pay for a product or service. Porter and Millar (1985) identify three ways that organisations can add value to their commodities or services (known as generic strategies for improving competitiveness):

- To competitors, but at a lower cost). This strategy allows a business to charge a lower price than competitors and make a larger profit by increasing market share, or charge the same price and make a higher profit per unit sale (Kling & Smith, 1995).
- *Produce a unique or differentiated good* (providing value in a product or service that a competitor cannot provide or match, at least for a period of time). It is hoped that customers consider the goods as being unique, and that they would be prepared to pay a premium price for this added value (Kling and Smith, 1995). If an organisation is the first to introduce a particular feature it may gain a competitive advantage over its rivals for a period of time. When another organisation matches that particular feature, it may not gain a competitive advantage but it will still be adding value, in the consumers' eyes, to its own products. Some ways in which information technology can be used to differentiate between products and/or services are (Sandy & Burgess, 1999):

  - *Quality:* This relates to product or service traits (such as durability) that provide a degree of excellence when compared to the products or services of competitors.
  - *Product Support:* The level of support provided for the product or service. This can include information on how to use the product, procut replacement; return strategies, and so forth.
  - *Time:* This works on the concept that buyers will pay more for a product that is provided/delivered quickly, or will chose a product of similar price and quality if it is available now over a competitor's product that is not currently available.

- *Provide a good that meets the requirements of a specialised market.* With this strategy, an organisation identifies a particular niche market for its product. The advantage of targeting such a market is that there may be less competition than the organisation is currently experiencing in a more general market. The specific characteristics used to target the niche market could be regarded as another type of differentiation.

Summarised, a competitive advantage occurs when a business achieves consistently lower costs than rivals or by differentiating a product or service from competitors (Goett, 1999). The three generic strategies are an integral component of other tools that Porter and Millar (1985) mention to help an organisation to gain a sustainable competitive advantage. The first of these, the value chain, consisted of "the linked set of functions or processes that enable businesses to deliver superior value to customers and therefore achieve superior performance" (MacStravic, 1999, p.15). The second tool, the five competitive forces model, recognises that competitive advantage is often only achieved defending the firm against five competitive forces (the bargaining power of suppliers, the bargaining power of customers, jockeying amongst rivals, the threat of new entrants to the industry and the threat of substitute products or services) or influencing the forces in the organisation's favour (Earl, 1989).

Are Porter's concepts, introduced some two decades ago, still useful today? MacStravic (1999) describes an application of the value chain to health care. Kling and

Smith (1995) use Porter's five competitive forces model to identify strategic groups in the U.S. airline industry.

In relation to Internet applications, Allgood (2001) uses Porter's three generic strategies to suggest how a stock broker offering Internet-based share trading is differentiating its services from a traditional stock broker and how stock brokers will need to use information technology in innovative ways to differentiate their services and maintain market share.

There are a number of critics of Porter's theories. The major concern is that these days it is impossible to achieve a sustainable competitive advantage, the likes of which Porter described in the 1980s. Businesses are more likely to use the theories to seize opportunities for temporary advantage. Whatever the length of the advantage, the basic theories can still be applied (Goett, 1999).

The authors contend that while businesses can still achieve competitive advantage though producing goods at a lower cost, and customers are willing to pay a premium price for differentiated goods that are better quality, delivered faster, provide extra support and so forth, that Porter's three generic strategies are as valid today as they were two decades ago.

Aspects of Course and Subject Delivery

There are two overall aspects to course and subject delivery, the educational and administrative components (Darbyshire & Wenn, 2000). Delivery of the educational component of a subject to students is the primary responsibility of the subject coordinator. This task is the most visible from a student's perspective. The administration tasks associated with a subject form a major component of subject coordination, but these responsibilities are not immediately obvious or visible to the students. However, if the administration tasks are performed poorly or inefficiently, the effects of this become immediately apparent to both subject tutors and students alike, and can be the source of much discontent. It is essential that all aspects of subject administration be carried out as efficiently as possible, so as not to distract the student from their goal, which is to learn.

In both instances, IT solutions can be employed to either fully or partially process some of these tasks. Given the complex and often fluid nature of the education process, it is rare that a fully integrated solution can be found to adequately service both aspects of subject delivery. Most solutions are partial in that key components are targeted by IT solutions to assist the subject coordinator in the process.

Administrative Tasks

There are a number of administrative tasks associated with subject coordination for which IT solutions can be applied in the application. These include (Byrnes & Lo ,1996; Darbyshire & Wenn , 2000):

- Student enrollment
- Assignment distribution, collection and grading
- Grades distribution and reporting
- Informing all students of important notices

Most universities have a student enrollment system administered at the institute level. However, at the subject coordinator level, there are often local tasks associated with enrolment such as user account creation and compilation of mail lists, etc. Some of these tasks can be automated, or in fact partially automated and user driven (Darbyshire & Wenn, 2000). Many academics implement their own solutions to many of these small tasks.

As the written assignment still remains the basic unit of assessment for the vast majority of educators, there have been many initiatives to computerize aspects of this task. Notable initiatives include Submit at Wollongong University, New South Wales, Australia (Hassan, 1991), NetFace[1] at Monash University, Victoria, Australia (Thompson, 1988), ClassNet[2] at Iowa State University, USA (Boysen & Van Gorp, 1997), and TRIX[3] at Athabasca University, Canada (Byrnes & Lo, 1996).

There are now many techniques to distribute student grades and other reports via the Internet. These range from e-mail, to password protected Web-based database lookup. Hames (2000) reports on the use of the Internet in distributing reports and grades to middle school students in Lawrenceville, Georgia.

Notice boards and sophisticated managed discussion systems can be found in many systems (WBT Systems, 1997). For example, products such as TopClass, Learning Space, Virtual-U, WebCT, Web Course in a Box, CourseInfo and First Class all contain such integrated messaging facilities (Landon, 1998).

Educational Tasks

Many of the tasks viewed as "educational" tasks can also employ IT solutions in order to try and gain perceived benefits. Some of these are:

- Online class discussions
- Learning
- Course outline distribution
- Seminar notes distribution
- Answering student queries

Just how many of these are actually implemented will relate to a number of factors, such as the amount of face-to-face contact between lecturers and students. However, using the Internet for many of these can address the traditional problems of students misplacing handouts, and staff running out of available copies.

Discussion management systems are being integrated into many Web-based solutions. These are usually implemented as threaded discussions, which are easily implemented as a series of Web pages. Other tools can include chat rooms or listserv facilities.

Answering student queries can take place in two forums, either as part of a class discussion via a managed discussion list, or privately. Private discussions online are usually best handled via an e-mail facility, or in some instances, store and forward messaging systems may replace e-mail.

Implementing IT solutions to aid in the actual learning process is difficult. These can range from Intelligent Tutoring Systems (Ritter & Koedinger, 1995; Cheikes, 1995), to facilitated online learning (Bedore et al., 1998). However, the major use of IT solutions in the learning process is usually a simple and straightforward use of the Web to present hypertext-based structured material as a supplement to traditional learning.

Using Internet Technologies to Improve Efficiency and Added Value

With the recent explosion in Internet usage, educators have been turning to the Internet in attempts to gain benefits by the introduction of IT into the educational process. The benefits sought from such activity depend on the driving motivation of the IT solution being implemented. This is further complicated by external influences on the various stakeholders. There are three main stakeholders in the education process (we consider only the education process at the university level in this case). These are: the student, the academic and the university itself.

Over the past decade or so, with economic rationalization there is an increasing tendency to view an educational institution such as a university, as a business. This seems abhorrent to some and educationalists resist it, arguing that education cannot be treated as a business lest we lose too many of the intangible results. Yet, external influences seem to be dictating otherwise. This has a definite influence on the roles of the stakeholders, and thus the value of the benefits gained from the introduction of IT technologies into this process. While many may not perceive a university as a business, it is nonetheless possible to match the current uses of the Internet in tertiary education with traditional theory related to the reasons why firms use IT.

Internet technologies in education, which are used for the learning process itself, target the student as the main stakeholder. While the motivation may be the enhancement of the learning process to achieve a higher quality outcome, we can loosely map this to the "support of managerial decision making" concept identified earlier. Such technologies allow educators to obtain a far more sophisticated analysis of individual student's learning progress, and thus provide them with decision support tools on courses of action to take to influence this process.

Technology solutions which target the academic as the stakeholder (Darbyshire & Wenn, 2000, Central Point) implement improvements or efficiencies that can be mapped to the "support of the business operation" previously identified. Improvements or efficiencies gained from such implementations are usually in the form of automated record keeping and faster processing time, ultimately resulting in lower costs in terms of academic time, and added value to the students.

By default, the university also becomes a stakeholder in the implementation of either of the above types of technology enhancements. Benefits gained by students and staff by such uses of technology also translates ultimately to lower costs for the institution or the provision of more and/or better quality information. The benefits of such systems can be mapped onto the "support of strategic advantage" concept (as Porter's low cost and differentiation strategies), previously identified as a reason for using technology in business.

Most of the reported advantages of this type, from online course and subject delivery, relate to cost savings in terms of efficiency, flexibility and/or convenience for the students. These represent the traditional added value benefits of lower cost and faster access to course and subject materials. While there may be some dollar costs savings to the university, this is usually only slight, and can be offset against other resources used in developing or purchasing the technology solutions. Most of the added value lies in the flexibility such solutions offer to both staff and students.

For postgraduate students, or for part-time students who have daytime jobs, online courses or courses supplemented with online material offer them the flexibility to study or obtain this material at a time convenient to them (Bedore, 1998). For the full-time students still studying in a traditional mode, the use of Internet technologies to supplement teaching also provides some convenience and flexibility in obtaining course materials, missed notes, or in contacting the course instructor and other students. There is also added value in terms of flexibility for academic staff in distribution of class material, sending notices and contacting students.

The successful implementation of technologies to provide IT support for both the educational and administrative aspects of subject delivery will clearly benefit the university. If these institutions are to regard themselves as a business, then the successful use of IT in subject delivery could give the university a strategic advantage over other universities which it would regard as its business competitors.

In terms of comparison with Porter's three generic strategies, if universities can lower the costs of their subject delivery then it can provide two advantages — they can offer their courses at a lower price to students or they can use the benefits gained from increased margins in other areas. In terms of differentiation, by being able to offer value added subject delivery as an "improved product," over their competitors, the client base can be expanded. The differentiation can come in various forms, such as flexible delivery times, increased access to resources and so forth.

BACKGROUND: VICTORIA UNIVERSITY

Victoria University is a relatively new university, being formed as recently as 1992 from the merger of two former Institutes of Technology. The recent addition of the Western Melbourne Institute of Technical and Further Education, which was previously involved primarily with industrial training, has made Victoria University into a large dual-sector university catering for a wide range of tertiary students from the central and western suburbs of Melbourne.

As a new and rapidly changing technological institution, Victoria University has been very conscious of a need to continually reevaluate its curriculum and to look for new ways of best providing for the needs of its students and the business community that it serves in a cost-effective manner. The Faculty of Business and Law, of which the School of Information Systems is a part, has pioneered a project, with which it hopes to encourage all faculty members to use Internet technologies to support the delivery of courses (refer to Current Challenges Facing the Organisation, later).

The School of Information Systems has a full range of undergraduate and post graduate courses relating to business computing. The school offers a range of subjects. Some are quite technical in nature while others are much more generally business-related. On average, students tend to mix electives across both ends of this spectrum.

CASE DESCRIPTION

At Victoria University, there are no "fully online" courses at the moment. There is developmental work currently proceeding on such courses, but adoption is slow. This is because there has been no centralized driving force behind the adoption of the Internet as a teaching tool in the faculty until the adoption of Central Point (refer below). There are, however, many good initiatives to supplement traditional on-ground courses with material available on the Internet. This material takes the form of lecture notes, useful links, supplemental material and examples, assignment distribution, managed discussions, assignment submission, notice boards and grades distribution.

Initially, such activity was centered on a few people, but now there are faculty-wide initiatives to get as many academic staff as possible online with supplemental material. A range of subjects within the School of Information Systems were selected to gauge the effectiveness of the initiatives in relation to business practice.

All of the subjects chosen in the case study use the Central Point System (Darbyshire, 1999; Darbyshire & Wenn, 2000) as the vehicle for providing the initiatives. The Central Point system is discussed briefly in a later section.

This case study incorporates two undergraduate subjects (Information Technology Management and Object Oriented Systems) and two postgraduate subjects (Management Information Systems and Building Small Business Systems). Each subject runs for one "semester" of 13 weeks duration (39 contact hours in total). Each subject coordinator has used Internet technologies in a different way to supplement their traditional teaching methods. Each expected a range of benefits to themselves and to students. For each subject, coordinators outline why they have implemented Internet technologies in such a way, and what types of benefits they perceive to have occurred from the implementation. When presented with the major reasons why business adopted IT, all agreed that their motivation was the equivalent of supporting their day-to-day business operations. Finally, students from each subject have been surveyed to determine if the 'added value' benefits from their viewpoint match the expectations of their subject coordinators. At the conclusion of each of the subjects chosen for this case study, students were asked to complete a standard survey form consisting of the following nine questions.

1.   Have you accessed details about your assignments from the subject Web site?
2.   Have you accessed your subject outline from the subject Web site?
3.   Have you followed an external link via your subject Web site?
4.   Have you used the Web site in class?
5.   Have you accessed the Web site outside of class?
6.   Have you contacted your lecturer by e-mail during the semester?
7.   Have you contacted your tutor by e-mail during the semester?
8.   Have you submitted an assignment by e-mail during the semester?
9.   Have you submitted an assignment online via the subject Web site during the semester?

For each question, the student could respond either "Yes," "No" or "N/A" if the feature in question was not applicable to the subject. If the student responded with a "Yes", then they were further asked to respond in numeric terms approximately how many times they used the feature, and then to gauge its effectiveness on a Likert-type scale of five ratings, using the choices "Very Useful," "Useful," "OK," "Not Useful" and "Useless." Provision was also made for the student to enter their subject code. These responses were converted to an "effectiveness" ratio by arbitrarily providing a score 4,

3, 2, 1 and 0 for the range of responses "Very Useful" to "Useless" and comparing it against the maximum possible score if every response was "Very Useful." To provide an example, three responses of "Very Useful," "Useful" and "OK" would return an effectiveness ratio of 75% (determined by [4+3+2=9] divided by [3 x 4=12]).

The survey was placed online at the following address–http://busfa.vu.edu.au/dbtest/page2.html, and the students were asked to complete this survey within a few days after the completion of the subject. Responses were hence fully anonymous, and upon completion the data from each survey was automatically e-mailed to the authors. The results of the surveys for the individual subjects in the case study are shown in the case study results. In some cases, if a particular feature was not used within a subject, the students' responses to that question would be N.A. These responses are not shown in the individual survey results.

Each of the four subjects involved in the case study and the survey results are now discussed.

Information Technology Management

Subject Description

This subject typically runs over two to four campuses at any one time, the number of enrolments ranging between 150 and 350 per semester. It is a first-year subject for students studying the Bachelor of Business (Information Systems).

This subject provides an introduction to information management and management of information technology. This is achieved by introducing concepts relating to:

- The changing nature of information and information technology (IT).
- Information Management in the Internet age.
- Managing the information technology resource and projects within the organisation (emphasising the smaller- to medium-sized business).

The main subject topics are:

- Information and Management. Organisational information and its interaction with technology. Technology developments.
- Information Management in Internet-worked Enterprises. Formulating information technology strategy and its linkage with business strategy.
- Application of models to link information technology and corporate strategy. Establishing information technology projects.
- Managing technological change and applications of information technology (small business context).

Subject Coordinator View

Figure 1 represents the homepage for the subject, Information Technology Management, for Semester 2, 2000.

In such a large subject, the major benefits from such a Web site are economies of scale. Lectures are conducted on multiple campuses at different times in large lecture theatres, so there is little flexibility available in subject delivery.

Figure 1. Information Technology Management homepage

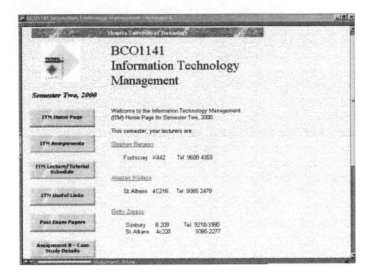

The main features of the Web site are:

• Provision of assignment details, past examination papers and a schedule of class times. One printed copy of each of these is handed out to students at the start of the semester. The idea of putting the information on the Internet is to reduce the need to reprint a copy when a student misplaces it, thus saving time and money for the institution.

• Links to extra information about the subject on the Internet. It was felt that some students may like to access extra materials to support their study as a means of improving their understanding of subject material.

• Lecturer contact details. Many lecturers travel between campuses and are not always directly accessible by telephone. This feature is placed on the Web site to encourage students to contact lecturers by e-mail, which is location independent.

Educational benefits expected were mainly in the area of flexibility. Administrative benefits expected were flexibility and cost savings (in terms of time), particularly in the area of assignment submission and collection, and distribution of material.

Student View

Student responses are tabulated in Table 1. In all, there were 55 responses to the survey from students at three campuses.

Summary

The features used by students generally received an effectiveness rating of Useful or Very Useful. Most of the students (more than 9 out of 10, on average) accessed the

Table 1. Student survey results: Information Technology Management

Internet Feature	Use	Effectiveness Ratio
Accessed assignment details	87%	82%
Accessed subject outline	75%	84%
Used external links	75%	74%
Accessed Web site in class	96%	84%
Accessed Web site out of class	93%	88%
Contacted lecturer by email	28%	85%
Contacted tutor by email	42%	91%
Submitted assignment by email	24%	87%

Web site during class (in tutorials) and outside of class, to perform information gathering (accessing assignment details and the subject outline) and to examine external links to gain extra information.

In such a large subject, it was always more likely that the students' first contact point would be with the tutor rather than the lecturer, with whom they would have a more personal relationship.

Interestingly, one in four respondents submitted an assignment during the semester. This was not instigated as a formal method of subject submission, but was allowed by tutors if the student requested it.

The use of the Web site by students "out of class" was very encouraging, as was the number of students that accessed external links. There were definite benefits by students and staff from the information provision services provided, as expected. The contact with tutors and lecturers by e-mail was also encouraging as it was not extensively promoted in class. Finally, the level of submission of assignments by e-mail was unexpectedly high, as it was not formally announced at all!

The challenge in the subject was in the encouragement of the adoption by the students of some of the features of the software. It seemed that while many were happy to read material placed on the Internet, most were reluctant to use the submission features of the software. When questioned, many said that they simply didn't trust the software to deliver their material to the academic coordinator. It seemed that the students still preferred to physically hand their submitted work to the coordinator, as they weren't sure that the coordinator actually received it when it was submitted electronically. This seems to be a challenge to the software designers to provide functionality that addresses these concerns, such as a facility for a student to see their submitted work online.

Object Oriented Systems

Subject Description

This subject runs over two campuses each semester, and the total number of enrollments ranges between 45 to 75 students per semester. This is a final-year subject for students studying the Bachelor of Business (Information Systems).

This subject provides an introduction to object oriented programming and design using the Java programming language. The main concepts covered during the course of this subject are:

- The nature of objects
- Object oriented analysis and design
- Object oriented programming

The main topics covered during this subject are:

- Objects,
- Object oriented design using UML
- Encapsulation
- Introduction to Java
- Java objects
- Java event programming
- Using Java GUI objects
- The AWT, using JDBC to interface Java to a database

Subject Coordinator View

The homepage for this subject is shown in Figure 2. This is not an overly large subject, but is taught across two campus locations and does contain some post-graduate and part-time students. The main point in supplementing the course delivery with Internet functionality in this subject was to gain flexibility for academic staff and students. The main features of the Web site include:

- *Detailed lecture notes and supplementary lecture material.* It was hoped that by including detailed notes here the students could print this out, and use the lecture to concentrate on the material rather than just a note taking exercise.

Figure 2. Object Oriented Systems homepage

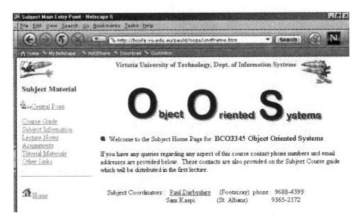

- *Provision of administration material, such as assignments, course guides and notice boards.* By placing such material here, timely notices can be sent to students, and assignments distributed at anytime throughout the week. Also, material such as course guides can be easily obtained by students at any time, and academic staff do not have to act in a pseudo-secretarial role by continually passing out such material.
- *Assignment submission and collection.* Using this Web site, students can submit their assignments either through a Web page, or via e-mail as an attachment. Thus, flexibility is introduced for all students, and particularly part-time students that don't want to make a special trip in just to drop off an assignment. Also, there is flexibility of staff in collecting the assignments as this can be done anytime online.

No real educational benefits were expected in this case, apart from flexibility of access to material placed on the Web. For the administrative benefits, it was hoped to gain flexibility for students and staff and there was also an expectation of some time savings to be gained in the administration side. For instance, the provision of administration material online and the assignment submission, offers a degree of automation that requires little or no academic coordination. Assignments would be time and date stamped so there was no dispute as to submission times.

Student View

For this subject, there were 22 survey submissions. The results are displayed in Table 2.

Summary

From the survey of this component of the case study, there was a very high percentage use of the Web site's features. However, all the features of the Web site were actively promoted by the subject coordinator. The lowest feature use was that of assignment submission through the Web site. It seemed that most students were more comfortable in submitting assignments as e-mail attachments, even though there was no immediate conformation of submission as there was with the Web site submission. There was also a very high user satisfaction of the features as evidenced by the effectiveness

Table 2. Survey results: Object Oriented Systems

Internet Feature	Use	Effectiveness Ratio
Accessed assignment details	100%	91%
Accessed subject outline	68%	90%
Used external links	73%	91%
Accessed Web site in class	73%	92%
Accessed Web site out of class	86%	95%
Contacted lecturer by email	95%	90%
Contacted tutor by email	95%	93%
Submitted assignment by email	95%	96%
Submitted assignment by Web site	59%	90%

ratio percentage (nothing under 90%). The highest feature use was in the area of assignment submission and contact with the subject lecturer.

The expected time savings did not eventuate. In fact, as a result of this case study we found that quite the reverse was true. With the Web site, students had much flexibility in contacting their tutor or lecturer. This generated many e-mails, particularly in relation to assignment problems when submission time was close. Many of these queries would not have been generated without electronic access to the tutor. In that case, due to the delay, the student would either get help elsewhere or save some questions for a face-to-face visit. With Internet-based access the students could quickly generate an e-mail for the tutor as the problems arose. This increased the workload to the tutors significantly.

The main challenge in this subject was the realization of some of the added value benefits in terms of time savings for the academic coordinator. There was certainly added value achieved from the students' perspective, in terms of cost, time and flexibility. However, the increased access to the coordinator by the student nullified any time savings benefits gained. Such problems could most likely be overcome with a more disciplined approach to the use of the technology

Management Information Systems

Subject Description

This subject typically runs over one or two campuses at any one time, the number of enrolments ranging between 100 and 200 per semester. It is a first-year subject for students studying the Master of Business Administration and an elective often chosen by other Master of Business students.

This subject aims to introduce students to a broad range of topics relating to the field of information systems, to give an appreciation of the advantages computers confer in the management of information and to provide an introduction to how information systems are built. Theoretical issues are reinforced through laboratory work that leads to the design and implementation of small systems. Upon completion of the subject, students will have a management perspective of the task of building and maintaining information systems applicable to any organisation.

The subject covers a selection of topics relating to: the effective management and use of technology, the concept of information and how it can be managed, how information technology can be used to assist in managing information and elements of systems development. Specific topics covered are:

- Management, information and systems
- Problem solving and decision making
- Process modelling, function charts, data-flow diagrams (DFD)
- Databases and data modelling (ER diagrams)
- Types of information systems
- Data access: building database front ends
- Project management
- Innovation and the management of technological change
- Use of IT in small business
- Data communications

Figure 3. Management Information Systems homepage

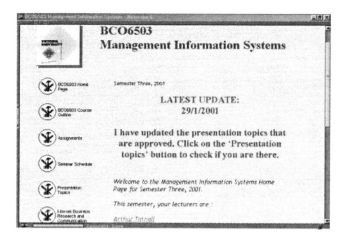

- Electronic commerce
- Strategic applications of IT

Subject Coordinator View

Figure 3 represents the homepage for the subject, Management Information Systems, for Summer Semester 2000-2001 (southern hemisphere).

This subject is also a relatively large subject, so the major benefits from the Web site are similar as those for Information Technology Management. The subject has typically been conducted in three hour seminars combining tutorial rooms and computer laboratories, so there is some flexibility available in subject delivery.

The main features of the Web site are:

- *Provision of assignment details, subject information, and a schedule of class and presentation times.* As with Information Technology Management, one printed copy of each of these is handed out to students at the start of the semester. The result is to reduce the need to reprint a copy when a student misplaces it, thus saving time and money for the institution.
- *Links to extra information about the subject on the Internet* (out of screen in Figure 3). For students that wish to access extra materials to support their study.
- *Lecturer contact details.* To encourage e-mail contact with lecturers.

Web Discussion List. As a means of introduction to the subject, the students' first assignment is to read two or more articles on a related computing topic and to answer a discussion question posted by the lecturer on a Web-based bulletin board. After this they are to read the contributions of other students and comment upon those. Students are broken up into groups of about four to six students for the assignment. Refer to Figure 4 and Figure 5 for typical screens from the Web Discussion List. The use of this feature has the advantage that students can read the articles and submit their opinions at any

Figure 4. Sample messages posted to the Web Discussion List

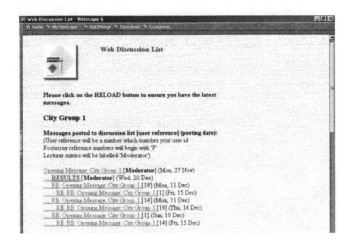

Figure 5. A sample message and reply screen for the Web Discussion List

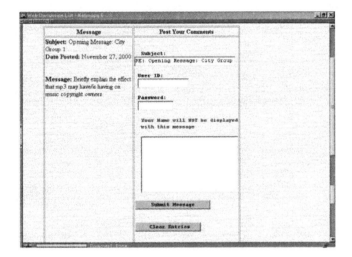

time, provided that they have access to the Internet. For the lecturer it provides a more structured form of dialogue to assess than the "live" version that could be conducted in class.

• Another use of the Web site is to use it for live demonstrations during lectures. The lectures for the topics of Internet Business Research and Communication and Electronic Commerce are conducted in computer laboratories, where students are encouraged to browse "live" to various sites as they are being used as examples. This provides the benefits of students seeing actual examples of what is being

Figure 6. Electronic Commerce "lecture" Web page for Management Information Systems

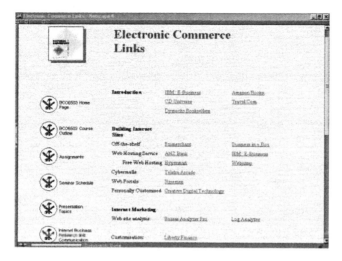

Table 3. Survey results: Management Information Systems

Internet Feature	Use	Effectiveness Ratio
Accessed assignment details	88%	68%
Accessed subject outline	75%	79%
Used external links	50%	81%
Accessed Web site in class	88%	86%
Accessed Web site out of class	75%	88%
Contacted lecturer by email	63%	90%
Submitted assignment by email	38%	92%

discussed, with the option of revisiting the sites at a later date if they require further reinforcement. The sample lecture screen for Electronic Commerce is shown in Figure 6.

There was one main educational benefit expected, and this was related to the stimulation of discussion expected from the use of the messaging system. It was hoped that some synergy would develop from emerging discussions, and further discussion would ensue. Flexibility in access to material was also an expected advantage. Some administrative benefits were expected, and it was hoped that some time savings would result as a consequence of the students using the technology for assignment submission.

Student View

For this subject, there were 24 survey submissions. The results are displayed in Table 3.

Summary

It is interesting that the effectiveness ratio for accessing assignment details was the lowest for all of the features over all of the subjects in this case study. This could be because the assignments were quite detailed, and only a summarised version of them was made available on the Web site. The fact that 88% of respondents accessed the Web site in class showed that some of them must have contributed to the assignment discussion list out of class. For a master's-level subject, the use of external links was disappointing, especially as some of these were to online databases of journals that could be used to gather material for the research paper. Most of the students were quite familiar with the use of e-mail, and many of them contacted lecturers throughout the semester.

The challenge in this subject was in getting the students to make use of the Internet features available. Many choose to not use the technology, and the effectiveness ration determination for this subject was low. More active encouragement was needed and in fact is planned for future semesters.

Building Small Business Systems

Subject Description

This subject runs on one campus at any one time, the number of enrollments ranging between 20 and 30 per semester. It is an elective subject for students studying the Master of Business (Information Systems) and the Master of Business Administration, as well as being an elective chosen by other Master of Business students.

This subject introduces the student to a broad range of topics relating to the field of information technology and small business. Topics covered include:

- Determining small business IT needs
- Selecting applications for small business: business processes
- Selecting hardware and operating systems for small business
- Networking for small business
- Building small business applications
- Office suite programming
- Sharing data with other applications
- Use of automated input devices
- Calling other office suite applications
- Automating applications across packages

Subject Coordinator View

Figure 7 represents the homepage for the subject, Building Small Business Systems, for Summer Semester 2000-2001 (southern hemisphere).

This subject is also a relatively small subject, and is typically conducted in three hour seminars combining tutorial rooms and computer laboratories, so there is some flexibility available in subject delivery.

The main features of the Web site are:

Figure 7. Homepage for Building Small Business Systems

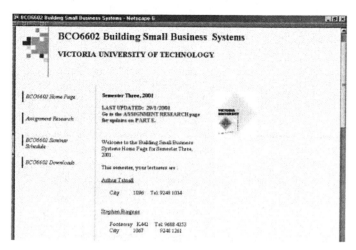

- *Links.* To small businesses research sources on the Internet for students to assist in the preparation of their research assignments.
- *Downloads.* There are a number of sample files used by students in the subject. This facility provides a convenient way for students to access these files as needed, in class or at home.
- *Lecturer contact details.* To encourage e-mail contact with lecturers.

No real educational benefits were anticipated. However flexibility was cited as an advantage. Some time savings was anticipated, and flexibility for the subject coordinator in distributing material and collecting assignments was cited as the main administrative benefits expected.

Student View

For this subject, there were only 12 survey submissions. The results are displayed in Table 4.

Table 4. Survey results: Building Small Business Systems

Internet Feature	Use	Effectiveness Ratio
Accessed assignment details	83%	92%
Accessed subject outline	75%	83%
Used external links	83%	75%
Accessed Web site in class	100%	88%
Accessed Web site out of class	100%	69%
Contacted lecturer by email	100%	100%
Submitted assignment by email	75%	100%

Summary

The use of Internet technologies was strongly encouraged in this subject. This is reflected in the levels of usage shown in Table 4. All students used the Web site in and out of class, and all contacted the lecturer at some stage in the semester. Three out of four students submitted assignments by e-mail. In the other subjects, there was little difference between the effectiveness ratios for accessing the Web site in and out of class. The results for this subject show that accessing the Web site outside of class was less effective. The reasons for this are unclear and need to be explored further.

The major challenge in this subject has been traditionally to get students to make use of the technology. This subject is part of the masters by coursework degree, and many of the students are full-time overseas students. There has always been a reluctance to use the technology, mostly based on a lack of familiarity, and to encourage this, use of components of the technology were made compulsory — compulsory in the sense that part of a students score was allocated to their use of some Internet features. This form of active encouragement seemed to give the students confidence in using the Internet for the non-compulsory features.

Case Study Summary

The findings from both the "pre" and "post" implementation views of the subject coordinators were in accordance with their expectations and had delivered the "value added" as expected. Student responses indicated that they had used the "compulsory" aspects of the Internet technology applications extensively, and the optional aspects to varying degrees. All of the applications were seen to be either "very useful" or "useful" by the students.

It seems apparent that Internet technologies can be used to supplement traditional subject delivery in a number of ways, leading to a number of "business" type benefits for both institution and students. The subjects in the case study indicate that the use of the Internet technologies by students seems to be higher where it is openly encouraged by the coordinator and where the students are already familiar with the use of the technology.

CURRENT CHALLENGES FACING THE ORGANISATION: THE FACULTY OF BUSINESS AND LAW AT VICTORIA UNIVERSITY AND THE CENTRAL POINT PROJECT

The previous section indicated some of the benefits available from the use of Internet technologies to assist with subject delivery. There are two major problems that arise when considering the adoption of these technologies by Victoria University. The first problem has been that developments have been ad hoc. That is, they have occurred on the basis of individual interest by faculty members, not a coordinated approach. As such, there was never a standard interface to be followed nor templates to be used. Work was continually being reinvented. Formatting of information (such as course outlines) occurred in a number of different ways. There were faculty members that did not even know the technology was available, or did not know how to use it if they did know it was available.

The Central Point project began as a series of Web pages in 1996 to help two staff members coordinate the distribution of lecture material for subjects taught over multiple campuses. Since then, Central Point has evolved into an interface for subject Web sites (Darbyshire & Wenn, 2000). Central Point is the tool used to deliver the Internet functionality in all the subjects in this case study.

The original design of the Central Point site was to "alleviate pressure on the subject administrator." Central Point is now a series of Web pages and databases designed to provide basic administration functionality to academic subject sites that "hook into" it. This interface uses Cold Fusion scripts and Microsoft Access databases to provide this functionality.

Central Point was not designed to impose any type of format or structure on subject sites. Quite the opposite in fact, as academics generally tend to oppose the imposition of any such structure. Instead the academics are free to develop their own Web sites for their subjects, and then "hook into" the Central Point site. This is done by a Central Point administrator making a subject entry into one of the Central Point databases. The subject's Web site is then displayed as a "clickable" link on the Central Point homepage (see Figure 8), and the subject coordinator is then automatically set up with a Subject Notice board, Web-based assignment box for assignment submissions and access (Darbyshire, 1999), facility for recording subject grades on the Web, and the ability to manage subject teams and diversion of assignment submissions to tutors.

These facilities can be accessed by staff and students from the Central Point home page. Thus the students do not have to remember multiple home pages for different subjects, but just one central link. Hence the name: Central Point.

Over time, the Faculty of Business and Law saw the development of Central Point as an opportunity to coordinate the efforts of faculty staff in their use of Internet technologies to assist with subject delivery.

In Semester 2 of 2000, the faculty adopted Central Point as the standard Web tool to tie together all the subject Web sites being developed, and to provide the basic functionality mentioned above. The homepage of the Central Point System is shown in Figure 8, and is still being developed by the faculty to increase the functionality provided to subject Web sites, and to personalize the site for students logging into the system. The homepage is at http://www.business.vu.edu.au/cpoint/index.html.

There is also current research underway to increase functionality and hence achieve accelerated benefits as discussed by Parker and Benson (1988) by the application of agent technology (Darbyshire & Lowry, 2000).

The challenge for the Faculty of Business and Law is to demonstrate the "business" benefits to faculty staff of adopting Central Point as their gateway to using the Internet to support their own subject delivery. The case studies have shown that if the subject coordinator actively encourages the use of the technology, then the students will use it.

In addition to this, the use of Central Point does provide a single-entry point for students to access the Web sites of subjects within the faculty. Standard information, such as subject outlines, is provided in standard formats. Training courses in MS FrontPage are being run for faculty members that are not used to set up Web pages.

Figure 8. Central Point homepage

Once the adoption of Internet technology is realized for the majority of academic staff in the Faculty of Business, further challenges will soon present themselves. At this time we should begin to see requests for further development based on desire for accelerated benefits that occurs after the initial adoption of technology (Parker & Benson, 1988). If further initiatives are not explored at this time, the Faculty could miss an opportunity to capitalize on any momentum towards Internet technology acceptance. The adoption of further complex technology to assist in subject delivery will need to be handled carefully. Support mechanisms will need to be put in place to assist both faculty and students in the event of difficulty. This is an area often overlooked when large-scale adoption of technology is considered.

Currently, both governmental and Faculty polity dictate the duration and structure of the courses to some degree. As the move towards the adoption of Internet technology progresses, many benefits will be perceived in moves towards complete online delivery. Courses taught in this mode do not necessarily fair well when the same rigid structures are imposed on them as on-ground courses. At this point, major challenges will be faced in the flexibility of the faculty in accommodating any such moves.

REFERENCES

Akerlind, A., & Trevitt, C. (1995). Enhancing learning through technology: When students resist the change. *Proceedings ASCILITE'95*. Retrieved from http://www.ascilite.org.au/conferences/melbourne95/smtu/papers/akerlind.pdf

Alexander, S. (1995). Teaching and learning on the World Wide Web. *Proceedings of AusWeb'95*. Retrieved May 22, 1999, from http://www.scu.edu.au/sponsored/ausweb/ausweb95/papers/education2/alexander/

Allgood, B. (2001). Internet-based share dealings in the new global marketplace. *Journal of Global Information Management, 9*(1), 11.

Bedore, G. L., Bedore, M. R., & Bedore, G. L., Jr. (1998). *Online education: The future is now*. Socrates Distance Learning Technologies Group, Academic Research and Technologies.

Boysen, P., & Van Gorp, M. J. (1997). *ClassNet: Automated support of Web classes*. Paper presented at the 25th ACM SIGUCCS Conference for University and College Computing Services, Monterey, California.

Byrnes, R., & Lo, B. (1996). *A computer-aided assignment management system: Improving the teaching-learning feedback cycle*. Retrieved May 12, 1999, from http://www.opennet.net.au/cmluga/byrnesw2.htm

Cheikes, B. A. (1995). GIA: An agent-based architecture for intelligent tutoring systems. *Proceedings of the CIKM'95 Workshop on Intelligent Information Agents*.

Darbyshire, P. (1999). Distributed Web based assignment submission and access. *Proceedings-International Resource Management Association, IRMA '99*, Hershey, USA.

Darbyshire, P., & Lowry, G. (2000). An overview of agent technology and its application to subject management. *Proceedings International Resource Management Association, IRMA '2000*, Alaska.

Darbyshire, P., & Wenn, A. (2000). A matter of necessity: Implementing Web-based subject administration. In *Managing Web enabled technologies in organizations*. Hershey: Idea Group Publishing.

Earl, M. J. (1989). *Management strategies for information technology*. Cambridge: Prentice Hall.

Goett, P. (1999). Michael E. Porter (b. 1947): A man with a competitive advantage. *The Journal of Business Strategy, 20*(5), 40-41.

Hames, R. (2000). *Integrating the Internet into the Business Education Program*. Alton C. Crews Middle School, CS Dept. Retrieved January 26, 2001, from http://www.crews.org/media_tech/compsci/pospaper.htm

Hassan, H. (1991). *The paperless classroom*. Paper presented at ASCILITE '91, University of Tasmania, Launceston, Australia.

Kling, J. A., & Smith, K. A. (1995). Identifying strategic groups in the U.S. airline industry: An application of the Porter model. *Transportation Journal, 35*(2), 26.

Landon, B. (1998, April 10). *On-line educational delivery applications: A Web tool for comparative analysis*. Canada: Centre for Curriculum, Transfer and Technology. Retrieved October 10, 1998, from http://www.ctt.bc.ca/landonline/

MacStravic, S. (1999). The value marketing chain in health care. *Marketing Health Services, 19*(1), 14-19.

McNaught, C. (1995). Overcoming fear of the unknown! Staff development issues for promoting the use of computers in learning in tertiary education. *Proceedings Ascilite '95*. Retrieved from http://www.ascilite.org.au/conferences/melbourne95/smtu/papers/mcnaught.pdf

O'Brien, J. A. (1999). *Management information systems, Managing information technology in the Internetworked enterprise* (4th ed.). Irwin McGraw Hill.

Parker, M. M., & Benson, R. J. (1988). *Information economics: Linking business performance to information technology*. Prentice-Hall.

Porter, M. E. (1980). *Competitive strategy*. NY: Free Press.

Porter, M. E., & Millar, V. E. (1985). How information gives you competitive advantage. *Harvard Business Review, 63*(4), 149-160.

Ritter, S., & Koedinger, K. R. (1995). *Towards lightweight tutoring agents*. Paper presented at the AI-ED 95-World Conference on Artificial Intelligence in Education, Washington, DC.

Sandy, G., & Burgess, S. (1999, November). Adding value to consumer goods via marketing channels through the use of the Internet. *CollECTeR'99: 3rd Annual CollECTeR Conference on Electronic Commerce,* Wellington, New Zealand.

Thompson, D. (1988). *WebFace overview and history*. Monash University. Retrieved January 2, 1999, from http://mugca.cc.monash.edu.au/~webface/history.html

WBT Systems. (1997). *Guided learning — Using the TopClass Server as an effective Web-based training system* (WBT Systems White Paper). Retrieved from http://www.wbtsystems.com

ENDNOTES

[1] http://www.educationau.edu.au/archives/olcs/case25.htm

[2] http://classnet.cc.iastate.edu/

[3] http://oc1.itim-cj.ro/~jalobean/CMC/resource9.html

Stephen Burgess is a senior lecturer in the Department of Information Systems at Victoria University, Melbourne, Australia. Stephen lectures widely to undergraduate IS and master's degree students, teaching subjects such as Information Technology Management, Building Small Business Systems and Information Management. His research interests include the Internet for use in small business, as well as small business use of technology and the application of Internet technologies. His current research project is in the provision of an automated tool to help small businesses decide on a Web based implementation strategy.

Paul Darbyshire is a lecturer in the Department of Information Systems at Victoria University, Melbourne Australia. He lecturers in object oriented systems, C and Java programming, and has research interests in the application of Java and Web technologies to the support of teaching. His current research is into the use of the Web for university subject management and the use of AI techniques for the development of second-generation Web-based teaching support software.

This case was previously published in the *Annals of Cases on Information Technology*, Volume 4/2002, pp. 390-409, © 2002.

Chapter XXIII

Design and Implementation of a Wide Area Network:
Technological and Managerial Issues

Rohit Rampal, Portland State University, USA

EXECUTIVE SUMMARY

This case deals with the experience of a school district in the design and implementation of a wide area network. The problems faced by the school district that made the WAN a necessity are enumerated. The choice of hardware and the software is explained within the context of the needs of the school district. Finally the benefits accruing to the school district are identified, and the cost of the overall system is determined.

BACKGROUND

The organization is a reasonably small school district with about 2,700 students, in a town named X that is comprised of four villages (A, B, C and D). There are five schools in the district, the Board of Education that oversees those schools, and the Bus Garage. The buildings that house these seven entities are spread over the four villages, and the distance between locations is more than ten miles in some cases. While the school district is not exceedingly wealthy, it does have access to state and federal grants for bringing schools online. What isn't covered by the grants has been funded out of the municipal budget, with virtually no opposition from the Board of Finance, setting a stage whereby the town maintains an aggressive approach to deploying technology in its schools.

SETTING THE STAGE

The Town of X needed to connect the computers and systems at the seven locations in order to share information and resources. The district is spread out over the four villages that make up the town (A. B, C and D) and its seven locations within that area are separated by as much as ten miles. The faculty, staff and students within the school district needed the ability to communicate with each other and with the rest of the world via electronic mail. Accessing information via the World Wide Web was needed to keep up with the changing world, both for the faculty and the students. They needed to share files without printing and distributing hard copies. There was a distinct need for getting access to the Internet into the classroom. The schools in the school district had computer labs, and computer-based education was a priority, but the computers could not connect to the Internet. The students, teachers and administrators in the school district needed to run collaboration applications like groupware and various administrative applications that required connectivity to a central database. Given the geographical distances involved, a wide area network seemed to be a good solution, that would take care of most of the needs of the school district.

An example of this need for connectivity and bandwidth is the Phoenix database, a school administration package used in the school district for town X and many other districts. Phoenix is a large and complex database written for Microsoft Access 2.0. It is clunky and painful to deploy in many respects, and it does not run on the newer versions of Access. So Access 2.0 must be installed for the database to run, but package serves its purpose and is popular. Phoenix has modules for maintaining student records (attendance, grades, discipline, etc.), class and bus schedules, payroll for faculty and staff, cost accounting and a number of other functions needed to run a school district. The ability to run this software product across a network is indispensable in this environment, as it allows each school to handle its own administrative tasks independently. In order to enable the proper running of the Phoenix software over the network, a good bandwidth is a requirement.

When the project was envisioned, direct fiber over the distances involved was not a viable option (the distances between two consecutive schools could run to more than 10 miles). A wide area network, using some form of leased transmission lines was considered the most efficient way to connect the school district and create a cohesive enterprise. Such an enterprise could communicate efficiently and act in concert, providing connectivity and services to the faculty, staff, students and even the citizens of the Town X school district.

CASE DESCRIPTION

A Look at the Network

The Town X wide area network is described from two perspectives. The first perspective gives an overview of the network and the services that it delivers. This view primarily describes the software. The second perspective explains how these services are delivered. This perspective provides a discussion of the hardware and the network protocols used in the wide area network deployed.

Network Services

Just as a desktop computer requires an operating system (OS) to run, a network needs a network operating system (NOS). For the town X school district, the NOS chosen was Microsoft's Windows NT Server 4.0. There were a number of reasons for choosing Windows NT 4.0, but the overriding reason was the fact that the network administrator was familiar with the software. The NOS resides on a single computer known as a Primary Domain Controller (PDC). The school district network is configured as a single NT domain network — that is, a user needs to log on only once to the network in order to reach all resources he/she is allowed access to. The NT Server OS can also be configured to run as a stand-alone server to host other services on the network. There may be a number of such servers in a domain. In the case of the Town X school district WAS, there are two other NT servers that serve as hosts for additional services.

The Role of Domain Controller(s)

While the PDC, of which there can be only one, hosts the Security Accounts Management (SAM) Database, it is the Backup Domain Controllers (BDC) who maintain a copy of this SAM Database and actually perform the authentication of logins to the Domain. So when a user turns on a client machine and enters a user name and password, the "request" is directed not to the PDC but to a BDC, provided there is BDC present. For the Town X school district, each location has a BDC. At login the user name is compared against the SAM Database, and those privileges and accesses that were assigned to that user's group or groups are granted to that user. (A user may belong to one or more groups). Some of the pre-defined groups in the NT domain include "domain users," "administrators," "print operators," or any of a wide variety of others. In a LAN running Windows NT, normally access is granted by group membership and not on an individual basis. This practice makes management of the domain a lot easier. In addition a user may be assigned a home directory on the server, access to which is normally restricted to the user only.

The process of creating accounts in the NT domain occurs on the PDC and then is copied to the BDCs by replication. The BDC also mirrors most of the other functions of the PDC, so that in the event the PDC fails, a BDC must be "promoted" to PDC status until the primary machine is fixed or replaced. This redundancy enables fault tolerance and is an essential part of smooth network operation.

Internet and Proxy Service

At the Town X High School, the BDC also hosts two other critical packages, namely Microsoft's Internet Information Server (IIS) and the Proxy Server. This IIS connects both to the Internet and to the internal network. It also hosts the Web site for the Town X school district. In order to limit the number of IP addresses needed, the network in the Town X school system uses private IP addresses. What that means is that these internal IP numbers cannot be routed to the Internet. So the IIS server needs to have two network cards, one with a private address to connect to the internal network, and one with a public address to connect to the Internet. In case the IIS server had only the public IP address, the way for the computers on the internal network to reach the Internet server would have been to go out to the Internet via the proxy server and reach it via the public IP address.

Proxy service allows the administration and monitoring of Internet usage as well as logging of inbound and outbound traffic. Also, more importantly the Proxy server allows the filtering out of objectionable materials from the office or academic environment. The school district was so concerned with this issue of restricting access to objectionable material that they augmented the filtering function by the addition of a third-party product called Web Sense. Web Sense is a subscriber service that downloads, on a daily basis, an updated list of URLs from which the objectionable materials issue. Web Sense then works closely with Proxy Server to block access to these URLs.

Once the Proxy Server has been installed on the network, the client machines are configured to "point" at it for their access to the Internet. The server caches a configurable amount of content (100 MB-200 MB is common) as it is accessed from the Web and compares its cached material to the remote material periodically, keeping it updated. This way the frequently accessed material is kept readily available locally, speeding up Internet access for the user, and freeing up bandwidth for other more desirable uses.

DHCP and WINS

The Town X school district network uses DHCP and WINS to locate computers on the network. Dynamic Host Configuration Protocol (DHCP) is a dynamic TCP/IP addressing method that assigns each machine a TCP/IP address when it gets on to the network. Windows Internet Name Server (WINS) provides a NetBIOS host name address resolution for WINS clients. WINS increases network efficiency by reducing the number of network broadcasts needed to resolve a NetBIOS host name to an IP address. A NetBIOS name is a unique computer name that is entered in the identification tab in the network control panel of a computer. This is the name of the computer that would show up in the Network Neighborhood. Names are generally mnemonic, and so identify either the primary user of that computer (smith, librarian) or the location of the computer (Library12, Lab3), etc. On a TCP/IP network, however, the DHCP server assigns an IP address (which is a number) to the computers. A WINS server maintains a database that is updated periodically which matches the NetBIOS name with the IP address for each computer on the network.

Mail Services

The e-mail service in the Town X domain is provided by a dual-server MS Exchange 5.5 approach, while the e-mail client software used is MS Outlook 98, a 32-bit Windows application.

The two-server configuration was chosen to enable greater security by separating student and administrative accounts while providing fault-tolerance though information store replication. So in case one e-mail server goes down, the second can take over till such time the first server comes back up.

The protocols standardized for the students, teachers and administrators to send and receive electronic mail are SMTP and POP3 respectively. SMTP is the de-facto standard for sending mail, while POP3 was chosen as the e-mail client protocol as it limits the amount of incoming mail that stays on the e-mail server. One of the reasons for choosing MS Exchange as the e-mail server was the ease of use and management tools available. As with other Microsoft products (like MS Proxy Server) which are part of

Microsoft's Back Office suite of products, MS Exchange provides a number of tools for tracking usage and monitoring performance. Since e-mail tends to become indispensable in an enterprise, keeping it online requires proper allocation of resources, and monitoring is critical to this process.

Server and Desktop Management

The network management software chosen for the Town X school district was the MS Systems Management Server (SMS), yet another member of the Back Office family. The software is powerful enough that with the client portion of the software installed on the workstations, the network administrators can remotely control and manage desktops and servers as well as troubleshoot. SMS can be used to inventory and collect data, distribute software and upgrades, and monitor server and network traffic, all from remote management stations.

Desktop Configuration

A conscious decision was made to configure each desktop in the domain in a standard fashion to present a consistent look and feel throughout. A standard computer has the MS Office Professional suite, a World Wide Web browser and MS Outlook 98 for e-mail and group scheduling tools. Other software as needed for the individual users is installed on a case-by-case basis.

All the network clients are configured to use DHCP and WINS servers as well as MS Proxy server. Finally each machine is configured to log on to the NT domain and access appropriate logon scripts, user profiles, home directories if applicable, shared applications and other network resources.

THE DESIGN OF THE WIDE AREA NETWORK

The Town X domain involves servers and desktop computers spread over seven buildings at five sites within the town. The Board of Education (BOE) and Town X High School, which are adjacent to each other, are connected by a fiber optic backbone allowing transfer of data at 100 Mbps. The Main Distribution Frame (MDF) at Town X High School is the network core. The remaining sites are connected by means of a wide area network implemented over frame relay connections at T-1 speed (1.54 Mbps). Another frame relay connection at T-1 speed links the network to a local Internet Service Provider (ISP) which then links it to the Internet.

Frame Relay vs. Point-to-Point

A considerable amount of thinking went into the decision to use Frame Relay as the connection technology for this WAN. Traditionally, WANs were created using point-to-point technologies. The difference between Frame Relay and point-to-point is that point-to-point is implemented over a dedicated copper line from one site to a phone company digital switch and from that switch to a second site. In order to connect the entire Town X school district network in this fashion, each site on the network would need to have five such connections, one to each of the other sites. Since point-to-point connections are leased on a per-mile monthly fee (about $40 per mile in this case), the costs add up pretty quickly.

Frame relay, on the other hand, requires only one connection per site. That connection is leased at a set fee and goes to what is called a "frame relay cloud," a group of digital switches which on request forms a connection to another such cloud where the object of that request (the machine that is being accessed by your network) is connected. So the T-1 speed in the case of the Town X school district WAN is actually "burstable to T-1" speed. If there are simultaneous connections to more than one other site on the school district network, the bandwidth is then split among them. In spite of that, frame relay was found to be the more cost-effective and efficient way to connect at the time the network was designed. This was true even when compared to some of the lower cost technologies like ISDN or modems. The problem with ISDN or simple modem technologies was that neither of these was sufficiently robust to carry the kind of traffic generated by the school district. At one time in the past, the school district network was actually connected by 56 kbs frame relay, but the administrative database program, Phoenix, which is extensively used, required a much higher bandwidth so the connection was increased to a T-1.

The Main Distribution Frames

At the heart of each of the Town X school district WAN's school is an MDF built around 3Com's Corebuilder 3500 Layer 3 switch. The "Layer 3" refers to the fact that the Corebuilder provides for not only 100 Mbps port switching but also routing at "wire speed." (The backplane of the Corebuilder routes packets as quickly as the network electronics can deliver them. Cisco's router did not have this capacity; as it could only route at the speed of its processor). The Corebuilder design is also fully scalable by means of plug-in "blades," each with fiber or Ethernet ports. So this design also allows for expandability using technology not even available yet.

The servers at each site connect to the network through the TX (Ethernet) blades, while the Ethernet Switches, both at the MDF and the various IDFs (intermediate distribution frames), connect by fiber to the fiber blades.

An added feature of the 3Com Layer 3 product is that its routing capacity allows for the creation of Virtual LANs (VLANS), by segmentation of a LAN into IP subnets. Since routing is a managed activity, this feature is used to provide extra security on the network by blocking traffic (via a process known as "packet filtering"), between the student and administrative subnets. For each LAN, three VLANs were set up so that students and administrators could be denied access to each other's areas, without denying them access to common resources such as the servers themselves. In the absence of Layer 3 switching this packet filtering can also be done at the routers, but would slow down the network.

Switched Port Hubs

The Town X school district network computers connect to their respective distribution frames (DFs) via 3Com's Superstack 3300 24-port Ethernet Switch. In the High School alone, there are 15 of these. (The Appendix has full details on hardware, software, labor and costs for the entire project.) The reason switched hubs were chosen for this implementation is the speed. One advantage of using switched port hubs is that the bandwidth is not shared, unlike in ordinary hubs. A second advantage of a switched port hub is that packets are delivered to the port the destination computer is attached to, instead of being broadcast to every computer connected to the hub.

Connecting to the WAN

The Town X High School LAN connects to the WAN through a Cisco 2500 Router and a Motorola CSU/DSU. A router passes data packets from one IP subnet to another, and the CSU/DSU acts somewhat like a modem. It translates frame relay protocol from the telco loop (telephone company line) into TCP/IP (Transmission Control Protocol/ Internet Protocol). In this case there are two connections through the router to two separate CSU/DSUs. One connection comes from the Internet server and routes a genuine Internet IP address to the ISP. Another comes from the TX blade on the Corebuilder and routes a private or internal IP address to the other LANs in the Domain. There it is picked up by the LAN's respective CSU/DSUs and routed to their respective Corebuilders. There too the IP addresses are private and are not routed to the outside world.

TCP/IP Addressing in the Town X Domain

The IP addresses of machines directly connected to the Internet never start with a 10 as that number is reserved for private networks. All the machines at the middle school have an IP address starting with 10.0.10. At the high school and BOE the addresses are 192.168.*.*. These are also IP addresses reserved for private networks. At the Mem school they are 10.0.9.*, and so on.

The routers meanwhile have addresses of 10.0.11.*, 10.0.8.* and so on. They use a convention where the .1 is on the WAN side and the .2 is on the LAN side. It helps the network engineers to keep track when they are configuring the network.

The only actual real-world IP address in the whole domain is the one that gets routed from the IIS server to the ISP. The ISP has actually given the Town X school district network a quarter of a class C address: in this case the IP addresses 209.R.S.128 through 209.R.S.191 with the .128 and .191 being the network and broadcast addresses respectively. The number of IP addresses available to the school district was not sufficient, thereby making the use of private IP addresses for computers on the internal networks a necessity.

Problems in Implementation

Problems in the implementation of an IT solution can come in many forms, all the way from technical/technological, to financial and political.

The wide area network was designed and implemented for the school district of a small town. The cost of networking the individual locations averaged about 125,000 (for a total cost of about 750,000). While the WAN had a fixed initial cost, it had a recurring monthly cost. The WAN cabling was done by the local telephone company who brought the frame relay connections to the individual locations at a cost of $1,200 each. There was also a monthly line charge of $400 per location. As we said, Town X has six such connections. For a small municipality that was a considerable financial burden.

One of the things that helped in Town X was a program of state-sponsored infrastructure grants, a program established specifically for the purpose of bringing schools online. There was also federal grant money available for these purposes. The Town X school district spent a lot of time and effort to make sure that they made a very good case for getting the state and federal funds. Approximately 90% of the cost of implementing the WAN was covered by the State and Federal grants. What wasn't covered by grants was funded out of the municipal budget, with virtually no opposition even from the Board of Finance. Even though the remaining 10% of the cost was a significant amount of money for a small town, the Board of Education kept the Board of Finance involved and informed at all points in time. So in spite of the ever-present danger of financial problems, the WAN implementation managed to get through the implementation stage without any major hiccups. This was primarily due to a lot of work up front by the BOE members who made sure that everyone on the town council was committed to the project.

The WAN project was overseen by an implementation team, which included a representative from the Board of Finance, the district technology coordinator and a Superintendent's representative, who was also the grant writer for the Board of Education. These three spent a lot of time with representatives of the contractor to set up the implementation plan. Good planning ensured that the contractor was aware of the school schedules, and had the appropriate resources available when needed during the implementation. The implementation team, along with representatives from contractor, both administrative and technical, met regularly to coordinate their efforts and keep the project on track. The meetings were more frequent during those phases of the implementation when there was a lot of activity, and less frequent during the slower phases. The meetings helped ensure that all the parties involved were aware of the situation at all points in time, and any deviations from the plan were discussed and then agreed upon. This approach worked quite well and the project was completed pretty much on its schedule, roughly two months for the cabling of the various locations and configuring the network electronics. Most of the work was done in the summer or after school hours with the school custodians providing the cablers and technicians access to the buildings, often until 11:00 at night. On the administrative side the major headaches were in the accounting and coordinating of the grant and municipal funding and the allocation of costs to each budget as appropriate. The contractor had prior experience working with school districts, and on projects based on similar state and federal grants. So the contractor's representatives were able to provide inputs into the process that helped properly administer the grants.

There was one unexpected snag when asbestos was discovered in one of the wings of the high school. The work at the high school had to be halted while the asbestos was removed. However, due to the existence of a comprehensive implementation plan, resources could be shifted to other school sites while the asbestos abatement took place. The fact that the affected area in the high school didn't include any of the wiring closets helped, as it did not impede the phased completion of the project significantly.

Benefits to the District

Some of the benefits in this case are immediately apparent. Running administrative applications like Phoenix from a central location is not just a convenience. It represents significant savings in man-hours, and therefore money, to the town. The fact that each school can now provide inputs into the scheduling system, maintain student records and other administrative tasks, without having to send over hard copies of data to the BOE, is a significant improvement in efficiency. Packet filtering at the routers and Layer 3 switches restricts access to the administrative side of the network while allowing for the use of less sensitive services by others, thereby increasing the overall utility of the project.

Another part of this benefit is centralized Internet access. Having this access through a central point gives the network administrator the ability to establish an Internet use policy and enforce it without having to do so on a location-by-location basis. As mentioned before, a third-party product called Web Sense is used to implement URL filtering of potentially offensive materials (the use of this filtering is a politically charged issue, but it is common in a number of school districts).

The use of the Internet in the classroom is becoming more prevalent. The students and teachers use the Internet to research current topics, and get access to information that was not readily available before the WAN was set up.

With Internet access the district is beginning to make many services available to the citizens in the community at large. School schedules, lunch menus and library card catalogues are all posted for access through the Internet. Students and faculty can retrieve their internal and/or external e-mail, through the Web, whether on or off site. Eventually other services, such as distance learning, may be implemented in Town X.

CURRENT CHALLENGES/PROBLEMS FACING THE ORGANIZATION

The Town X school district is considering expanding the scope of the wide area network. One of the locations in the school district, the Bus Garage, was not connected to the WAN due to lack of planned funding. This has resulted in perpetuating some of the inefficiencies that the WAN was supposed to eliminate. Since the garage is not networked, the bus availability and the bus schedules have to be transported manually from the garage to the BOE offices and back.

A number of classrooms in the schools are still not connected to the network. So it generates a feeling of inequity among the teachers and students who are assigned the non-networked classrooms.

While the connectivity is good in most locations, the high school seems to hog most of the available bandwidth on the WAN. The ability to access information from the Internet seems to slow down during the time the high school is in session. There seem to be some problems with the filtering service, as there are cases of objectionable material being found on some of the computers in the high school labs. Some of the lab computers have been used to serve up copyrighted material, while others run bandwidth-intensive applications that bog down the network.

ACKNOWLEDGMENT

This case was researched on site in Town X and through many hours of conversation with the senior systems engineer with the contractor and the designer of this particular network.

FURTHER READING

Stallings, W. (2001). *Business data communications* (4th ed.). NJ: Prentice Hall.
Comer, D. (1999). *Computer networks and Internets* (2nd ed.). NJ: Prentice Hall.

APPENDIX: PROJECT COSTS

High School and Board Of Education (BOE)

Cabling	Qty	Each	Total	Purpose
Classroom Drops (4 data/1 voice)		450	28800	Connect classrooms to each other and the internet
	64			
Office Drops (2 data/1 voice)	30	400	12000	Connect offices to each other and the internet
6 strand fiber between MDF and 5 IDFs	5	5200	26000	Connect all wings of the school together
Cabling Total			66800	

Network Electronics	Qty	Each	Total	
3-Com Superstack 3300 24-port Ethernet Switch	15	2,20	33000	Connect servers to classrooms and offices
3-Com Corebuider 3500 Layer3 Switch	3	8000	24000	Layer3 switch to connect fiber and add network security
3-Com Corebuider 3500 6 Port Fiber Blade	7	4800	33600	Connect all wings to servers
3-Com Corebuider 3500 6 Port TX Blade	6	3000	18000	Connect all servers to the network
Network Electronics Total			108600	

Additional Network Hardware				
4 foot Category 5 patch cables	316	3	948	Patch rooms to network electronics
15 foot Category 5 line cord	316	4	1264	Patch computers to network
APC 1000 Net UPS Rack Mount	6	829	4974	Protect servers from faulty power and power outages
Fiber Patch Cables	36	62	2232	Connect fiber to network electronics
Additional Network Hardware Total			9418	

Network Servers	Qty	Each	Total	
LAN & WAN File Servers	2	8500	17000	NT server/primary DNS, mail server, backup domain/web server

Built-IN POP & SMTP Mail
Built-IN DNS service
Built-IN WWW server
Built-IN Appleshare File & Print Services

Network Software	Qty	Each	Total	
MS BackOffice	3	1400	4200	Server operating system - NT, Mail, Web, Management
Windows Backoffice Client License	300	60	18000	License for clients to access servers
Network Software Total			22200	

Labor - Hardware & Software Installation, Training & Preparation				
Cabling	Included			
Network hardware & hub installation	30	100	3000	Installation and configuration
Network server install & configuration	60	100	6000	Installation and configuration
Desktop configuration	316	75	23700	Installation and configuration
Training	40	75	3000	Administration of network
Total - Labor			35700	
High School and BOE Total			**259718**	

Middle School

Cabling	Qty	Each	Total	Purpose
Classroom Drops (2 data)	68	$200	$13,600	Connect classrooms to each other and the internet
Cabling Total			$13,600	

Network Electronics	Qty	Each	Total	
3-Com Superstack 3300 24-port Ethernet Switch	6	1,450	$8,700	Connect servers to classrooms and offices
3-Com Corebuider 3500 Layer3 Switch	1	$8,000	$8,000	Layer3 switch to connect fiber and add network security
3-Com Corebuider 3500 6 Port TX Blade	2	$4,800	$9,600	Connect all servers to the network
Network Electronics Total			$26,300	

Additional Network Hardware				
4 foot Category 5 patch cables	136	$3	$408	Patch rooms to network electronics
15 foot Category 5 line cord	136	$4	$544	Patch computers to network
APC 1000 Net UPS Rack Mount	3	$829	$2,487	Protect servers from faulty power and power outages
Additional Network Hardware Total			$3,439	

Network Servers	Qty	Each	Total	
LAN & WAN File Server	3	$8,500	$25,500	
Built-IN POP & SMTP Mail				
Built-IN DNS service				
Built-IN WWW server				
Built-IN Appleshare File & Print Srvices				

Network Software	Qty	Each	Total	
MS BackOffice	3	$1,400	$4,200	Server operating system - NT, Mail, Web, Management
Windows Backoffice Client License	136	60	$8,160	License for clients to access servers
Network Software Total			$12,360	

Labor - Hardware & Software Installation, Training & Preparation

Cabling	Included			
	10	$100	$1,000	Installation and configuration
Network server install & configuration	60	$100	$6,000	Installation and configuration
Desktop configuration	136	$75	$10,200	Installation and configuration
Training	30	$75	$2,250	Administration of network
Total - Labor			$19,450	
Middle School Total			**$100,649**	

Mem School

Cabling	Qty	Each	Total	Purpose
Classroom Drops (4 data/1 voice)	27	$450	$12,150	Connect classrooms to each other and the internet
Office Drops (2 data/1 voice)	15	$400	$6,000	Connect offices to each other and the internet
6 strand fiber between MDF and 2 IDF's	2	$5,200	$10,400	Connect all wings of the school together
Cabling Total			$28,550	

Network Electronics	Qty	Each	Total	
3-Com Superstack 3300 24-port Ethernet Switch	5	2,200	$11,000	Connect servers to classrooms and offices
3-Com Corebuider 3500 Layer3 Switch	1	$8,000	$8,000	Layer3 switch to connect fiber and add network security
3-Com Corebuider 3500 6 Port Fiber Blade	1	$4,800	$4,800	Connect all wings to servers
3-Com Corebuider 3500 6 Port TX Blade	1	$3,000	$3,000	Connect all servers to the network
Network Electronics Total			$26,800	

Additional Network Hardware				
4 foot Category 5 patch cables	138	$3	$414	Patch rooms to network electronics
15 foot Category 5 line cord	138	$4	$552	Patch computers to network
APC 1000 Net UPS Rack Mount	3	$829	$2,487	Protect servers from faulty power and power outages
Fiber Patch Cables	4	$62	$248	Connect fiber to network electronics
Additional Network Hardware Total			$3,701	

Network Servers	Qty	Each	Total	
LAN & WAN File Servers	3	$8,500	$25,500	NT server/primary DNS, mail server, backup domain/web server

Built-IN POP & SMTP Mail
Built-IN DNS service
Built-IN WWW server
Built-IN Appleshare File & Print Services

Network Software	Qty	Each	Total	
MS BackOffice	3	$1,400	$4,200	Server operating system - NT, Mail, Web, Management
Windows Backoffice Client License	108	60	$6,480	License for clients to access servers
Network Software Total			$10,680	

Labor - Hardware & Software Installation, Training & Preparation

	Qty	Each	Total	
Cabling	Included			
Network hardware & hub installation	20	$100	$2,000	Installation and configuration
Network server install & configuration	60	$100	$6,000	Installation and configuration
Desktop configuration	138	$75	$10,350	Installation and configuration
Training	40	$75	$3,000	Administration of network
Total - Labor			$21,350	
Mem School Total			**$116,581**	
Cen School				

Cen School

Cabling	Qty	Each	Total	Purpose
Classroom Drops (4 data/1 voice)	34	$450	$15,300	Connect classrooms to each other and the internet
Office Drops (2 data/1 voice)	15	$400	$6,000	Connect offices to each other and the internet
6 strand fiber between MDF and 2 IDF's	2	$5,200	$10,400	Connect all wings of the school together
Cabling Total			$31,700	

Network Electronics	Qty	Each	Total	
3-Com Superstack 3300 24-port Ethernet Switch	7	2,200	$15,400	Connect servers to classrooms and offices
3-Com Corebuider 3500 Layer3 Switch	1	$8,000	$8,000	Layer3 switch to connect fiber and add network security
3-Com Corebuider 3500 6 Port Fiber Blade	2	$4,800	$9,600	Connect all wings to servers
3-Com Corebuider 3500 6 Port TX Blade	2	$3,000	$6,000	Connect all servers to the network
Network Electronics Total			$39,000	

Additional Network Hardware				
4 foot Category 5 patch cables	166	$3	$498	Patch rooms to network electronics
15 foot Category 5 line cord	166	$4	$664	Patch computers to network
APC 1000 Net UPS Rack Mount	3	$829	$2,487	Protect servers from faulty power and power outages
Fiber Patch Cables	4	$62	$248	Connect fiber to network electronics
Additional Network Hardware Total			$3,897	

Network Servers	Qty	Each	Total	
LAN & WAN File Servers	3	$8,500	$25,500	NT server/primary DNS, mail server, backup domain/web server

    Built-IN POP & SMTP Mail

    Built-IN DNS service

    Built-IN WWW server

    Built-IN Appleshare File & Print
Services

Network Software	Qty	Each	Total	
MS BackOffice	3	$1,400	$4,200	Server operating system - NT, Mail, Web, Management
Windows Backoffice Client License	136	60	$8,160	License for clients to access servers
Network Software Total			$12,360	

Labor - Hardware & Software Installation, Training & Preparation				
Cabling	Included			
Network hardware & hub installation	20	$100	$2,000	Installation and configuration
Network server install & configuration	60	$100	$6,000	Installation and configuration
Desktop configuration	166	$75	$12,450	Installation and configuration
Training	40	$75	$3,000	Administration of network
Total - Labor			$23,450	
Cen School Total			**$135,907**	

Cen School

Cabling	Qty	Each	Total	Purpose
Classroom Drops (4 data/1 voice)	34	$450	$15,300	Connect classrooms to each other and the internet
Office Drops (2 data/1 voice)	15	$400	$6,000	Connect offices to each other and the internet
6 strand fiber between MDF and 2 IDF's	2	$5,200	$10,400	Connect all wings of the school together
Cabling Total			$31,700	

Network Electronics	Qty	Each	Total	
3-Com Superstack 3300 24-port Ethernet Switch	7	2,200	$15,400	Connect servers to classrooms and offices
3-Com Corebuider 3500 Layer3 Switch	1	$8,000	$8,000	Layer3 switch to connect fiber and add network security
3-Com Corebuider 3500 6 Port Fiber Blade	2	$4,800	$9,600	Connect all wings to servers
3-Com Corebuider 3500 6 Port TX Blade	2	$3,000	$6,000	Connect all servers to the network
Network Electronics Total			$39,000	

Additional Network Hardware	Qty	Each	Total	
4 foot Category 5 patch cables	166	$3	$498	Patch rooms to network electronics
15 foot Category 5 line cord	166	$4	$664	Patch computers to network
APC 1000 Net UPS Rack Mount	3	$829	$2,487	Protect servers from faulty power and power outages
Fiber Patch Cables	4	$62	$248	Connect fiber to network electronics
Additional Network Hardware Total			$3,897	

Network Servers	Qty	Each	Total	
LAN & WAN File Servers	3	$8,500	$25,500	NT server/primary DNS, mail server, backup domain/web server

    Built-IN POP & SMTP Mail
    Built-IN DNS service
    Built-IN WWW server
    Built-IN Appleshare File & Print Services

Network Software	Qty	Each	Total	
MS BackOffice	3	$1,400	$4,200	Server operating system - NT, Mail, Web, Management
Windows Backoffice Client License	136	60	$8,160	License for clients to access servers
Network Software Total			$12,360	

Labor - Hardware & Software Installation, Training & Preparation				
Cabling	Included			
Network hardware & hub installation	20	$100	$2,000	Installation and configuration
Network server install & configuration	60	$100	$6,000	Installation and configuration
Desktop configuration	166	$75	$12,450	Installation and configuration
Training	40	$75	$3,000	Administration of network
Total - Labor			$23,450	
Cen School Total			**$135,907**	

Rohit Rampal is assistant professor of management information systems at the School of Business Administration, Portland State University, OR. He received his PhD from Oklahoma State University. He has previously worked in the College of Business Administration, at the University of Rhode Island. His areas of research include telecommunications, information systems in manufacturing, virtual enterprises, forecasting, and neural networks. He has previously published in the International Journal of Production Research *and the* Encyclopedia of Library and Information Science.

This case was previously published in the *Annals of Cases on Information Technology*, Volume 4/ 2002, pp. 427-439, © 2002.

About the Editor

Mehdi Khosrow-Pour, D.B.A, is executive director of the Information Resources Management Association (IRMA) and senior academic technology editor for Idea Group Inc. Previously, he served on the faculty of the Pennsylvania State University as a professor of information systems for 20 years. He has written or edited more than 30 books in information technology management. Dr. Khosrow-Pour is also editor-in-chief of the *Information Resources Management Journal, Journal of Electronic Commerce in Organizations, Journal of Cases on Information Technology,* and *International Journal of Cases on Electronic Commerce.*

Index

Symbols

3G mobile license 128
3G mobile systems 151

A

academic and research network 177
Adalat 283
added value 372
administrative services (AS) 30
administrative tasks 374
aerospace research information 224
analyses 68
ANX® (see automotive network exchange)
application integration 309
application server 43
application service provider (ASP) 145, 349
architecture 211
Asian Flu 244
assessment of threats 42
Association of South East Asian Nations (ASEAN) 244
asynchronous transfer mode (ATM) 21
automotive industry 17
Automotive Industry Action Group (AIAG) 18

automotive network exchange (ANX®) 17, 21
automotive network exchange overseer (ANXO) 21

B

B2C (business-to-commerce) market 347
background of developers 332
bar code 128
basic telecom operator (BTO) 91
Baycol 283
Bayer Corporation 283
British Telecom (BT) 129
Brno University of Technology 210
building small business systems 388
business and operational support systems (BSS/OSS) 311
business model 13
business operations 372
business partners 6

C

campus wide network 157
"cash cow" drugs 282
cellular mobile telephone system (CMTS) 248

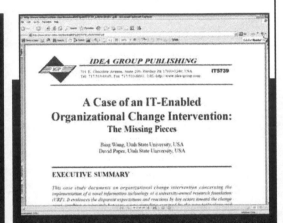